The **Absolute** and **Fundamental Law** of **Order** and **Truth**

"ANY ORDERED SYSTEM, INCLUDING THE UNIVERSE, MUST HAVE A FIXED REFERENCE AND BE OF INTELLIGENT DESIGN."

This law is fundamental for the development of life on our planet and any ordered system that we undertake.

John Constantine Capleton

Copyright © 2024 **JCLSM Publishing**

All rights reserved. No part of this publication may be reproduced, distributed, or transmitted in any form or by any means, including photocopying, recording, or other electronic or mechanical methods, without the prior written permission of the publisher, except in the case of brief quotations embodied in critical reviews and certain other noncommercial uses permitted by copyright law. For permission requests, write to the publisher, addressed "Attention: Book Rights and Permission," at the address below.

Published in the United States of America

ISBN 978-1-958518-26-7 (SC)
ISBN 978-1-962730-88-4 (HC)
ISBN 979-8-89395-955-0 (Ebook)

JCLSM Publishing
222 West 6th Street
Suite 400, San Pedro, CA, 90731
www.stellarliterary.com

Ordering Information and Rights Permission:

Quantity sales. Special discounts might be available on quantity purchases by corporations, associations, and others. For details, contact the publisher at the address above.

For Book Rights Adaptation and other Rights Permission. Call us at toll-free 1-888-945-8513 or send us an email at admin@stellarliterary.com.

TABLE OF CONTENTS

ACKNOWLEDGEMENTS..iv
FOREWORD 1
MY EARLY YEARS 2
INTRODUCTION...16
THE ABSOLUTE LAW OF UNIVERSAL ORDER AND TRUTH............17
THE MYTH ABOUT EVOLUTION 22
'ORDER NEEDS CONSTANT MAINTENANCE'.................................26
ORDERED SYSTEMS FOUND ON EARTH 52
PHYSICS 77
CHEMISTRY 89
INTRODUCTION TO THE GOD HEAD... 106
DEFINITIONS AND CONCEPTS.. 136
CONCLUSION 157
AUTHOR'S BIO.. 159
REFERENCES 160

INTELLIGENCE DEFINES ORDER

"IT IS ALL ABOUT REFERENCE"

ACKNOWLEDGEMENTS

To my wife, Cheryl, for the extra time I had, between chores, so that I was able to focus on writhing this book.

Thanks to our three daughters Lisa, Stephanie and Marissa for taking the time to review the draft, giving me feedback to help ensure it made logical sense and that the basic concept would be readily understood.

Lisa R. - For encouraging me to write the book and offering to assist me in the review of the draft.

Jenn - Thanks for your advice on rearranging the topics in order of technical complexity, making it easier for the reader to be gradually introduced to the various concepts.

Pete - You were always willing to help me find the Bible passages referred to in the narrative to better show the obvious relationship between scripture and science.

Dan Caffese -Thanks for honing my understanding of who GOD is and how he relates to us.

Jack - Your prophecy about the message you received from the Holy Spirit and your assistance in reviewing the draft, gave me the confidence to write this book.

FOREWORD

Born in Oracabessa, a small port town on the northeast of Jamaica, just thirteen miles east of Ocho Rios, John was aware of the world beyond his door from a young age. He attended boarding school in Kingston as a boy, opening the doors for later studies in the United Kingdom, where he received his master's in engineering. Following emigration to the United States, he worked as a Risk Control Consultant until he retired in 2014.

John values working with his hands and has often been found completely absorbed in his projects. As a child, John enjoyed drawing and painting as well as designing and building models. In his teenage years, he developed an interest in electronics and moved on to making amplifiers and preamplifiers. John thrived on gaining knowledge and applying what he had learned to develop his craft. He later combined those skills and made enclosures for speakers, seeing his projects through from conceptual stage to completion.

An insatiable curiosity about how things work has encouraged an almost lifelong pursuit of the answers to what initiated the universe and the forces needed to maintain order. Adamant in his belief that all natural occurrences are governed by logic, John always knew there must be a rational explanation for existence.

Having raised three daughters, John is currently enjoying retirement with his wife and their two dogs in Arizona. His passion for creating has remained with him, and he still paints and designs & builds furniture.

MY EARLY YEARS

My hometown of Oracabessa is small port town on the northeast coast of Jamaica. For those who are familiar with the popular vacation spots on the island, it is thirteen miles east of Ocho Rios.

The name of the town evolved from the original Spanish name Cabeza Ora, or Golden Head, which the Spanish named it, because of the beautiful sunsets over the bay. The sun sets on the side of the bay over the projection of the coastline. In English, the adjective comes before the noun and so the name became Ora Cabeza and spelled, 'Oracabessa'.

Jamaica was colonized by the Spanish in 1494. The British captured it in war and colonized it in 1655. However, many of the towns in Jamaica still have Spanish names. Jamaica gained its independence from the British in 1962.

Oracabessa was a very popular banana shipping port in the 1950s and 1960s. You may have heard of 'The Banana Boat Song' by Harry Belafonte. This is the operation he was singing about as bananas were checked before being loaded onto ships in the harbor for export to England and other foreign countries.

It was a popular destination for tourists visiting the island and staying at the nearby resorts. People would come from all over the world to see this operation which would go on for two to three days, every week.

It was a day to look forward to as it brought out some of the very interesting characters in the town and nearby villages to display their unique talents, some in return for money from the tourists.

One of my first memories of Oracabessa was looking at the vast ocean, from our front yard, and seeing the blinding reflection of the sun on the water, as it always did in the early afternoon.

One day, as I gazed at the water, I said to myself, "I am really alive. What I see is real". The reflection was so blinding that one could not look at it for more than an instant without being forced to look away. I believe it was from that time I developed the desire to try to understand not only the force behind the universe we live in but also the reason for life itself.

(Above was the view of the port operations from our front yard)

This was also the town that Ian Fleming chose as the site for his home in Jamaica. He wrote many of his James Bond books here. One of his books was named after the property, Golden Eye. The property was previously named, Rock Edge, as I would hear my family refer to it, when it was owned by a friend of theirs.

My grandparents had a small property surrounded by his property and we had the privilege of using his private beach for all our summer vacations from school, along with our friends. We had a lot of fun spending time at the beach, as he was never there for more than two months of the year and never during the summer months. We took over the beach the rest of the year.

I remember my first and only encounter with Mr. Fleming. I was about 6 years old and at my grandmother's house. I saw him walking down his driveway as he sometimes would on his exercise walks and to survey his property. He was always only in a pair of shorts, sandals and smoking a cigarette.

I had seen him before and recognized him as the owner of the property surrounding my grandmother's. He always came about the same time every year. This was where he could relax and write his books without any interruption.

This time he stopped and asked me if I could call 'Mrs. Simmit'. I knew that he was referring to my grandmother, 'Mrs. Smith'. He called her Mrs. Simmit because that was how the local people pronounced the name 'Smith' and he had obviously been told her name by one of his property employees.

I had no idea why he wanted to speak with my grandmother but I went to call her and she immediately came to see what he wanted. As I stood beside her, he first asked if she was Mrs. Simmit to which she responded affirmatively.

My grandmother was a retired school teacher, well respected in the town and well spoken. She spoke perfect English and demonstrated her command of the English language in her response to him.

What he had come to complain about was that the fence at the property line, between the two properties, was broken and there was no longer a clear demarcation. I think he wanted it repaired and she agreed to have it done. However, at the end of the encounter she corrected him concerning the incorrect pronunciation of her name. She told him her name was "Smith" and not "Simmit". In addition she scolded him on his source of information, indicating that if he wanted accurate information about her and her property he should ask her and not his employees.

I was proud of how she stood up to 'Commander Fleming' as he was known by the people in the town. You see, he was an Officer in the British Navy during World War II and was well connected among the British aristocracy. Some of the people who stayed at the property were Princess Margret, Sir Winston Churchill and Sir Anthony Eden, another of Britain's Prime Ministers. When they were there, we felt safe, as a security guard was posted at the gate 24/7. We would give them snacks and meals as well as shelter on the veranda, when it rained.

That said, I must commend Mr. Fleming on his handling of an issue with the property title on his last visit to Jamaica. My grandfather had bought the property from a family friend, the same person who sold the property to Mr. Fleming. Unfortunately, there was no proper separation of the two properties and no separate title drawn up for my grandparent's property. As a result, she had no legal claim on the property and my grandfather was now dead.

She got legal council requesting that Mr. Fleming sign over her section of the property to her. He could have easily refused or fought it in court. But he did not! On his next visit to Jamaica he signed her section of the property over to her. That was very gracious of him and I have never forgotten this and respected him for that decision. That was his last visit to Jamaica as he died in England soon after.

Memories of my early life are now somewhat spotty but I do remember those incidents that made a permanent impression on me. Looking back to when I was about five years old, I was in the living room alone one day listening to the radio. I started wondering, how could someone be talking, also singing and music coming from this little box on the table. There had to be some little people and a band somewhere in there. I then got a knife from the kitchen, turned the radio around and started unscrewing the back cover. I could feel the heat being generated by the vacuum tubes, as in those days they did not yet have transistors. Vacuum tubes are not only hot but they operate with moderately high voltage. I looked inside and did not see the man or the band so I decided to try to move some of the components aside that might have been blocking my view. Somehow I touched what must have been the power supply and experienced my first electric shock. After recovering from the 'shock', I carefully replaced the back cover but told no one about what had happened.

Throughout my life I have always had a passion for designing, making and testing things to verify if my design was successful. I have built model cars, planes and boats. I liked boats because I grew up with boats in my home town, Oracabessa.

I once challenged myself to build a full sized boat from pine wood. This was one with a keel, starting with a wooden frame and adding the panels to finish it. It was 19 feet long and about 4 feet wide.

I had just finished high school but had not yet decided to go to college. I needed something to do and this seemed to be a good project. I got the raw materials from the local hardware store. Since I was not working, my mother provided the funding which she did without complaining even though she could hardly afford it. She was concerned about me and would have done anything she could to help me get through this rather difficult period of my life. She opened an account with the local hardware store and whenever I needed tools, lumber or nails I would go and get the necessary supplies. With all of this, I did not finish the boat as I left Jamaica to go to college in England before completing it. A local fisherman bought it from my mother after I had left.

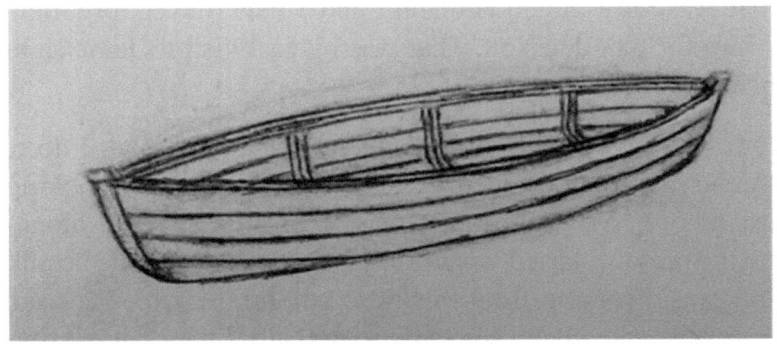

MY PINE BOAT

It was at this time in my life that I also had time to practice my oil painting. I painted a portrait of my sister from a black and white photograph that she had sent to us from London where she was attending law school. (See portrait below.)

MY FIRST OIL PORTRAIT

I was then about 19 years old and needed to make some important decisions about my life. I was the 'black sheep' in the family and going nowhere.

It was during my last high school years that I started to explore various theories or philosophies of life including Eastern Religions and even 'The Occult'. I needed to find a philosophy that satisfied my basic understanding of life at that time.

One afternoon during the time I was experimenting with occultism, I had just come from school and started reading one of my books. I came upon a chapter that described the steps on how to teleport one's spirit. According to the book, your spirit would leave your body and go anywhere you desired. I followed the steps indicated and something did happen, but not what I expected.

I was laying on my back in bed and going through the steps when I realized that I could not move. I was fully conscious and awake and so was beginning to get really scared. No matter how I tried, I could not move.

I was totally aware of everything going on around me. I could see the curtain by the window slowly rising and receding. I could hear the wind as it slowly moved the curtain and then die into the ambient outside background noise of the birds and other random sounds. I began to panic but could not call for help as I had also lost my voice. I have no idea how long this lasted and it may have been only a few seconds, but it appeared to last forever. With adrenaline building up in me from my state of panic, I managed to move one arm, and then the other and finally my whole body.

That was the last time I dabbled in 'The Occult'. I felt so out of control and did not ever want to feel that way again.

This period of my life was very disappointing to my parents, particularly my father, who was an attorney and was of the firm belief that a good education was essential to one's success in life. He wanted me to go to college immediately after high school but I was definitely not ready.

I remember him telling me that I had no ambition and that artists never made enough money to adequately support themselves, much less a family. He would give me extra lessons in Latin as he thought a person was not fully educated if he or she was not familiar with the language from which most modern languages were derived. Classes were held on our veranda where all my friends and neighbors could see. They would make fun of me by making faces from a distance as they watched me squirm when I could not answer a question on the day's lesson.

My father was self taught after leaving high school. He taught himself law, getting all the necessary courses and exams by mail from England. He was very self motivated and expected his son to do even better than he. I was a disappointment to him for most of my teenage years but after attending engineering school and getting a postgraduate degree, I think I might have redeemed myself in his eyes. He even offered to pay for my Masters course but I had already been granted an extension of my government scholarship and did not need his help anymore.

Some time before he died, he told me that he thought engineering was such an interesting occupation and he always wondered how a massive, steel oil tanker floated on water. I was glad for the opportunity to explain

'Archimedes Principle' of flotation to him. I don't think he really understood, although he believed what I told him.

(Archimedes principle states that an object will float on water when it displaces the volume of water equal to its weight. Water weighs 62 pounds per cubic foot and so it would support an object weighing 62 pounds if that object displaced one cubic foot of water when placed on its surface. The oil tanker, therefore, displaces the volume of water equal to its weight. This principle is true for any liquid)

I remember my mother being a very nice human being. She would give anything she had to those in need and was well liked in the town. I needed someone like her to keep me balanced. I was very close to her and felt as if I could talk with her about anything and she would not judge me.

She was the secretary at the United Fruit Company, one of the banana shipping operations, the main industry in the town. Her job included administrative work such as typing documents and making out checks to pay the banana suppliers who sold their produce that was harvested weekly.

I would sometimes watch her as she typed checks for some of the suppliers and could not help noticing the amounts on some of the checks. These people were making tens of thousands of 'British' pounds per week and in some cases hundreds of thousands. This was a lot of money in the 1950's.

Whenever I think about the wonders of word processing, I remember my mother in those days of carbon paper. Typed copies were made using carbon paper. If you made one mistake you had to stop and erase the mistake on the original and all the copies, of which there were sometimes 5 or 6. It was not only time consuming but it was also messy. Although she made few mistakes and was good at making the corrections, I could still see the look of frustration on her face when she made a mistake, as she knew what she now had to do.

One plantation owner, for whom she would sometimes do special secretarial work, had one of the larger properties with hundreds of acres of bananas and so he got a healthy weekly check. On 'Banana Days', I would see him come into the office, from his chauffeur driven Mercedes Benz, to see how things were going and to talk with the office staff. He would then go out on the landing in front of the office overlooking the port to watch the shipping operations. He also owned two launches powered by Mercedes Benz engines, used for fishing and for pulling the banana boats out to the ships. A Mercedes Benz mechanic would visit annually to perform routine maintenance on the boats.

It was on one of these launches that we were promised a Saturday afternoon ride for the office staff and their families. I was then about 8 years old. I had looked forward to this day. It was a beautiful afternoon with the sun shining brightly on the ocean. We waited and waited but he never came. There must have been a good reason why he was not able to come but since there was no telephone at home at the time, we did not know what had happened. This was probably one of the most disappointing experiences of my childhood and I still think about it. I believe he was really sorry for disappointing us as he would give my mother the annual calendars and catalogues from The Mercedes Benz Company to pass on to me. It worked because I always looked forward to getting my annual catalogs and calendars. My mother always gave me the United Fruit Company magazines with pictures of the fleet of all their ships. That was a real treat for a young boy.

I thought Mercedes were the most classy cars. One day a traveling salesman came into town and stopped at the office. Can you guess what he was selling? Battery operated toys which were the latest upgrade from push and wind up toys. He had fire trucks and cars but there was this red 'Gullwing' Mercedes Benz sports car that really caught my eye. It was not cheap but somehow I persuaded my mother to get it for me.

After she bought it for me I was completely preoccupied with this toy. I played with it from the time I got it until I met with my friends that afternoon. I was looking forward to showing off my new car. They passed it around so each could examine it and play with it. Then it came to one boy's turn.

You know, in every group there is always one that you wonder whether or not he is really a genuine friend. It was now his turn to play with my car. He picked it up, looked at it and then dropped it. I did not know whether or not it was on purpose but the car stopped working.

I went home with my broken car and took it apart in an effort to fix it but could not get it to work. I then took it to my friend, Maxie, to see if he could fix it for me. Maxie was like a big brother to me and my mother liked and trusted him. All I had to do was tell her that I was going over to Maxie's and she would ask no further questions.

We worked on the car for several hours but still could not get it to work. We tried everything but nothing worked. Eventually, we had to give up but I have never forgotten the disappointment and the regret for having shown my car to that friend.

I believe this was when the seed was sown for my interest in engineering. I made every effort, at every opportunity to learn about how things work and how to design and build things.

In the early years my models were built from wood, metal, and plastic. Later, I used fiberglass and carbon fiber. These are typical manufacturing materials and so I would experience the same challenges working with these materials as one would in a manufacturing facility.

My Education

The system of education in Jamaica started with kindergarten, then elementary school at about the age of six years. We were taught the basics- English language, English literature, Arithmetic, Geometry, Algebra and some basic Science. In addition, there was Home Economics for girls and Woodworking for boys.

At the age of about ten years, you were now prepared to take The Scholarship Examination which one had to pass in order to be accepted into High School.

We had very good teachers who would spend extra time with students after school to help them prepare for the scholarship exams. I was a very dedicated student and remember doing 100 math problems daily, under test conditions, in preparation for the scholarship exams.

I had now taken the scholarship examination and was awaiting the results. The morning when the scholarship results were published in the Jamaican news paper, The Daily Gleaner, when I scanned the list, I did not see my name. I thought I had done well enough to pass so I was extremely disappointed. I was very worried as I knew that my parents, especially my father, would not be happy and I would have a lot of explaining to do. It wasn't too long before one of my friends who was also scanning the news paper for his name said 'Si yu name ya'. This is Jamaican dialect for 'Your name is right here'. I was so relieved! The reason I could not find my name was that I was looking for it in the wrong place. I had been awarded a special scholarship and my name was printed in a separate place with some other students who had also been given special recognition.

Everyone now had great expectations of me in high school but I performed as an average student and definitely not up to the level of expectation of my father. After the first two years at the first high school, he transferred me to a boarding school in Kingston, as he thought this would be a better work environment for me and a place where I could focus on my studies.

I hated boarding school and continued to be an average student but did pass my Ordinary and Advanced level examinations in preparation for college. The last two years of high school, I stayed with an aunt and uncle who lived in Kingston. I would no longer be attending boarding school.

After leaving high school, for the first year, I stayed at home in Oracabessa and explored my hobbies including oil painting, woodworking and electronics. Electronics also became one of my interests as I have always liked music and would listen to the radio to the popular music of the time. I built amplifiers and preamplifiers, from circuits published in Popular Electronics magazines, to enhance the fidelity of the music.

Moving forward to when I was now about 20 years old, I got a job as a teller in a bank. This was probably not the best choice for me as in those days the bank employees were known for their drinking and partying and I was now one of them.

I worked at the bank for less than two years, then got another job as an auditor for the Courts Offices. This lasted only a few months as I soon realized that I needed a change and applied for and got a government scholarship to go to college in England to do engineering.

I left Jamaica for England in 1970 to continue my higher education. I spent my internship with a company named Metro-Cammell. Here, they built buses and trains including those used in the London Underground Transport System (Tube). They also built buses and trains for the Commonwealth countries, including Jamaica and India. My training alternated between six months there and six months in college, for my first degree.

At Metro-Cammell, I was trained in each department, initially at the Training Center where I learned to operate production machines such as the lathe, milling machine, shaper and surface grinder.

Before I was allowed near a machine, I had to learn to use my hands, with great precision, filing a piece of steel until it was perfectly square in all dimensions. This was a small piece of mild steel, commonly used in manufacturing, rough cut from a plate about 1/8" thick, 6" long and 2" wide. I was given a hand file to transform this rough piece of steel to a tolerance of 5/1000" on all dimensions. It seemed to me like an impossible task but there were other trainees who went before me and had their finished work displayed as proof that it could be done.

You may be wondering how was I to know when I had achieved success? In addition to the steel and the file, I was given a reference, called a square, Fig R1a. This is a reference tool that is perfectly square and perfectly smooth and flat on the reference surfaces. It is 'L' shaped. The perpendicular section is at a perfect 90 degree angle. This was my reference for squareness and flatness. This is one of the reference tools that is still invaluable even today. To measure the dimensions, a vernier caliper was used. This is a highly accurate measuring tool with an accuracy to less than 1/1000th of an inch.

To use the square as a reference, when you held it up to the light against the work piece, if you saw any light coming through between the contacting surfaces, then you had to continue filing because the side you were working on was either not flat or not square. To achieve success, each of the four corners had to be square relative to the adjacent corners; as well as the thickness, relative to the larger flat surfaces; See Figs. R1a and R1b through R3a and R3b below.

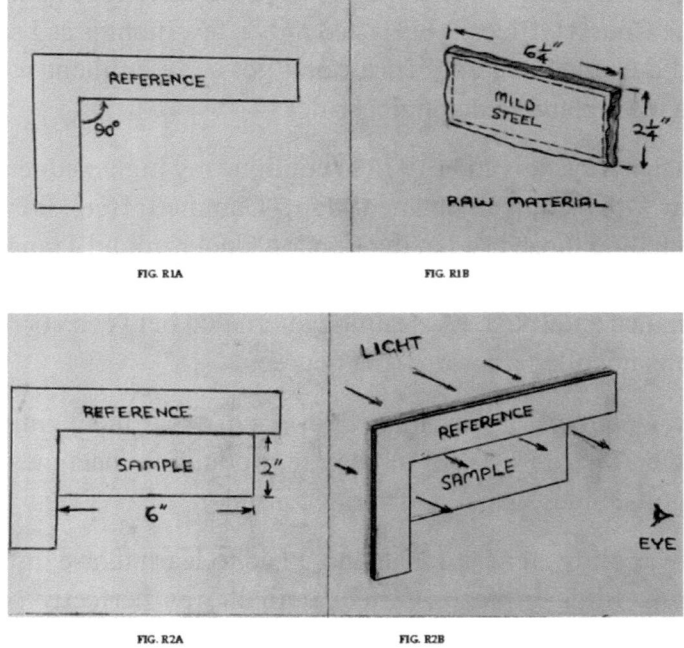

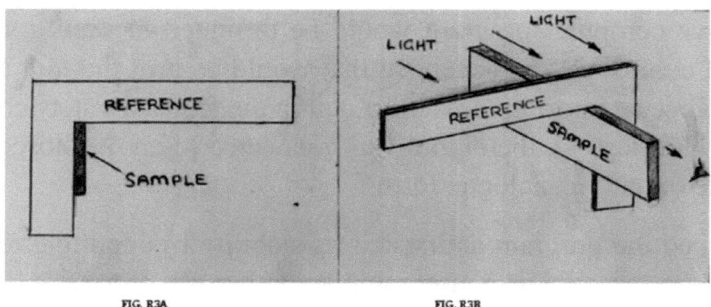

FIG. R3A   FIG. R3B

I think it took me two to three weeks to achieve success but, once completed, it did give me a sense of accomplishment.

I did not spend all of my time doing this project as that would have been too much. My muscles would get fatigued holding the small file, trying to accomplish the task. To break the monotony, I was given other projects to do but was expected to eventually complete this project.

I could not have achieved success without a reference with known accuracy. I had to complete this project before I was allowed to learn how to operate the machines, so it was important.

After the training center, I spent time in the Design and Manufacturing Departments and wrote reports on what I had learned in each of the respective departments. Research and Development worked closely with Design and Manufacturing and I had the opportunity to test new designs and verify tensile and compression strengths under different load conditions. I even did some simple designs myself.

I then moved on to the administrative departments where I also wrote reports on their operations. Training was therefore hands on and the best way to become familiar with the design, manufacturing, and administrative processes.

The hand filing exercise taught me to develop patience and an appreciation for what it took to achieve accurate results in the manufacturing process.

My Masters thesis in engineering was to develop a computer program that would aid in the design of a stable control system for a production machine in order to produce accurate parts. The title was 'Computer Aided Control Systems Design'.

The active computer program would be designed to send a signal to a proportional control servomechanism that would control the tool during the machining process. The output would result in the tool accurately copying the input signal. For testing, the output was connected to an oscilloscope which displayed the output in analogue form.

I developed the program and tested its stability by inputting test signals such as a sine wave, step response, ramp and other waveforms.

The first test of the program resulted in failure. We later found that all calculations in the program were correct except for the scale on the visual display, on the oscilloscope, which initially showed a flat line with every input signal. I controlled each input signal and knew the expected output, but the output did not match the input.

Then, my professor got the brilliant idea to change the scale on the 'Y' axis. This made all the difference because the proper graph now appeared on the screen.

This indicated to me that every step in the journey is critical. The entire sequence has to be accurate and complete before the final goal, and hence success, could be achieved. There has to be a set sequence of events starting from a fixed reference with all of the steps accurately completed before success can finally be achieved. This is not a random process.

Success does not come randomly. It takes a determined, focused effort. You have to know the desired outcome and take the sequential steps and make the necessary effort to achieve it. Time is not the most important element even though time is a necessary component. It became apparent towards the end that even if all other steps had been correct it was just as important for the final step also to be correct.

Developing this program made me realize that the slightest error could result in failure. Every step had to be correct for the whole system to be successful.

After completing my education in England in 1975, I went back to Jamaica and worked in the manufacturing industry for about two years before immigrating to the United States, in 1977.

I got married in England before returning to Jamaica. Our first daughter, Lisa, was born there, less than a year before I graduated from The University of Birmingham. After returning to Jamaica, our second daughter Stephanie was born. It was fifteen years later that our last daughter, Marissa, was born in Glendale, Arizona where we currently live.

I had never been truly religious but always believed there was a unifying force in the universe. This force existed but, to me, it was always far away, somewhere up there, out of reach, in the heavens.

When I was a young boy, I went to church on Sundays because my parents required that I learn about GOD. To them, it was the right thing to do. When I was on my own for the next thirty years, I only attended church at Christmas and occasions such as weddings and funerals.

I still believed that there was a unifying force but it was still out of my reach. To that force, I was insignificant and unimportant or possibly, nonexistent. I only thought of GOD occasionally when I was in a reflective mood.

I was over fifty years old when things dramatically changed in my life. A series of revelations concerning life and the process of creation made everything begin to become clear and make rational sense.

When I look back at my early life, I can see how it has influenced my entire life. This included my choice in career and my chosen hobbies. It is as if these were chosen for me, and all I did was to fulfill the purpose. This purpose was not of me, so it must have been from a source outside of me; a higher source, the source we refer to as God.

INTRODUCTION

We are all faced with the same questions. How did this all begin? What is the purpose of our existence? What part do we play?

We are born into a world that is initially foreign to us. We have to learn everything about this material and spiritual world. We all start at the same point, birth, and then gradually grow through childhood, adulthood, and finally, old age, if we are blessed with an average lifespan.

Some believe that this is a random existence with no real purpose. Others believe there is a real purpose for our being here and our lives are a fulfillment of this grand purpose.

In this book, I will look at both of these beliefs in an attempt to determine which best fits how our world manifests itself and the possible source of its existence. What we will come to find is that there is nothing random about the order that we see and experience. Order and randomness are mutually exclusive. The laws that control our material world are fixed and never change. Together, these generate universal order.

As we discuss these laws, we will see that they work together toward one purpose: to maintain harmony in the universe. This is an indication that the source of these laws is also fixed and never changing. This is the only conclusion as only such a source could generate and maintain such laws. We will determine that only God has the attributes to be the source of our existence.

To help guide us through this life and help us determine our purpose here, we should take into consideration where we were born and the gifts or talents given to us, as well as our experiences from childhood to adulthood and old age. These shape our lives and teach us about the world. I will first share with you some of my experiences, from birth through adulthood and retirement, as examples of how these experiences shaped and directed me toward my purpose in this life.

THE ABSOLUTE LAW OF UNIVERSAL ORDER AND TRUTH

"Any ordered system, including the Universe., must have a fixed reference and be of Intelligent Design"

This law designed the universe to be an ordered system. Without this law there can be no order. If this law is broken, disorder will be introduced into the system and there will now only be partial order. In addition, once disorder is introduced, order will be affected and will start and continue to erode. This is the Universe in which we now live.

As indicated above, when we look at our universe, we see both order and disorder.

Let us first examine order. If we look at any ordered system, we see that there are certain attributes that we recognize for us to conclude that the system is ordered. Some of these are listed below:

Symmetry
Harmony
Consistency
Theme
Balance
Stability
Fixed Cycles
Fixed Sequencing
Aesthetics
Structure
Predictability
Pattern
The presence of purpose
The presence of limits
The ability to be defined

An ordered system may be recognized by having one or more of these above characteristics.

We see these characteristics all around us. They are an indication of how order manifests itself.

When we try to develop an ordered system for ourselves, there are certain things that have to be taken into consideration. We first develop a plan with a fixed reference, start, and then build the system until it is complete. Depending on the system we are creating, there are specific sequential steps that must be taken. We also have to recognize when the system is complete and functioning as designed. This is not a random or chance process but takes intelligence. It is our intelligence that defines the system. There are no exceptions to this process.

The basic requirements for the development of an ordered system are a fixed reference, and intelligence to define sequencing. Intelligence fixes the reference and defines the sequential steps needed to develop the system. This is true for any ordered system made by man or other intelligent living creature.

To accomplish this, the intelligence must be external to the system so that it has an objective perspective. Otherwise, it will be unable to develop an independent, definable system. If we are a part of a system, we are influenced by that system and thus unable to look at it objectively. Before an ordered system is initiated, it must be planned. Planning can only be initiated from outside the system.

If we now look at disorder or randomness, we find that neither can be accurately defined. If something cannot be defined, it cannot be reproduced. At the nano level (fundamental level), atoms and molecules can be accurately defined and so are ordered systems. This means that the building blocks of all matter are ordered. From this we can conclude that they are of intelligent design. All ordered systems must, therefore, incorporate fundamentally ordered matter and components so that the complete system may be ordered.

By the same reasoning, the universe must be of intelligent design because it is an ordered system. Intelligence is that which defines order. As far as matter is concerned, both disordered and random matter are the raw materials for creating an ordered system. It takes intelligence to use these raw materials and transform them into independent ordered systems. It also took intelligence to create the fundamental components as ordered systems (atoms and molecules).

We can confirm this truth by examining any ordered system. Order cannot be randomly achieved. Neither can it be achieved by chance. It takes intentional, determined acts to achieve order. Additionally, only the intelligent designer is able to recognize when the system is complete and functioning as designed.

If we examine spiritual order, the principle is the same. This means that the statement below is 'Absolute Truth'.

'Any ordered system, including the universe, must have a fixed reference and be of intelligent design.'

Now, let us look at The God of the Bible. God has the following attributes as well as many others.

- He is Eternal
- He is the only God and there is no other
- He does not change

These attributes would have been fundamental in initiating an absolute, ordered system.

1. In order to initiate an absolute ordered system, there has to be an Eternal God. He was not created. He always existed in the form of infinite energy and intelligence. We cannot understand the concept of eternity because we are trapped in a temporal system.

2. He must be the only God as there can be only one ultimate source of control in any ordered system, especially if there are attached subsystems. All the systems must operate in harmony and so must be under a singular control. Otherwise, there will be chaos and disorder.

3. God does not change. This is critical for the development of any ordered system, again, especially for the development of the most complex systems such as Heaven and the Universe. If the reference changes in an existing system, disorder develops. In the material world, a reference change manifests itself as material disorder. In the spiritual world, it manifests itself as Sin.

Order is the product of intelligence. This is how we identify an intelligent designer; by the order inherent in the product's design. Order is how intelligence manifests itself. There can be no order without intelligence. Intelligence defines order. This is Absolute Truth.

Since the fundamental elements of the universe are ordered, there must have been an intelligent designer. This intelligent designer is whom we refer to as God. There can be no other.

If there were another initiating designer, the fundamental elements would not be of similar properties and structure. There could not be harmony as there would be fundamental differences preventing harmony. In the spiritual world, there could be no peace or joy since there would be constant interference from non-harmonious forces, possibly as powerful as the order we know. The peace and harmony we have come to know is from a singular source, God.

If you cannot define something, you cannot reproduce it. If you cannot reproduce it, order and hence life as we know it would be impossible. Only intelligence can define, and so only intelligence can create order and hence, life. This is not a random process. It is fundamental that a reference be set for any ordered system to be initiated. This reference cannot change. God does not change and so He is the perfect reference for the creation of order and life. There is only one Truth as there is only one spiritual reference and that reference does not change.

If you are in a material or spiritually ordered system and you lose your reference, you become lost. Once lost, one cannot recognize or locate the true reference from within the system. This is because you now have no true reference that you can use to navigate. Only some source, outside the system, who knows the true reference can direct you back to truth. This is why an ordered system can only be initiated by an intelligent source outside the system, using the reference that has not been contaminated with disorder (the True reference). The reference must be set and monitored from outside a system for it to stay true.

Perfect order is eternal. An ordered system only deteriorates if disorder is introduced into that system. Intelligence recognizes the difference between order and disorder and can choose either.

The entity with Absolute Intelligence has Absolute Knowledge and Control over all systems. We refer to this entity as God.

You may ask, what is an ordered system. Here are some examples:
- All living things
- All things man-made
- The things made by intelligent creatures
- An atom
- A molecule
- The Universe
- Anything of which we can make sense
- Anything that can be defined
- Anything that is or can be reproduced

What does The Eternal God say about Himself:

"I am God and there is no other"

"I am the same yesterday, today and forever"

"I am eternal"

The above are critical attributes for a creator. Only God defines spiritual and ethical order. Man is given authority to define material ordered systems.

Based on the above reasoning, we can clearly see that God has revealed to us the necessary clues for us to conclude that He is the One True God.

THE MYTH ABOUT EVOLUTION

The fingerprint of GOD is evident in the fabric of the universe. Here, I have discussed scientific evidence that shows, beyond a reasonable doubt, the existence of GOD.

Introduction

This book presents a 'singular' concept that is the fabric of the universe. 'Singular' meaning its origin is a single source. It manifests itself as a recurring theme throughout the universe as well as in our daily lives. It is so simple that we take it for granted, yet it represents a connection between matter and life. Both exhibit the same basic 'TRUTH' as to their origin, a truth that is reflected in everything we do, in the way we think, in fact in everything that may be defined as being ordered.

When you do an investigation of an incident such a robbery, you may not initially know who committed the crime. You must therefore start with the crime scene to give you clues as to what happened. The things you look for are fingerprints and now, DNA, some characteristics that are unique to a single individual. The 'MO' - modus operandi- mode of operation, is another of the ways of identifying the perpetrator. This shows the pattern in which different individuals do things in their unique styles. Also, you look for articles that can be traced back to a single source. If you can connect all the clues to a single person, then you will have found the perpetrator. We will now use this same reasoning to try to determine the origin of the universe.

All of nature follows fixed laws, the characteristics of which we have determined by careful observation and experimentation. We cannot change these laws so we have learned how to use them to our advantage by discovering ways to compensate for their effects, to achieve a specific outcome or to enhance an outcome.

Only ordered systems with fixed references can be used to describe or simulate the universe in which we live. This is an indication that the universe itself is ordered and has a fixed reference. By observation, we know that the universe is an ordered system. It is also evident that all references are connected to a singular source. In other words, we are all connected. This is evident in that we are fully aware of each other and can interact with each other and our environment.

Nature's laws include those relating to gravity and nuclear forces, thermodynamics and electromagnetism. We have even developed equations to define them mathematically and to determine the mathematical relationship between them, making it easier for us to use them in practical applications. These applications include research in physics, chemistry, biology and nuclear science and have resulted in all scientific inventions.

I think everyone should try to understand this concept. While reading this book, use your own reasoning, experience and judgment to make a determination as to whether you believe this to be 'TRUTH'.

If you are a believer (in GOD and JESUS CHRIST), this book should give you more confidence when sharing your convictions with others. If you are not, it should raise questions about evolution, if that is what you currently believe. If you are an atheist, it should give you some food for thought about the existence of the universe and how it came into being. If none of this matters to you, it should still be an interesting read.

Here, I have looked at a typical human lifetime and examined the human experience from birth. We all have the same experience of becoming aware of our existence and preparing ourselves for a lifetime in this world.

We are born with the tools necessary to interact, appreciate and learn about the world in which we live. We have a brain that makes us intelligent beings and five senses that allow us to interact with our environment. Our five senses provide information that is fed to the brain for interpretation. They are all monitored and controlled by one central controller, the brain. This was all given to us without any contribution on our part.

Have you ever wondered where you came from and the purpose of your being on this earth? I have thought about this all my life but without much success, until now. At this time in my life, I looked back to see if I could find a pattern or some indication as to my origin and purpose. I examined my life from my first memory through childhood, school, work and now retirement.

What were the most important experiences that shaped my life and made me into the individual that I am today? I believe everyone should ask themselves these questions. It would help you make informed decisions as to the direction you want to pursue in life.

At the end of your time here, will you be satisfied with the life you have lived? Is there anything you wanted to accomplish but did not? Do you still want to accomplish more?

After reading this book, you should have a better understanding and a rational interpretation of life's origin and the part each of us plays.

The Universe

Let us look at the universe and its infinite size as we observe it from planet Earth.

When you first looked up at the skies on a clear night and saw all the stars tracking in unison across the dark background, do you remember what thought came to mind? Mine was, "How insignificant I am in all this infinite expanse"! A time lapse gaze reveals a slow, choreographed movement with all the planets and stars moving in unison. Since it is the nearest, the moon appears to move against this backdrop, so massive, yet so silent.

Another significant feature of these objects is that they are all spherical in shape. This again manifests order and a common reference for all of them.

I now know that there were billions of stars in my gaze and understand the awesomeness of what I was viewing.

It was at this early age that I began to question, 'what is this all about'? One thing I did conclude is that it all manifested 'ORDER'. There is no chaos in all this giant expanse of large moving objects. It is not random because they follow cycles that have been repeated over billions of years and still continue to do so. At that time in my life I was only about 7 years old.

With this in mind, I think we can all agree that the planets and stars move in an ordered manner in space. Ordered, as opposed to random or disordered.

Now, let us look at planet Earth! As far as we know, it is the only planet, with these characteristics, in our galaxy or any galaxy we have been able to study. It has an abundance of plant and animal life that perfectly complement each other. We have a sun that makes life possible, both animal and plant. In fact, this is the only planet that we know of where life exists. This is an indication of how unique it is to have this combination existing in only one place, planet Earth.

Earth is in an orbit around the sun in a way that makes it ideal for the development and growth of life on our planet. It is as if its orbit was predetermined.

As I previously mentioned, the basic shape of a planet or star is spherical. If we examine its form, any given sphere has a fixed radius. The radius of a sphere is constant from the center, to any point on its surface. This is the simplest representation describing a three dimensional object and is, by far, the most common manifestation. The fixed reference is the center of the sphere and any straight line equal to the radius, from the center of the sphere, describes a point on its surface. It is also interesting to note that protons neutrons and electrons are described by the same formula, but at the nano level.

Life

Matter, as we know it in this universe, obeys the law of entropy, the Second Law of Thermodynamics. Entropy basically means that, if left alone, over time, matter will continue to deteriorate or lose energy. Things only improve if a source, with a positive influence, intervenes. This implies an ordered (nonrandom) source intervening. Entropy, on the other hand, tends toward randomness.

Because of entropy and decay, any man-made ordered system begins to deteriorate immediately after completion. Initially, the change is so gradual that we may not even notice it. However, it becomes noticeable over time. This change is the process of oxidation or decay and loss of energy or entropy. It will continue until a stable lower energy state is reached where the process becomes truly random. This is why we always need to maintain man-made ordered systems in order to extend their useful life.

'ORDER NEEDS CONSTANT MAINTENANCE'

Life is the only self-generating, self-preserving process that defies entropy. It takes disordered or random matter and converts it into ordered systems. Life is able to reproduce itself and has continued to do so over millions of years. This is a unique ability that is not consistent with the law of entropy that governs matter. Some even argue that life is not of this earth. One thing I think we all can agree on is that anything that has life is an ordered system.

My belief is that 'Any ordered system, including the universe, must have a fixed reference and be of intelligent design'. I challenge you to find an ordered system that does not have these essential components.

We have discovered that all living things, whether plant or animal, have a fixed design reference. This is the DNA of each species. Now, all we have to show is that an intelligent source created life.

Since we were not there to witness creation, we all have our own beliefs as to how the universe came into existence. We can, however, look at creation for evidence as to its origin since the 'fingerprint' or evidence of the characteristics of the creator is always present in any creation.

The first step is to look at all ordered systems to determine if we can find any that is not of an intelligent source. If we cannot find any such system, then we can reasonably conclude that the previous statement is true, that order is created by intelligence and therefore life is of intelligent design.

If we look at any ordered system that has been developed by human beings, we can conclude that, without exception, it was created by an intelligent designer. If we examine any other ordered system on our planet, we see that it was created by some intelligent source. A bee hive (honeycomb), a bird's nest and an ant hill. The organisms that make these all exhibit some form of intelligence.

Now, we have looked at the universe and concluded that it is ordered. Then, by this definition, it must have a fixed reference and be of intelligent design.

Physicists have shown us that our universe originated at a single, fixed point in time that they refer to as the 'Big Bang'. Since the Big Bang (the fixed reference) developed an ordered system, then it must have been the product of an intelligent source. Only intelligence can define and create order. The latter cannot be proven but can be shown beyond a reasonable doubt. Then, if this is true, there must be a GOD, the intelligent source.

Human Life

Human life begins with two living cells fusing, rapidly multiplying and growing. In a few weeks, the cells begin to specialize to form our various organs, our body frame, the shape of each bone, and everything else that makes us unique individuals. We have discovered that this process is at the instruction of our DNA, the blueprint of life. Genes further fine tune the instruction to include family similarities that are passed down from parent to child.

DNA is an intelligent structure of cells that even has the information on how to form and develop a brain which will perform arguably the most important function in our bodies. DNA is found in each cell of your body and helps control and coordinate its development and function. Since everyone's DNA is different, it can be used to identify each individual.

In the human body, throughout life, the DNA (and genes) continues to control cell generation and development. For each individual, all cells generated are uniquely yours. This is why your body operates as a unique unit. Each cell can be identified as yours and will reject any other.

For the body to operate as a unit (in harmony), there must be one control center, the brain. Through the central nervous system, it is connected to the organs and all other parts of the body. It is the fixed reference that the body needs to function normally and as one.

It cannot be overstated how finely tuned our bodies are to the extent that even a small variation from the norm upsets so many other functions within the chain of control. When the body operates normally, it is no less than a miracle. It is designed to self maintain, an ability that only makes sense if the designer saw the big picture, including possible environmental changes that could affect its normal function.

Before memory begins, we have developed no conscious reference and so it appears to us that life only started when conscious memory was 'switched on'. Our history becomes a series of events or memories starting from the first. The first memory is fixed and has to remain so in order for us to develop a knowledge sequence and always be aware of who we are relative to that first memory. This is our experience, which defines us. This includes our family, friends, likes and dislikes, and knowledge gained both at the conscious and subconscious levels. It is what makes us unique.

To help us navigate life, we are each given aptitudes in specific knowledge acquisition in which we excel without much effort. We sometimes use these as a guide in selecting our careers in life. These are referred to as 'gifts'. They may be a hobbies or personal interests but these all make us unique in a world with seven billion people. Each of us is one of a kind.

To define how we see ourselves, we use the reference fixed by our first memory followed by other memories in sequence (knowledge). This ability is also programmed into our offspring. This is only possible because we are intelligent beings. Intelligence gives us this ability.

'We Are' - Consciousness

At some stage in this development we realize that we exist. 'WE ARE'. This is at an early age, probably between two and four years old. This is when conscious memory begins. Before this age, we are consciously and subconsciously gathering information and gaining knowledge that will prepare us for our lives ahead. Even in the womb, as we are developing, we are gaining knowledge. This is the first indication that we are intelligent beings but still unaware that we exist.

In order to recognize our existence, we have to develop a fixed reference in time from which we start a sequence of memory events associated with the order of which we are now a part.

Then, it is as if a switch trips and we become conscious. We are now aware of our existence because we have a recorded reference on which we build with each experience. All of this is programmed or automatic. Up until this point, we have contributed nothing to the developmental process. The world is already ordered, a fact we will soon discover.

Plant Life

The entire cycle of the life of a plant involves growing from a seed to a mature plant, then producing fruit and more seeds that will eventually grow into other plants to keep the cycle going. Within this cycle they provide food for animals as well as other plants.

If we look at a plant, as it blossoms to bear fruit, there are usually numerous blossoms. Bees are attracted to cross-pollinate and fertilize them. Then there is a selection process. The plant selects the most healthy blossoms for continued development into fruit. It is clear that there must be a standard or threshold by which this is determined. This is preprogrammed and the plant knows which blossoms to reject based on the probability of survival through the developmental process. This is not a random process and indicates some form of intelligent design. If a fruit gets prematurely damaged, somehow the plant recognizes the problem and rejects it, if bad enough. Again, this indicates feedback and monitoring of the process to ensure healthy fruit. This is a programmed response which is so much more than a random process.

Earth

Earth is a living planet. If we look at the analogy between human and plant life, we see that water is the common element that supports and maintains life on earth. Just like blood and sap in animal and plant life, respectively, water continuously circulates on earth giving life to its inhabitants.

Rain and snow fall onto the earth's surface dissolving the nutrients in the soil so they can be absorbed by the plants. For better distribution of water, rivers and streams flow like arteries and veins supplying life giving water to all parts of the earth. Water evaporates from the surface of the earth and the ocean to form clouds which then precipitate as rain. The cycle then continues.

Keeping in mind that animals and plants consist mostly of water, when we look at the earth, most of its surface is covered with water. There is a consistency in the design of life and the life sustaining process which appears to indicate a singular source.

Anything that is not connected to the life giving source dies. We have to be in the cycle of renewal to stay alive. This is essential in all forms of life; animal, plant and our planet, Earth.

All this relates to life in the material world, but the same is true for the spiritual world. To gain eternal life we have to be connected to JESUS, the eternal life giving source.

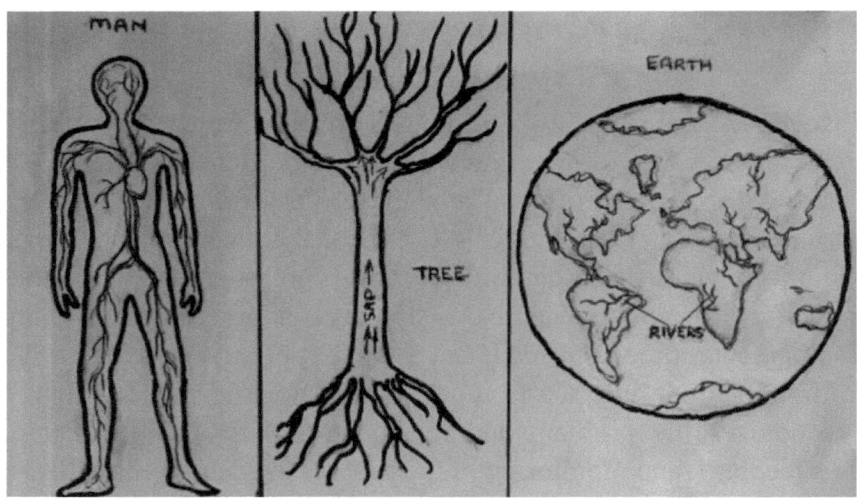

The diagrams demonstrate the similarity between the life sustaining source in the different forms of life, material and spiritual.

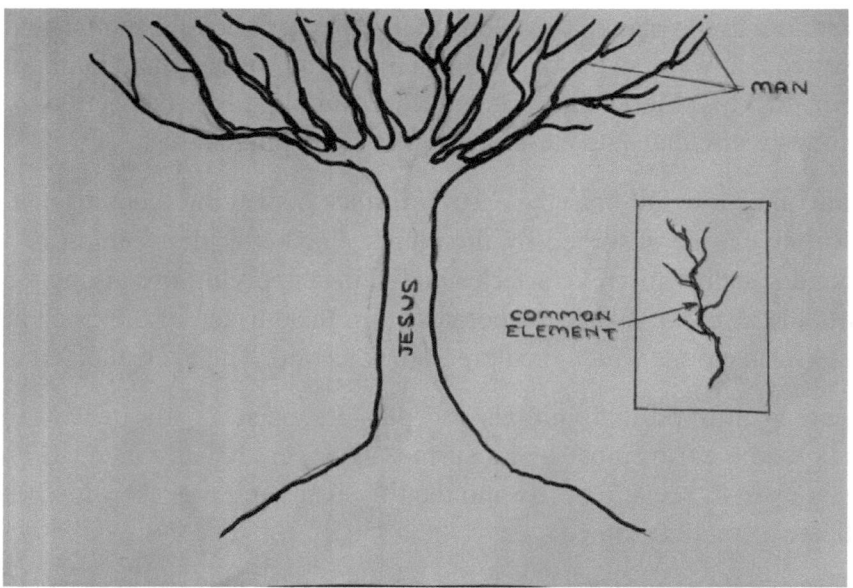

Nature's Cycles

Day and night, seasons and years are all examples of cycles. For every cycle, there must be a reference and this reference must not change otherwise behavior will be random and erratic.

We use day and night, weeks, months and years as references in time. We can only use them because they are constant and we can depend on them not to change. Wouldn't it be logical to assume that they all must have fixed references in order to remain constant over millions of years? By observation, we can conclude that this is the case.

Mathematics is used to simulate natural occurrences or events. We use this system to develop equations that closely simulate how nature operates within the confines of its laws. Mathematics uses waveforms to represent cycles.

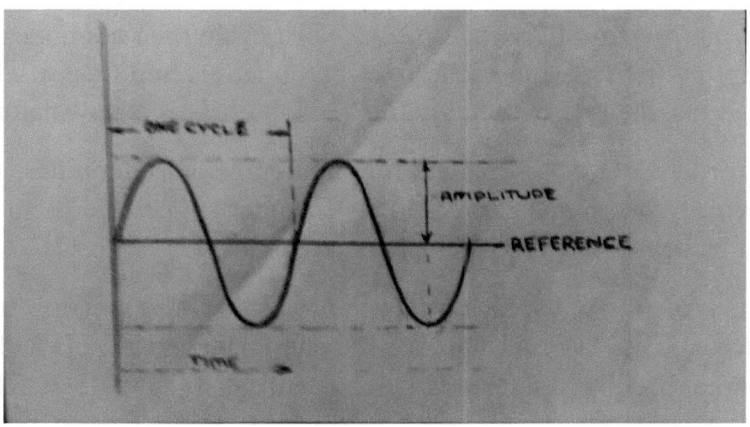

TYPICAL MATHEMATICAL WAVEFORM

Other examples of cycles include life and death and our daily activities. This even includes meal times. For a cycle to continue without change, it must have a fixed reference.

In the case of our solar system the sun is the reference. It influences all the planets that are in orbit around it as these are maintained in fixed orbits.

If we look at the big picture, we see that the universe is expanding. If there is a reference, there may be a point when it reaches maximum expansion and then begins to contract on itself.

But will it? If it was created from an infinitely intelligent mind there will probably be an intervention at some point. We have been told (in the BIBLE) that, this universe as we know it, was not designed to last forever.

Bible quote (The universe will be destroyed by GOD - Rev. 21:5, 2 Peter 3: 9-13)

At the nano-level, electrons orbit around the nucleus in fixed orbits. Other particles have been found in the nucleus which have specific functions controlling how the system works as a whole. However, the three most important components are the proton, neuron, and electron. Matter is consistent in nature in that it is made up mainly of these three particles. These are the building blocks of matter.

Life And Death

Life manifests itself in the cycle of life and death. Life first gives life to another living system after which it dies. The cycle then continues with the new life giving life to another, through reproduction, and then, itself, dying. Something must die in order to give life to another. This is how nature works.

Life is ordered. Order is created from randomness. Randomness is order's raw material. Death creates randomness. In order to recreate or sustain life, there must be death. The food we eat demonstrates this 'TRUTH'.

(This is probably why Jesus had to die for our sins to give us eternal life. Jesus is eternal. Only a GOD can give life eternal and so a GOD had to die to give us eternal life. Only such a sacrifice could satisfy GOD'S JUSTICE)

Bible quote -(1 Peter 2:24 ESV)

Anatomy And Physiology

The human body is the most complex living system know to man. It is a complete system with multiple subsystems that function as one. In order to function as one unit, each subsystem is connected to the central nervous system controlled by the brain.

Our body was perfectly designed. If we were to test a human body that was functioning to design specification, all systems operating within the design tolerances, it would operate perfectly.

Our body incorporates electrical, chemical, mechanical and mental systems all coordinated from a single point, the brain. Each system or subsystem is unique to an individual's body and does not function normally in anyone else's body. This is because the systems, in an individual human body, are custom designed for that particular body. The blueprint is the DNA.

A baby developing in a mother's womb has a separate blood circulatory system from that of the mother. This ensures that both human systems, although joined, operate independently. One reason being that they may have different blood types and there can be no cross contamination. In the same way, organ transplant patients must get anti-rejection medicine to prevent the body from rejecting the foreign organ.

DNA

When we look closely at our planet, we find order is being created from what initially appears to be randomly distributed elements. The soil which contains all the nutrients necessary for life is basically inert. These elements are continually being converted to order (life) and decay (entropy). From the elements of the earth, man is born, grows and reproduces.

DNA is the blueprint of all living things. It contains all the information (intelligence) needed to mold the elements into a tree, a flower, an animal, an insect and a human being. Such intelligence cannot be of the soil but from some external source. This is because soil, by itself, is inert. The DNA cannot develop without a reference because a reference is necessary to develop order from randomness.

Everything we are and will be is programmed into our DNA. It could not have happened by chance or randomly, as suggested by evolution. The probability of that happening is zero. The entire theory of evolution implies randomness or chance selection which is at odds with the order seen in ourselves and in every living thing around us.

DNA is designed to conjugate only with that of its own kind. It rejects all others. This is contrary to what would be expected from random behavior. In other words, DNA is consistent and is highly selective in its reproductive characteristics. Since DNA is so selective, it does not lend itself to random change, let alone a systemic change from a monkey to a human being which would suggest cumulative positive changes.

Although similar, the designs are uniquely different between these two DNAs. Any similarities only indicate a common designer (creator).

The only mutations we see in DNA are degenerative in nature resulting in malfunction, thus indicating damage to its structure. This may manifest itself in physical deformity or breakdown of the body's immune system, both of which are negative in nature. These occur because the DNA (or genes) is now defective and not functioning as designed.

We typically take our body function for granted without considering how finely tuned the body's design. The human body is a complex integrated system. Its organs work together as a unit with each performing its programmed function. It needs all organs working in harmony, performing their specific functions, for the whole to be efficient.

The body knows when something is wrong and immediately tries to correct the defect to compensate for any adverse effects. This is automatic and indicates intelligence.

If we get an infection, the body immediately sends antibodies to the site to fight it and, in most cases, is successful. This is a continuous process in our bodies of which, most of the time, we are unaware.

It would have been miraculous if life had developed from a random process, but even more miraculous if something as complex as the human body was constructed randomly or by natural selection. Without intelligence, how would the body know that the design of its organs was now optimal and changes should stop? How did it know that a convex lens in our eyes produces an inverted image and so the brain needed to invert it for the image to be accurate? How would it have known that the tissue of the lens needed to be transparent and invisible to us? We know that anything other than transparent are defects, such as cataracts.

Even the process of growth is highly complex as the entire body has to grow in harmony. How many attempts did it take before it recognized that this harmony was necessary and how could it tell subsequent reproductions to make this change?

How did it evolve a brain to control all body functions correctly without knowing the entire design and processes of the body beforehand? It even developed a skull to protect the brain realizing that it was a very delicate structure that needed to be protected. Was this also an evolutionary process?

Why can't we find a skeleton without the skull if the body with a skull evolved and survived by natural selection?

Which came first, the chicken or the egg? Which is more confusing? Is it if we say the chicken evolved from the egg or was it the other way around? If we consider it to be some external intelligence that created it, there would be no confusion.

The egg contains all the information needed to make a chicken. It is an ordered system and so could not have developed randomly. The entire egg would have had to be fully designed before it could be closed in the shell, with no other changes needed. It would have had to have known that the information needed to make a chicken was now all there. There had to have been a reference and external intelligence. This could not have been random.

All of these questions point to an external intelligent source that designed and created living things. They could not design and build themselves because there is never enough information or knowledge at any stage of development to make these positive changes and fix them in future reproductions. The process of evolution would have needed to know the 'big picture', what was right and what was wrong and have the ability to select the sequence of positive changes. THIS TAKES INTELLIGENCE! INDEPENDENT INTELLIGENCE!

DNA not only designs the body but instructs it throughout its life on how to grow and reproduce. What degree of intelligence does that require? Certainly not some random process. This takes complete knowledge and understanding of the raw materials and how to use them. The raw materials being the dust or elements of the earth- the food we eat. It seems apparent that all of this was designed and created by an external intelligence. Not only did this source create it all but it was planned down to the very last detail.

Life is a continuous process. There is no time to stop and change design. All changes have to be made while the process is ongoing. It is much more difficult to try to make changes under dynamic conditions. It is much easier to stop and make the change and then restart. Life does not have this luxury and so it would seem more rational or logical that the basic design was complete before it was initiated. Any changes resulting from environmental interaction were also built into the original design (adaptation).

The Eye

When you look at one feature of the human body, the eye, do you recognize the degree of intelligence it requires to create such an organ?

We live in a three dimensional world as far as matter is concerned. (Time would be the fourth dimension but, for what we are considering, the first three are adequate). We have two eyes which are ideal for three dimensional viewing. If the number of eyes we have was randomly selected, we could have had one or three eyes or more. It would be viable for us to have any number of eyes. Remember there is no fixed reference, so there could be any number of eyes. But humans have only two eyes and have always had only two eyes. If this were not the case, we would have found human skulls with one or multiple eye sockets, not just two. This indicates that the creation of eyes was by design. This is the only logical conclusion.

Darwin developed the theory of evolution after observing differences or mutations in species, which resulted in their selective survival, in a changing environment. This is adaptation, which is built into the DNA and genes because the designer was aware of these possible environmental challenges giving the genes the ability to make these adjustments.

How can one explain life, let alone life responding to light, to form an eye. Then realizing we need two to be able to judge distance or increase viewing angle. There was no trial and error as we see no evidence that this was the case. All human skulls show that we have always had two eyes and two ears, located at the most ideal places on our bodies. This is also true for other mammals, birds and fish.

Evolution, by its current definition, is a random process. Only the end result is selective. From a random process one cannot expect to develop an ordered system. Even if nonrandom, this does not translate into being ordered because, in this case, nonrandom defines the ultimate result and not the process itself. Only the resulting elimination or survival of a species is nonrandom (this is more fully explained under the section relating to 'EVOLUTION').

The Blood

Apart from the brain, which is the control center of the body, the most important organ is the heart. The heart is responsible for pumping blood around the body which is how life is sustained. The blood carries oxygen and food to all cells in the body. This is essential to life. It also carries the antibodies that fight foreign matter that attack the body. It keeps all parts of the body alive and healthy, including the brain.

Man has always known the importance of blood to the body. This has given rise to superstitious beliefs that blood has some innate spiritual or supernatural power. Some tribes drank blood and sacrificed their people to appease the gods so that they may gain favor with them. They realized that blood is precious and essential to life.

As indicated above, blood flows continuously through the body as it carries precious food and oxygen to each cell in the body. The sap in plants perform the same function as blood in an animal.

The function of the blood and sap demonstrate that the process of life needs continuous renewal to grow and maintain itself. It is constantly fighting against entropy and cannot afford to put its guard down.

Blood is pumped by the heart but the heart needs blood to survive. The brain also needs blood to survive. The most important organs and all others need blood to survive. This means that blood is the source of life. It needs to be circulated continuously to sustain life. In a plant, sap needs to be circulated continuously to sustain the plant. Earth needs water circulating continuously to sustain life on earth. Do you see a theme developing here? Life needs continuous sustenance otherwise it dies. This is life as we know it on earth. It is a continuously maintained, ordered process. There is a fundamental interdependence of all living elements in the loop.

Death is an integral part of the life process. At a certain point our defense system begins to break down. This is a function of age. Death of natural causes will occur at some point in our lives. The average age is between 70 and 80 years. However, there are diseases that are capable of fatal attacks on the human body resulting in premature death.

(Jesus died on the cross giving his precious blood so that we might live)

Bible quote - (Mark 15:25)

Eventually life will fail, because death is inevitable, but the reproductive process produces new life and the cycle continues.

Design of the Human Body

When you look at the human body, do you think you could improve on its design? I would imagine that there are many people who would say they could. Let us look at examples of possible improvements.

The Hands

Most of what we have to do with our hands requires only one. Two hands are the optimal number for lifting heavy weights and for balance. But then, if there were three hands, where would you attach the third? If it were placed next to an existing hand it would be redundant. If placed in the middle of the chest it would be in the way.

When we look at the design of the body, as a whole, it is the optimal combination and location of all the body parts. Now, should we be led to believe that the design came about by evolution without the designer knowing and understanding the intended ultimate function.

The Heart

We may also say we should have two hearts, since the heart is so important. But one is enough if we take into consideration that it is protected by the rib cage and is designed to last the lifetime of the human body. If there were two hearts the designer would have had to take into consideration blood flow issues which becomes more complicated with two hearts. The fact is, the body's design is adequate, efficient and optimal for its purpose.

Medicine

In medicine, the reference is the healthy human being. One of the first things a doctor does with patients is to take their temperature. For a healthy person, the temperature should be about 98.6 degrees Fahrenheit. If it is too high, the person has a fever. That means that something is wrong. If too low, that is also a problem and has to be corrected. We should keep in mind that 98.6 degrees Fahrenheit is the average acceptable normal temperature but small variations above or below this temperature are considered within tolerance.

What we have learned is that the human body operates within certain parameters. If it goes outside these parameters we know something is wrong, we are ill.

Other parameters include blood pressure, blood composition, urine composition, skin color and texture, body weight etc. As we gain more knowledge about the human body, we develop a more comprehensive list of parameters that are the norm. Any abnormal variation needs to be addressed.

Since the human body is an ordered system, there are certain norms that we look for in a healthy body. These remain fixed and so are the fixed references.

Having reviewed the order we see in nature and the universe, we will now look at evolution to see if its theories are compatible with the principles of order.

Evolution

I will define the evolution of man using the two interpretations or theories with which most of us are familiar. Evolution, in general, also follows the same principles.

1. Darwin - The changes, over time, in species as a result of genetic differences giving one an advantage over the other in a changing environment, with the eventual disappearance of the weaker or disadvantaged species. This change may be random or nonrandom.

2. Primordial - The random development of the life process, from what we call 'primordial matter', to life as we know it in its current manifestation.

Note: Neither of the above theories suggests how life began. Life can only develop from something already living.

Interpretation 1. - Darwin's theory of evolution.

Darwin seems to suggest that DNA gets more complex with time, with no explanation as to the mechanism of this change. In the process of NATURAL SELECTION, as Darwin postulates, the inferior or disadvantaged species eventually die out as a result of adverse environmental conditions which they are unable to overcome, whereas the more resilient species are able to survive and multiply. This would also imply that the change is cumulatively positive if we look at fossils of lifeforms connecting our human species to its current form. Species are defined by DNA and so, when species die out, they become extinct.

Interpretation 2. -The primordial theory.

The primordial theory is even less plausible since it not only does not explain the mechanism of the change but also suggests that there were billions of cumulative positive changes that seem to randomly occur.

Let us first look at the nonrandom process of evolution.

Nonrandom Evolution

We know, by observation, that life is not a random process. It is the result of an ordered process controlled by the DNA of each living system. We also know that healthy DNA resists change.

Nonrandom evolution may be said to be due to changing environmental conditions resulting in the elimination of some species and the survival of others based on DNA or genetic differences, giving one an advantage over the other. This would be considered a lateral change as this is only about survival and not necessarily considered an improvement. It is, however, selective, hence nonrandom. This is the only evidence of nonrandom change, as the mechanism of the internal change would be considered random, based on the Darwinian or the primordial theory, since the basic change mechanism is not defined and there is no explanation as to how or why it occurred.

The diversity we see among the species indicates change based on genetic differences. If genetic changes occur first and then the environmental changes follow, this may result in mutation or elimination of the weaker species.

But what would initiate a single change in the DNA or genes of a living organism, much less several positive changes, to develop a significantly superior species?

For this to occur, there would have to be an intelligent source setting or recognizing the fixed reference (DNA) and influencing these changes, in a sequential manner, in harmony with the reference (DNA). This cannot be a random change mechanism because it results in cumulative positive changes. Time also does not explain this change as the longer the period over which this takes place, the higher would be the probability of a truly random outcome. Over time, a random process does not improve on itself and, the more time elapsed, the more the process would tend towards truly random.

The believed evolutionary change from a gorilla to a human being could not have occurred without an intelligent source influencing the change, using the fixed reference (DNA) of that species as a guide in order to ensure a cumulative, positive outcome. The only connection between the gorilla and the human being is that they both have the same designer. They are actually two separate designs by the same creator.

Throughout this entire process, intelligence, whether built in or with direct intervention by an external intelligent source, must have been involved.

If the initial reference is not actively included, there is no guarantee that changes will be compatible with the system or go in a positive direction. Intelligence is built into DNA, but how did it get there?

How would the process have known when the goal had been met and no other systemic changes were necessary? Human DNA appears to have known that what it built was good and no further changes in the structure were necessary. Healthy DNA only allows changes compatible with its design and resists other changes. This shows a form of intelligence.

If this intelligence did not originate in the structure, it must have come from an external source. As far as we know, matter cannot develop intelligence by itself and so we can conclude that this intelligence is from some external source.

The original life process could not have been random because it developed an ordered system. Randomness can only develop order if instructed to do so by an intelligent source. Evolution would have to be an ordered process to develop or complement an ordered system.

Now, an ordered process must be of an intelligent source. Intelligence defines order. Evolution, as it manifests itself, must therefore be of an intelligent source. It cannot be defined any other way as it enhances an already ordered system.

However, neither of the above theories includes intelligence as the guide to the development of the evolutionary process. Nonrandom evolution, as it is currently defined, has no intelligence, it occurs by 'natural selection' of the species with the advantage in a given environment. It cannot reason. It is, therefore, not feasible for such a process to successfully initiate or continue to enhance an already ordered system.

Evolution, as currently defined, cannot explain how life was initiated. It also cannot explain the cumulative positive changes in the gorilla's DNA to become human unless it was guided by intelligence.

On closer examination, we would conclude that the amount of error associated with the design and creation of the human body is infinitesimally small or nonexistent. There are checks and balances inherent in its design to ensure minimal error. Where error is found, it is a result of an adverse interaction with the external environment and possibly from the system itself as a result of 'GOD'S CURSE' that has affected ordered matter, in various ways (see Bible quote below). One should also keep in mind that there is one reference in the Bible were JESUS said that a man was born blind, not because of SIN (or the CURSE) but because GOD wanted to demonstrate his power over sickness by JESUS performing a miracle to make the man regain his sight (see Bible quote below). We were originally, perfectly designed, but we now live in an imperfect world. Our design reflects an intelligent creator.

If we look at genetic defects such as Down syndrome, this occurs when an abnormal cell division causes extra genetic material from chromosome 21 to be formed. This is not how it was originally designed, but there has been some form of malfunction which results in extra genetic material being generated.

Based on the above reasoning, defects may either be from adverse interaction with the environment or a defect of the system as a result of the general effect of SIN on matter in an ordered system. Matter still conforms to the laws of nature, but SIN has not only affected our mind but also matter based on its effect on an ordered system.

Bible quote (Genesis 3 - GOD curses creation)

Bible quote (John 9:5- Jesus heals the blind man)

Random Evolution

Here, it is also implied that changing environmental conditions result in the elimination of certain species based on DNA or genetic differences making one more vulnerable than the others. If we consider this, it is also selective, as the weaker or disadvantaged species is eliminated. Only the internal change process could be considered random and so the same argument applies as was discussed in nonrandom evolution.

But again, what would initiate these internal changes? If these changes are randomly initiated there is an equal probability of each resulting in a positive or negative outcome. In the long term, things would not improve or result in a cumulatively positive outcome. Species would therefore not improve from their initial form. Also, as the system gets more complex, the probability of getting a positive outcome is continually being reduced as there is now a decreasing chance of getting a positive outcome. The reason is that the process would now be getting more selective.

The only way to ensure cumulative positive outcomes is if there is an intelligent source in coordination with the fixed reference (DNA), of that species, influencing the changes. Otherwise, changes may be detached and not compatible with the rest of the system. Changes must be overseen by an intelligent source to ensure that they are complementary. This could not be a random process.

Entropy works against life. Yet, life is designed to combat most of the day to day encounters that threaten it. These defenses are built into the system. This could not be a random process because life has a systematic way of dealing with environmental or internal attacks. It appears to develop a logical and rational line of defense, not a random one.

Based on what we see in the design of the human body, the designer would have to have had a plan. Only a source with intelligence can create a plan. After developing the plan, it had to have been executed in a step by step manner until completion. Each step would have to have been precisely followed.

When we, as humans, design and make things, we develop a plan and must use some form of reference so that we know where to start and the direction in which we should go. We have learned, from experience, that this is the only way this can be successfully done.

We are using the same building blocks, in the things we design and make, as those used in nature, as these are all we have available to us. We, therefore, must obey the same laws that govern matter. Nature and man must then follow the same fixed guidelines to achieve the same results (an ordered system).

To begin, there has to be a fixed reference, with all other references defined by the initial reference. This is the only way to coordinate the points or components in an ordered system with multiple subsystems. It takes knowledge of the complete system as well as sequential execution of the plan to accomplish this task. This could only be done by an intelligent designer.

A random process is the opposite of an ordered process, with a diminishing chance of a continually positive outcome. A purely random process cannot improve on itself. If it is biased towards the positive then it cannot be defined as being random.

Every ordered system has a fixed reference and follows a specific sequence of development. In addition to sequence, the designer must have some concept of the end product. Otherwise, how could it know when the system was complete and operating as designed? In other words, it must have intelligence. This principle is universal in its application. This is evident in creation and in the way we function, being made in the image of GOD.

Bible quote man made in GOD'S image (Genesis 1:26-28, Solomon 2:23)

In every material ordered system, there is a degree of tolerance, with extremes on either side of the normal distribution (Bell curve - tolerance is the allowable amount of error). This measuring standard was set by us as we have found that many natural occurrences follow this pattern. The extremes of the distribution, although not normal, are considered natural. An example of this is autism, where, although brain function is not normal in a person with autism, with proper treatment, the brain can be made to perform within normal parameters. This indicates that we are designed to function within certain parameters (ordered) and outside of this we do not operate in harmony with each other.

There is no evidence that the process of evolution, as we understand it, recognizes, much less uses a fixed reference and so it would be unable to create an ordered system.

Evolution can either be influenced by environmental (external) changes or internal changes. Evolution is 'LOST' without a fixed reference. It appears to have random references or no reference at all. Without a fixed reference, how is it able to know which direction to go? How can it, therefore, design a complex structure like the human body. Such a system requires the coordinated sequence of complex, positive changes which can only be accomplished by an external intelligent source using a fixed reference.

In the case of the human body, it would take infinite intelligence to design, create and incorporate self-maintenance, as well as the ability to reproduce itself.

If we look at our daily lives, our ultimate intent is to create order. Disorder and randomness are nonproductive. In order to accomplish a goal, we first need to have a plan as to what we want to accomplish. The next step is to determine the starting point, then take sequential positive steps until completion. This is not a random process and can only be accomplished by an intelligent mind.

How can someone examine the human body and conclude that it was put together by a process that was unaware of the end product but ended up with something that, even with our intelligence, is far beyond anything we could have conceived much less created. There is also the fact that it can reproduce itself.

Any ordered system that we encounter or create is planned and then executed. This is not in line with the principles of evolution as we know it. In fact, evolution is the only process that we know which is said to have developed a complex ordered system with no plan and no concept of its intent.

How could we expect that evolution could start from a primitive living organism (primordial) and eventually change into a human being, by a random or nonrandom process without intelligence. All odds are against this type of change as entropy is one of the main processes that would prevent this from happening.

This would mean understanding the laws of nature that govern matter. Evolution is not rational or logical. Some outside intelligent source must have intervened.

To create a human body, it would take billions of sequential, cumulative, positive developments which could not result from a random process. It would take a positively biased process with intelligence to accomplish such a feat. Evolution, as currently defined, exhibits none of these characteristics.

To believe that life evolved by the process of natural selection does not explain what we see in nature today. What we see are the completed products of an ordered developmental process (ordered systems). Natural selection is only the process of elimination. What we see are ordered systems both in life and in the manifestation of the universe. No random or nonrandom selection process could have developed either of these systems. It took infinite imagination with a detailed plan, executed in flawless sequence, monitored at every phase, and still is, to the most intimate degree.

This takes infinite intelligence and infinite knowledge that could only be accomplished by a GOD.

Creation was planned for man.

Bible quote -(Jeremiah 29:11, Ephesians 2:10, Philippians 1:6)

A Bird's Flight

Evolution implies that birds evolved wings that gave them the ability to fly. This conclusion does not make sense because, for this to happen, there had to have been an intentional effort with successful positive outcomes to develop a viable aerodynamic design.

Firstly, why was it necessary for any living thing to fly? There is no evidence that this was necessary and so it must have developed randomly as the process of evolution would probably suggest.

Now, if this were a random process, over a period of time, for every positive development there would be a negative, and the longer the period, the closer the process would approximate to truly random- say millions of years. At any point in time during its development, there would only be a few positive feasible events and billions of negative events that could halt the process. The process would have to have been guided by a plan as to which direction to go and what sequential steps to take to attain flight.

Some of the things that would have to have been successfully worked out would have been the following:

- The need for wings
- The wing design
- Location of wings
- Balance of the two opposing wings
- The design of the bones in the wings
- Where the joints needed to be
- What motion would be needed for flight
- The location and strength of the muscles
- The need for feathers

The design of feathers

The material needed for the feathers

The size and shape of the body relative to the wings

(Wings are so unique we copied the aerodynamic features of the design for our own airplanes.) We could not improve on them.

These are just a few of the obvious things that would have to be considered for a successful flight. If one of these was wrong or in the wrong sequence, flight would not be possible. Even after evolution got everything right, the bird would still then have to learn to fly and communicate this information to its offspring.

If we look at birds flying thousands of miles and finding their way to the same destination year after year, how do they manage such navigation? Even if we do not find out exactly how they do it, we know that they must have some form of a fixed reference. There has to be something that remains constant that they can use as a reliable guide.

What I am saying about evolution is that it got all these things right by trial and error. This is not possible unless the end product is known and there is an intelligent mind to guide the process. Evolution is blind without a reference. It could not have accomplished such a feat.

Below, I have discussed a possible explanation for the changes we see in fossils of humanlike bones and those of other species, as the species evolved.

Alternate explanation of Evolving humanlike and other Fossils GOD could have set creation in motion so that all the evolutionary changes would occur in sequence, over time. Like the development from a child to an adult and then old age. Sudden and dramatic like those changes at puberty and a chrysalis turning into a butterfly.

From 'homo erectus' to 'homo sapiens' as well as all the previous evolutionary changes associated with human beings. These could have all been programmed. One certainty is that these changes had to have been guided by an intelligent designer.

Initially, all that He needed to do was to create the basic DNA structures for each living thing and program changes to occur after predetermined periods of time, to adapt to changes in the environment. The basic structures would include man, animals, fish, birds, reptiles, plant life and micro-organisms. Programmed changes would create new species, with some becoming extinct over time and others multiplying and continuing to survive to what we see today.

There was one creation but it is probably much more complex than how it initially appears. GOD could have programmed man's DNA to change at specific points in time, to increase the capacity for the acceptance of more sophisticated knowledge processing and to change in appearance to what we see today.

This would mean that creation is much more complex and dynamic than what first meets the eye. This would present a much more complex picture of GOD and His power to create and maintain an ordered universe, but all within his capabilities. He is 'ALL POWERFUL' and 'ALL KNOWING'.

But, this should be apparent when we examine creation. The changes in a human body from birth to maturity and then old age. These changes are all programmed to occur at set times. Eventually, life ends in death.

GOD could have planned that at certain thresholds, systemic changes would occur in the structure of the DNA, so that a new and improved form of life was established. This would continue for a specific season, after which another change would occur to further improve that life form. Still, there may be some that are not programmed to significantly change such as the cockroach.

The average human life time is about 70 years. But, if we look at the creation of the universe, its life time may have been programmed to be billions or trillions of years. What we see could be the unfolding exactly as GOD planned. All of which appear to be evolutionary changes could be programmed changes like the chrysalis to the butterfly, the egg to the chicken, and the human body transformation at puberty. The cycle, however, is much longer.

GOD is capable of all this and much more. We cannot even begin to imagine His power and majesty. Anything you can imagine cannot even begin to describe our GOD. Our imagination does not have the scope to comprehend the power and intelligence of GOD.

We see these patterns everyday in our lives and in everything around us. This is all in line with GOD'S power and capabilities.

Adaptation

Adaptation is built into an already ordered system to adjust for environmental changes that threaten the species. It was designed by an intelligent source that anticipated these changes because the source could see the 'big picture'. This is another manifestation of planning.

Adaptation is more in line with the small changes we see in living organisms to compensate for environmental changes that would otherwise threaten their existence. This has been mistakenly extrapolated to represent the systemic changes that are necessary to develop a living ordered system.

Adaptation may be feasible for small changes but not feasible for systemic, positive changes while still maintaining an ordered system. Even so, it is of an external intelligent designer and is built into the system design.

Adaptation did not create order. Evolution did not create order. Intelligence defines and so must have created order.

Daily Human Activities

If we look at our daily lives, our ultimate intent is always to create order. In order to accomplish a goal, we first need to have a plan of what we want to accomplish. The first step is to determine the starting point, then take sequential positive steps until completion.

Your daily activities start with getting out of bed. First, we have to recognize our position- that we are in bed, in a horizontal position. We have to be aware of the location of our feet and where they are positioned. This seems very basic but what we are doing is using references, being always aware of our current position and what next to do. This is all automatic. We know what muscles to move and in what sequence to move them in order to get to the upright and then standing position. What we do next will depend on our subsequent plan or intent.

We are continually aware of where we were, where we are and where we want to be. Only an intelligent mind with memory can recognize and understand or interpret this sequence. If we are unable to do this, we are lost.

If we look at an object, we are only able to recognize it because we have a prerecorded reference. Our brain works in an ordered manner with a series of references to which it has multiple or what some might call, random access.

When we have a thought, it may have been initiated by something we saw, heard, felt, tasted or smelled. It may also have been a thought with an origin we are unable to specify. In any case, the thought must have a reference. Once that reference has been recognized the brain makes the association with the connecting information already recorded. Otherwise we are unable to interpret the thought or act on it.

Reproduction

The human body can only reproduce itself. It was designed using the DNA blueprint which was specifically designed for humans. Similarly, DNA for other living things can only reproduce themselves. This it true for all species.

In the same way, when the architect designs a building and draws the plans, it is only for that or buildings of the same design. Another plan has to be drawn for a building with a different design. If it is subsequently modified, this modification has to be done by an intelligent mind using the initial blueprint (reference) as a guide.

An analogy to this would be annual improvements made in the design of an automobile. Each year the model changes and improvements are made. The improvements do not occur by themselves. Design improvements are made by the design engineer, the intelligent designer. Automobiles evolve but only at the instruction of the intelligent designer.

The reproductive process follows the instructions of the DNA of the species. It does not randomly change.

Parent And Child

There are unique similarities between parent and child. These similarities include both physical and mental traits. In some cases, the visible resemblance is so striking that it almost guarantees a close relationship between the two subjects. This similarity is also sometimes evident in the extended family.

This is an example of how nature works. It is how things manifest themselves indicating close relations and connections, based on these similarities.

The only evidence we have to work with is nature, the universe and some understanding of how it all works. Now, wouldn't it be rational to conclude that, what we observe in nature shows evidence as to its origin. This means that there are traits in nature and the universe that manifest their origin. I think the most fundamental of these is basic 'ORDER'. Before these traits can be reproduced, there has to be an ordered transfer of information.

In electrical waveform amplification (an audio amplifier) the goal is to amplify the signal without introducing distortion. The amplifying circuit is carefully designed to accomplish this goal. But, even though this is the goal, there is always some evidence, in the amplified output, of the characteristics of the amplifying circuit. In this case, it manifests itself as distortion or some unique characteristics of the parent circuit that are evident in the output, even to a very small extent.

The laws of nature are also similarly designed to reproduce and maintain a specific theme or themes, but leaving some indication or evidence as to their origin. The product is intentionally designed for a specific purpose but contains evidence of its source or origin.

An example of this is a product made on a lathe. If we look at the macro structure of the surface, we see the grooves made by the cutting tool in the form of a close spiral on the finished part. This indicates the process used for making the product was 'turning', on a lathe.

ORDERED SYSTEMS FOUND ON EARTH

We will now examine a number of ordered systems that man has developed as well as ordered systems of other intelligent sources. To reinforce this theory, each of these systems must have a fixed reference.

Here are some examples of ordered systems of intelligent sources other than human:

> Birds flying in formation
>
> A school of fish swimming in unison
>
> Ants on a mission
>
> A swarm of bees in search of a new site to build a new hive
>
> Bees in the process of building a hive (honeycomb)
>
> Animal pacts living and hunting together

They all work together as a team, as one. This is order. It demonstrates intelligence. When we look at our world, we will see how references apply in every aspect of our lives.

Firstly, we must remember that to be ordered, events must occur in a set sequence. To set the reference and monitor the sequence, there has to be intelligence.

A Bird's Nest

Let us start with a bird's nest. The end product is a system that will offer protection and a home for its offspring. The bird starts by gathering twigs and leaves, beginning at the base and interweaving them to form the sides until it is complete. The twigs are interwoven to form a circular interior, starting from a single point and connecting the whole together.

All birds have a similar plan which they execute in sequence. The nest must be of adequate size to hold the eggs and the mother bird while she sits on them until they hatch. She must take into account the number of eggs that she may lay and the maximum size of the birds that will occupy the nest, before they leave. The nest is necessary to keep the eggs together in a safe place for them to hatch and the young birds grow, until they are ready to take flight and begin life on their own.

The nests of each species are consistently similar. In some cases it is possible to look at a nest and identify the species of bird that made it. Each nest is made from an instinctual plan and executed sequentially from start to finish.

Such a system requires a fixed reference because there are certain critical features such as the shape and the size that need to be consistent. The bird must decide where the base will be and where the sides will begin to determine the internal dimensions of the nest. It must therefore have a fixed reference from which to construct these dimensions relative to each other. In order to construct such a system, the bird must have the degree of intelligence, first to select a suitable spot for the nest, find the twigs and leaves, which will be the raw materials to build the nest, and then begin the process of building. It must also know when it is complete and ready to provide a suitable home for all the eggs.

The interior of the nest is basically circular. If we look at the geometry of a circle, we see that in order to make a circle, we have to have a fixed reference which is the center of the circle. The distance from the center to the perimeter is the radius and this is constant for any given circle. The bird has a mental image of the center and the radius or diameter and constructs the circle with these kept constant or relatively constant, as there are allowable tolerances for such a structure. In other words, some error is allowed.

It takes intelligence to construct a nest. It also takes considerable skill to weave such a structure. If you try to make such a structure you would find it very challenging and you may never get a finished product of the quality and structural integrity of a bird's nest. Yet, we have hands with opposable thumbs and a much larger and complex brain.

If you saw a bird's nest by itself and you did not know its connection to a bird, would you think that it was constructed by some form of intelligence? If we consider it an ordered system, then by definition, it would have to be of an intelligent source. By observation, we know it is an ordered system and by looking at the design, we see that it has a fixed reference.

A Bee Hive (Honeycomb)

Let us look at another ordered system- a honey comb. If one saw a honeycomb in a tree and did not see the bees that made it, we would believe that it was of an intelligent or ordered source. It takes intelligence, even at this level to put such a structure together.

Now, look at it from the bees' standpoint. They are given the task of making a symmetrical structure from a single material, which they produce, in order to provide a suitable habitat for their offspring and to store honey. In order to make a honeycomb, they have to start somewhere. This is the reference. Once they set this reference they must stick to it in order that each cell is in geometric symmetry with the adjacent cell.

There are many bees participating in this project and so the reference information has to be accurately communicated to all of them and they cannot afford to make a mistake. They have to work as a team. If they changed this reference while in the middle of building, the hive would be deformed and would not come together properly. It takes just one single change in the reference for this to occur and so, all subsequent cells, after the first, must be made with reference to the first. This is only a simple example but the principle is critical in the development of any ordered system.

Birds Flying In Formation

For bird's flying in formation, the lead bird is the reference. For bird's traveling long distances, the lead is rotated as this position requires significantly more energy exertion, whereas the other birds fly in the 'slip stream' made by the lead bird. Flying in the slip stream requires less exertion, making longer flights possible. The lead is rotated so that the lead bird gets to rest. When we observe this formation, for geese, this is a perfect 'V' indicating it is ordered, thus exhibiting intelligence on their part.

These are a few examples of intelligent sources, other than human, creating ordered systems.

Now we will look at the human being and how we have developed ordered systems for communicating, teaching, manufacturing and navigating, as well as some other familiar ordered systems we have constructed.

Any ordered system we have developed, in every case, it was done to simulate, describe or communicate aspects of the ordered system of which we are a part.

Remember, if we now look objectively at the big picture, in our effort to try to determine creation's origin, we must look at it like a detective examining a crime scene. We look for 'fingerprints' or patterns in creation that point to the character of the creator.

I will now review several disciplines, with which most of us are familiar, to highlight each of their fundamental components. The theme that will become apparent is that each, fundamental component has a fixed reference and complies with physical laws that do not change. This is an indication that there is some reference, of an intelligent source, keeping these laws constant or ordered. Things only remain ordered if they have a fixed reference. The reference must be set by an intelligent source that understands sequencing. This is the only way to orderly develop and grow from the initial reference. But first, I will start with man made ordered systems.

MAN MADE ORDERED SYSTEMS

Communication/ Language

Since we come into this world by way of our parents, we are directly linked to them. They display an innate desire to love and protect us. They take care of us when we are unable to take care of ourselves. They provide us with the necessities of life and are the first to teach us the basics for survival. This trait is seen throughout the animal world of which we are a part.

After we gain awareness of being alive, one of our first instincts is to communicate with others. As a baby, we instinctually use the universal language which is crying, meaning, 'I am unhappy', a smile which means 'In am happy', and so on. We then begin to learn the common language. This is the first ordered system that we learn for detailed and accurate communication.

The universal body language is the way we first communicate without the spoken or written word. We all, as humans, understand a smile, a frown or other facial expression. Similarly, gestures like beckoning with one's arm means 'come this way'. Body language has meaning and we immediately recognize an expression of friendliness or rejection without the spoken or written word. Communication requires intelligence on the part of both parties involved. The universal reference for communication is body language but, as we get more sophisticated, we develop the spoken and written language.

The spoken language incorporates specific sounds in a set sequence and with specific meaning that is always consistent. Then, we go to school to learn how to read and write the language. Later, we learn mathematics, the sciences and the arts. This is the accepted practice. This is how we are prepared to take our place in society in the pursuit of knowledge and to perform a responsible role. Through all of this, we are trying to understand ourselves and our world.

If we examine language, we see that it is ordered. We develop specific sounds or combinations of sounds to communicate specific thoughts. Everyone in a community must use the same combination of sounds in the proper sequence to express the same thought, request or emotion.

With time, we learn how to communicate fluently. The better we communicate the better we will be able to accurately exchange thoughts and ideas. This is important as we need to operate as a community. Each of us is a part of that community and to make our best contribution, we must be in harmony.

Then we progress to the written word, since it is sometimes necessary to communicate with someone out of ear shot. We develop common references such as an alphabet, words and sentences. The alphabet represents specific sounds. A word is the combination of alphabet symbols representing these sounds, in sequence. The sentence is a combination of words expressing a thought, a wish, a question or a command.

In any specific language, these 'words' must remain the same for each specific communication or meaning that they represent. In English, a 'girl' is a young female human being. A 'dog' is a specific type of animal, as is a 'horse' or a 'cow'. We have to keep these constant otherwise we would be totally confused and language would lose its purpose.

Each nation has its own native language or dialect. Within that language, the words must have the same meaning. We take this for granted but what we have done is created a communication standard or reference for all in that group. In addition, the reference has to be fixed to avoid confusion.

As I have mentioned earlier, from the time we are born we begin to be programmed by our environment. We are gathering information and knowledge about our world. The only way we are able to do this is because we are intelligent beings. We are born with intelligence. We have a memory so we are able to recognize sequence and learn by a process of information or knowledge accumulation.

Information cannot be stored randomly, otherwise we would not be able to use it efficiently or possibly even access it. There must be a reference to which all individual bits of information are connected and logically stored and accessed. In other words, it has to be ordered. This is the only way we would be able to access, store and retrieve this information accurately and efficiently. We have also learned that information storage in the brain is redundant which makes it easier to access.

We can now see how important it is for consistency, in order to maintain accurate communication when using language.

Societies are continually inventing new words to describe specific new inventions, concepts and thought processes as we grow in knowledge and change. However, we still have to be consistent in their application to avoid confusion. Sometimes words disappear from our vocabulary if they are no longer relevant or necessary in our changing environment or society.

In all communication, consistency is critical as in a common meaning of a particular word or expression so that we accurately convey our thoughts. Accurate verbal communication is extremely difficult as sometimes we never know how our words are interpreted. Cultural differences (accents) can result in different interpretation of words, even if this is not the intent. For the most part, this is the best we have and so we should constantly hone our language skills.

In our world, there are several languages which make it difficult for us to communicate verbally with everyone. Wouldn't it be nice if we all had a universal verbal and written language. Then we would have a universal communication reference. Different cultures could still have their ethnic languages or dialects that are unique to them.

Our 'heart' is that which accurately represents our thoughts. Language, at best, is only a close approximation of the thoughts we are trying to convey to others. Sometimes we say things that are the opposite of what we are actually thinking. This occurs when we practice deceit. GOD knows our 'hearts' and so He cannot be deceived.

Mathematics

We take learning mathematics for granted. It is expected of us. But why is this necessary? What we are doing is simulating natural events or occurrences. We need a standard or reference to communicate what we see in nature- two oranges as opposed to one. How about 100 oranges or more! How do we communicate this to someone else? We use language and symbols (numbers) to represent mathematical principles so we can communicate linear or even continually changing conditions as in the case of calculus. We learn to count. But first we have to have a reference and that reference must be fixed. We call this reference 'ZERO'.

If some people changed this 'zero' reference to 'one' and everyone else kept it at 'zero', all their calculations would be incorrect with reference to all the others in the group. So we can see how important it is for us to keep the reference fixed for everyone. We cannot be in harmony unless we have the same reference. Harmony means ordered as opposed to random or disordered. We are only able to simulate natural occurrences with this mathematical system because nature is also ordered. Nature came first with order and mathematics was designed to simulate it. Only order can simulate order.

As we progress from simple linear calculations to hyperbolic functions, the principle still holds. Hyperbolic functions, such as sine and cosine, simulate wave forms such as light and sound waves. Exponential functions simulate population growth and other real life occurrences that start slowly but dramatically increase over time or vice versa. (see diagrams below)

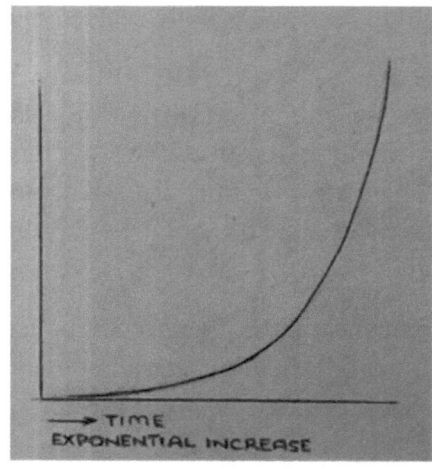
EXPONENTIAL INCREASE

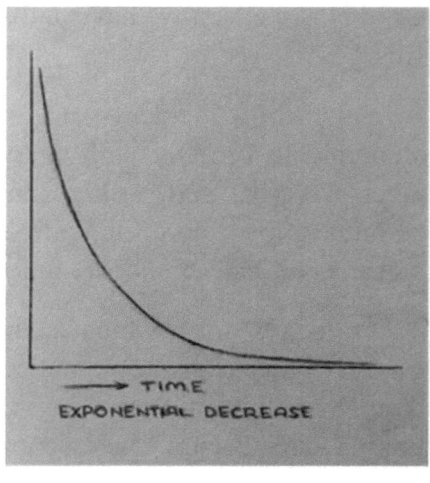
EXPONENTIAL DECREASE

Population characteristics are represented by the 'bell curve' (on page 63). This is discussed later including standards such as mean, and standard deviation. In every case, we use these equations to simulate nature. Also, in every case, we need a fixed reference. Intelligence is a given in setting these references.

Foot Note

The following is a simple example of how we calculate unknowns from an equation with other known factors. This is a typical application of a linear mathematical equation.

Look at the simple equation below:

$$C = A \times B$$

To determine the value of C we must know A and B. To determine how C changes with respect to A or B we need to keep either A or B constant and vary the other. If we make B= 2, then C= 2A. In this equation C and A bear a linear relationship. C is always twice the value of A.

If we plot these values on a graph, we get a straight line with a gradient or slope of 2. The line also passes through the origin since when A is equal to 0, C is also equal to 0.

This is very basic but the concept is true in all such calculations.

In defining nature's behavior, we develop equations like this that approximate what actually occurs. There may be some errors or small variations but these are negligible based on the generally predictable characteristics obtained from the results.

We say one and one equals two. This is hypothetically true. If we were to apply this to an actual situation, we would say one egg plus one egg equals two eggs. However, no two eggs are the same. Technically, one egg may be larger than the other which would make them not equal. Or there may be color or shape variations. But for all practical purposes, one egg plus one egg equals two eggs.

When we design something, we take into account that there will be variables that we will not be able to fully control. So we develop 'tolerances' to see how much error can be tolerated without affecting the intended function of the product. Tolerance is discussed later on in this book.

Manufacturing

I am a Manufacturing Engineer by training. I obtained my first degree from the City of Birmingham Polytechnic and my Masters Degree at the University of Birmingham, England. My major was Manufacturing Management and my job in industry was to coordinate the various processes and disciplines in a manufacturing plant to ensure that the finished product was made on time and to the design specifications. These disciplines included Mechanical Engineering, Electrical Engineering, Civil Engineering and Electronics. This gave me the opportunity to work in and also visit a variety of manufacturing facilities. In so doing, I became familiar with many manufacturing processes, from General Manufacturing to Medical, Technology and Aerospace. In addition, I was employed for 35 years with an insurance company as a Risk Control Consultant. This gave me access to many manufacturing facilities looking at quality control procedures from a product design and product safety standpoint.

All manufacturing operations have one main thing in common, detailed manufacturing procedures including quality control at all stages of manufacture. This is essential for producing high quality products in a predictable, repeatable manner.

Have you ever made something, anything? Do you understand the precision it takes to produce mechanical parts or parts for furniture so that the finished product is functional and aesthetically pleasing?

I have always enjoyed several hobbies including model making, oil painting, electronics and woodworking. With this combined experience, I know what it takes to design and make a product. What it takes to select and shape raw materials into a finished product; an ordered system.

I am now retired and have been for more than six years. I therefore have the time to review the knowledge and experience that I gained throughout my life and put it all in perspective.

I have learned that the materials must be right for the application in order to make a quality product. You have to understand the properties of the various materials used in a project so that they are compatible in the finished product. It also takes the relevant knowledge of production processes and the proper sequence throughout the component manufacturing and assembly processes. If the final step is incorrect, the entire project is at risk even though you have done everything else right.

It takes intelligence to keep track of the sequencing so that it is carried out in the proper order. In some cases, even the timing is important- not too fast and not too slow. The process is far from random. In fact it cannot be random as, by definition, it is an ordered process.

Frustration occurs when you are working on a project and it is not progressing in the way you anticipated. This happens because your production procedures are incorrect or you are not following nature's laws.

If you are fabricating something from steel, you must use cutting tools that are harder than the material being cut and also not brittle that it breaks under the cutting force. Cutting speed is also important, otherwise, the tool may overheat and lose its hardness and become ineffective. Cutting speed may also affect the tolerance and, if this is critical, we need to control these variables as much as possible to obtain consistent results.

What I am trying to say is that we have to be consistent in order to have predictable and repeated success. This process is not random but follows a set sequence that must be closely followed to produce the desired finished component. The same procedure is required for all components. The final assembly of the product is an added dimension where there is an additional possibility for error, even if all components are within the required specifications. This is because things like alignment and compatibility must now also be taken into consideration.

To produce several accurate components that make up the finished product, a system for quality control must be in place. This is critical to ensure that quality products are consistently made.

Instead of machining, an alternate method would be to melt the steel and cast it into the desired shape. This is a different ordered process to achieve a similar end product. The desired end product is known and the manufacturing sequence is formulated to get it done. This takes intelligence. Throughout the process, accuracy is continually being monitored in order to produce a quality product. We are also aware when the end is reached and where the process must stop.

Similar rules apply when working with various other materials including wood, fabric, plastics, chemicals etc, which all have different properties.

There are always set guidelines with which one needs to be familiar in order to succeed. If any one of these steps is missed or selected out of sequence, you will be unsuccessful.

Looking at the big picture, anything that is made or produced is done using set guidelines from start to finish. In all cases, only the raw materials available to us are being used. These are the same materials, with the same properties, used to make the universe.

Some people may be gifted in design and fabrication and are therefore able to make an original design and finished product without much effort. This is not the norm. It takes talent and experience to accomplish such a task. One should never underestimate the challenges such a task presents.

CNC Machining (Computer Numerical Control)

One uses the same principle of planning and setting a reference in CNC machining, to accurately make machined parts for any application. The machine tool control program must be given a fixed reference point typically(0, 0, 0), before it can accurately machine a part. These are the coordinates of the fixed reference point. The cutting tool must always know its location with reference to this point. This is the only way it can recognize its current location or coordinates and know where next to proceed.

This is also the technology used in 3D Printing. Instead of using a cutting tool as in CNC machining, the program now controls a type of spray nozzle that ejects the product material onto a flat surface (grid). Based on the dimensions of what is being produced, the nozzle ejects the material to the corresponding dimensions starting at the base and building up in the vertical direction, to exactly duplicate the three dimensional coordinates of the product being made. What it does is build the product slice by slice, until complete. This is the process of integration as is learned in mathematics. The smaller each incremental step, the more accurate will be the finished product.

Before CNC machines were invented, standard references were used to measure dimensional accuracy as the machining process approached the dimensions of the finished product. The machinist needed to have these standards at his or her machine to measure the accuracy of each part as it was machined. These standards are made from a very stable material to very high dimensional accuracy. They are, therefore, suitable for use as references when measuring to high degrees of accuracy. These standard references include, micrometers, vernier calipers, and similar high accuracy measuring instruments.

The first straight edge could have been a container of water with the surface as a reference. The surface of the water would be perfectly flat under gravity and so would be a good reference. Another reference could have been a string suspended with a weight on the end or a string pulled tautly between two fixed points. We had to start somewhere, but once we established the basic references, we were able to improve on them to obtain more accurate results.

Tolerance

Every material ordered system must have some limits or degree of tolerance for error. Within these limits, the tolerance is considered acceptable and the product will function as designed. However, outside these limits the product will not function satisfactorily and would be considered defective.

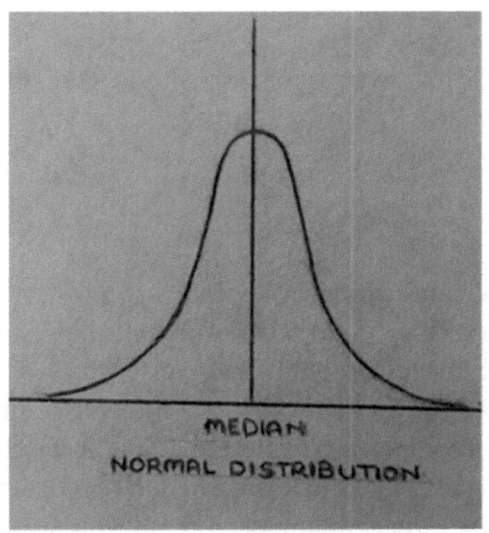

NORMAL DISTRIBUTION

A defect is an indication that something is wrong and needs to be corrected. Tolerance is allowed around a fixed reference (the drawing specifications). If we look at this mathematically, this is typically represented by a normal distribution, around the median or reference (see diagram above) for multiple samples of similar dimensions or characteristics. Most samples are close to the median in the form of a normal or equal distribution on either side of the reference. We can then look at the distribution data and calculate the standard deviation. Based on the acceptable tolerance, one standard deviation may be the acceptable tolerance on either side of the median. Any reading outside one standard deviation would be unacceptable.

The diagram below shows two standard deviations, one on either side of the median. This would represent the typically accepted tolerance in a manufacturing process. Any component within this range is acceptable and anything outside would be rejected.

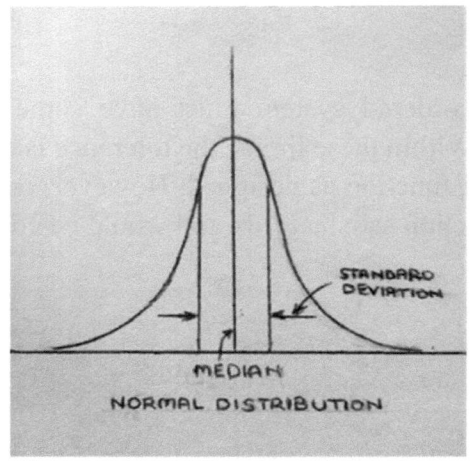

NORMAL DISTRIBUTION/STANDARD DEVIATION

An example of the application of tolerance is a piston that must fit into a cylinder. If the radius of the cylinder is six inches, the piston must be smaller in order to fit inside the cylinder. It must also have a certain amount of clearance so that it will move freely in the cylinder. We must remember that the cylinder also has a tolerance. A six inch cylinder would have a tolerance of say, plus or minus .005". Therefore, the smallest acceptable diameter would be 6" minus .005". The largest diameter of the piston must be this value (6 - .005)", less the required clearance between the piston and the cylinder, so that the piston will move freely inside the cylinder. The tolerance value must be fixed by the allowable gap between piston and cylinder so that the there is a good fit. If the fit is too loose, there will be too much play and if too close, the piston will not fit or be too tight. One can now see the importance of getting the right tolerances to make a successful product.

Mass Production

The process of mass production involves timing, sequencing and repetition. This is important when we need to make a product, at a high production rate, with a continuous production rate. To set up such a system, it is important that all stages of production be coordinated including raw material availability from suppliers, warehousing for raw materials and finished products, and layout of the manufacturing process as far as sequence and quality control are concerned. Typically, there are quality control procedures at each component production stage.

On any production line, there are sequential stages of production. If we look at the complete system, we need to take into consideration everything from design to completion and even use, as one will need to consider product safety and reliability based on the history of the product during use.

Parts made must come to the assembly line when needed for the assembly process. This has to be planned and monitored closely to ensure that the process progresses smoothly.

The process starts at the concept or idea for the design, with research and development and then manufacture, starting with the prototype. It is also necessary to have customer feedback so the necessary changes can be incorporated in future designs that will better satisfy consumer needs.

If we look at our natural world, we see that there is production or shall we say, reproduction of living things. Since the materials used all follow the same laws both in our production lines and in the processes we see in nature, do you think it is reasonable to assume that a similar process was used for the design and reproduction of living things? There had to have been an idea and a plan, the source of which must have been intelligent. In addition, there had to have been sequencing and timing which further complicates the production/reproduction process. Again, this can only be done by an intelligent mind. No ordered system can be created or produced without following these sequential and timely steps, as well as knowing when to stop.

In the natural process, our food is the raw material that builds and maintains us. All the necessary elements used in the design of our physical bodies are in the food that we eat and air we breathe. The DNA (also genes) does the design, manufacturing and quality control.

The initial design must have been done by an intelligent designer as was the sequential developmental process.

Robotics

Robots are designed to simulate human body motion. Those who have designed robots know that this is a very challenging prospect as human motion is very complex.

Let us look at a simple production line, where it is necessary for human motion to be simulated in areas such as filling and packaging. Where the human being is replaced by a robot or automatic process, everything has to be precisely located. There is very little flexibility whereas, with a human being, flexibility is almost limitless. The difference is that the human being knows what needs to be done and when faced with changing circumstances, outside the norm, is able to compensate.

The robot has to be taught every movement and is unable to do anything outside the scope of what it has been programmed to do.

If a bottle to be filled on the production line accidentally falls or for some reason becomes skewed, it will not be properly filled and the robot has to be programmed to stop, if this happens. The human being would immediately recognize the problem and correct it.

The robot is given a fixed reference from which to start and all movements are referenced from that point. If the reference is moved, it will not work properly. If the reference is unchanged, the coordinates will be accurately located, but if the bottle to be filled is not in the right position, the liquid will spill unless the robot and production line are programmed to stop when such a situation arises.

The robot is designed to do highly repetitive work with great speed and accuracy, but only if conditions remain consistent with those for which it has been programmed.

Note- Robots could be designed with cameras that provide feedback similar to what our eyes provide to us. This would be an improvement on the basic design. However, to replicate the reasoning ability and emotions of a human being is still out of reach in this technology.

Computer System Data Storage

Everyone knows that inherent in the design of computers is accuracy. How can they achieve such accuracy? First they have to be given a fixed reference.

A computer stores data in binary form to the base eight. This means that it only recognizes two conditions, 0 and 1(binary). These are used in various combinations using 8 pieces of descriptive content made up of this binary format. Eight was chosen because it allows a relatively infinite number of combinations for representing each symbol stored or processed. Symbols meaning numbers and letters of the alphabet, etc.

Each symbol is made up of eight bits using only 0's and 1's. This is how simple it is made. In electrical language, this means zero voltage (or current) or one unit of voltage (or current) the value of which has to remain constant for it to be accurately recognized. This may be represented by only a fraction of a volt so low voltages can be used as long as it is consistent and recognized as such by the computer circuit. It only has to recognize the difference between 0 and 1. In computer circuits, the power supply voltage has to be highly regulated and stable so that it provides a stable, accurate reference for the computer circuit voltage.

Each symbol has its unique combination incorporating the eight bits. Symbols representing our alphabet, our numerical system and other symbols such as commas, periods, colon and semicolon are all represented. When the eight bits in the combination representing a symbol are input, say to the memory, the circuit recognizes that the information representing that symbol is complete and the next eight will represent the next symbol. They are stored in units of eight. Each of the eight spaces is filled with zeros and ones in the specific combination representing each symbol. See example below.

00000001, 00000010, 00000011 represent 1, 2, and 3, respectively

There is a specific combination for letters and for numbers so each is readily recognized. Each number or letter has its own combination, starting from the right hand side of the array. Each symbol is therefore clearly represented leaving little or no room for error. The only problem would be if the computer mistook a zero for a one, or vice versa. Care is therefore taken to ensure the accuracy of the voltage (or current) used to represent 1 and 0 and to prevent interference from outside the system.

As long as these are consistent, accuracy will be assured.

We can therefore store a name or a number, a sentence or paragraph or a book using these combinations. The computer is very fast because computations are done at about 1/100 the speed of light, at the speed of electrons flowing in a conductor.

It is therefore clear that order is critical and to make this possible, it is necessary to have fixed references for each operation to proceed in an ordered manner.

Computer Data Entry/Processing

If you have worked with computers you know how precise you have to be in inputing data. This is an indication of the precision needed for the system to accurately interpret the input so that an equally accurate output can be guaranteed. By now you must have heard the term ' Garbage in, Garbage out'.

Do you think you could randomly program a computer? To the contrary, every bit of information has to be accurate and in the proper sequence. For the same reason, the process of evolution could not have programmed a human being. It had to have been an intelligent designer.

Every command needs to be precise and in the proper sequence for us to even hope for an accurate output. This is an indication of how the data input into nature had to be precise to ensure that life was made possible and even more so, the creation and development of a human being.

When you think about this, do you believe that life was created from a random data entry process or was it an ordered process by some super intelligent mind?

Semiconductor Integrated Circuits

Anyone familiar with the manufacturing process of a semiconductor knows that the process requires the ultimate in quality control. There can be no contamination of the 'wafer' except for the intentional introduction of additives that modify the characteristics of the finished product, based on its intended function. This extreme level of quality control is also necessary to ensure replication to very high tolerance or low error rate. The level of rejects can sometimes be extremely high in order to maintain the desired very high quality standards.

This is what it takes to make one of the critical components in a computer. This is the only way to ensure accuracy and reliability when the product is incorporated into a computer circuit. It is far from a random process. It is totally ordered.

Television Data Transformation

The image we see on our flat screen TV is transmitted in the form of a matrix or pattern using radio frequency waves, as the carrier, which are part of the electromagnetic spectrum. This frequency is longer than that of visible light.

In order for the picture to truly represent the image in front of the camera, when it gets to the viewer's end, the bits in the matrix have to be sent to the screen in accurate sequence. The output must truly represent the input. There is only one way to do this and that means accuracy down to the last bit. Otherwise, the picture we see will be a distortion of the original or there would be no picture at all. To ensure accuracy, there must be a reference that stabilizes the signal so that the sequence cannot change. To achieve this, each station transmits each signal at a specific frequency so when we tune to this frequency we can receive the signal. The picture and sound data are superimposed on the carrier frequency in the same sequence in which they are recorded and transmitted in the same order. The sound and video information are then separated in the receiver and sent to the respective circuits to amplify the picture and sound information before being sent to the picture monitor and the audio output sections, respectively. The bits in the matrix, as we see in the specifications for the TV monitor display, for example- '1080' per square inch, each has a specific place on the screen and there is no room for error. There is no room for randomness.

Our interpretation of the picture we see, as far as continuous motion is concerned, is dependent on what is called 'persistence of vision'. The same phenomenon applies in our interpretation of a motion picture. When we look at an object, the picture we see makes an imprint on the optic nerves (retina) but does not immediately disappear. There is a delay of the imprint on the retina. This delay is short but is enough for us to perceive the entire picture as well as continuous motion even though the picture on the TV screen is being sent as individual bits of information sent in rapid sequence. The bits are being sent at about 1/100 the speed of light (2,000 miles per second) and so there is a lot of information being sent, per unit time. As the picture changes, it appears continuously in motion even though it is a series of individual bits sent in rapid sequence.

The brain interprets the information transmitted from the eye in such a way that we see color, contrast, motion and distance or depth. Do you have any idea of the degree of order it takes to develop the eye and the brain working together for us to have such vision? Do you think it was randomly achieved through a process of trial and error, without a fixed reference? Do you think such a system could be obtained without input from an intelligent designer?

If this is the type of accuracy it takes for success at the data transmission level, how much more accuracy would it require to develop a system as complex as the human body?

If we went to another planet and saw a television set in operation wouldn't we believe something intelligent must have made it? We can probably use anything as an example such as a car, a plane, a house, etc. Yet, some of us believe that the human body was made from randomness, as well as put together randomly, over a long period of time.

Would you describe the human body, with all its complexity, as a random or as an ordered system? If you say random then we would have to use another definition for random. If ordered, that would make more sense. Remember any ordered system has to have a fixed reference. This reference cannot change and must be of an intelligent designer. In the case of the television set, man is the intelligent designer.

Radio Transmission

Like video transmission, radio transmission also uses radio frequency waves which are part of the electromagnetic spectrum. These are longer than those used for television transmission. You must have heard of AM and FM. These are amplitude modulation and frequency modulation, respectively. AM uses the lower frequencies of the radio wave spectrum and FM the higher frequencies, for transmission.

Sound waveforms are superimposed on an electromagnetic carrier (radio wave), in one case modulating the amplitude and the other, modulating the frequency. The sound information can then be transmitted at the speed of light and decoded at the receiver to get the sound information only. This is done using electronic filters. After decoding, the sound wave, in the form of an electrical current, is then amplified and fed into a speaker which vibrates the air at the frequencies fed into it, thus reproducing the sound.

We now transmit in digital format but this format must accurately represent the analogue format and be converted back to analog as this is the only way we would be able to hear or interpret sound.

When you tune your radio, it must be to the electromagnetic carrier frequency you are trying to receive, otherwise you will not find it. The associated electronic circuits must be precisely tuned to that frequency. This is the reference frequency.

Construction

Starting from basics, if we want to construct a building, we must first decide on a unit of measurement. This may be the inch, foot, meter or whichever we choose to use. However, once we chose a unit of measurement, we must stick with it or use some accurate form of conversion. This is now our reference unit of measurement.

In that chosen system, for any dimension to be accurately represented, it must be described relative to the origin. The origin is therefore the fixed reference. This is necessary in order to communicate accurate dimensions on the system 'blueprint'. One must be able to accurately interpret the magnitude and direction from that reference point to any point on the blueprint.

(What is referred to as the nominal value on a blue print or other engineering application is one which is acceptable and within the desired tolerance.)

As in any three dimensional system, we need X, Y and Z axes for it to be accurately represented. The X and Y axes represent the plan view and the Z axis the elevation. Here again, we must choose a fixed reference, which is typically given the coordinates, 0,0,0. We originate from zero on each axis. Each axis represents the magnitude and direction in each dimension. The distance from the reference is the magnitude. We need to select this reference and this cannot change after we start. If it is changed, the representation will be in error.

So, where did the reference come from? It came from a source from outside the system, the designer. It takes an intelligent decision maker to determine what is required including the starting point (the fixed reference).

Here again, tolerance is important especially in the structural members of a building. The minimum strength on a structural member must be met for it to be acceptable. In other words, the lower end of the tolerance must be at or above the minimum strength requirement. The required load specification would also include any safety factor.

Architecture

This is a combination of art and engineering. In order to create a structure that is aesthetically pleasing and also functional, we combine the two disciplines, art and engineering.

Art is both an emotional and spiritual expression of ourselves. To express this in material form we must obey material laws.

A building is designed with functionality in mind. In some cases, the design is to the individual taste of the client. After the aesthetic design is agreed upon, the next step is to calculate the structural properties to ensure that the structure can be safely constructed for the intended use. This is the job of the structural engineer.

At this point, the calculations are done on the loading of the structure including the footing, floor, walls and roof- in the case of a building. These calculations take into account the forces that the building will be subjected to including a safety factor, which is an additional load factor over and above the basic design loads. These factors are calculated to withstand static and dynamic loads. This takes into account weight (force), and distance or length (moment). These elements are critical in the calculations for the design strength.

Again, we see that fixed references are critical. All the specifications involved are predetermined and the minimum/maximum quantities set or fixed. The building can now be safely constructed to these design specifications.

Art/Painting

In painting, the primary colors for pigments are magenta, cyan and yellow. In order to achieve the brightest color, one can mix no more than two of the primary colors. I typically use the most brilliant of these two primaries such as chromium yellow and red for a brilliant sunset.

I cannot use blue in the mixture otherwise the color will be muted or greyed and so less brilliant. By adding blue I will have added the third primary which grays or neutralizes the color. This is the reason why when we clean our brushes, in water or turpentine, the color of the solvent eventually becomes muted or dull. At some point in our painting, we will have introduced the three primaries to the mixture causing the color to become muted. You see, with pigments, the three primaries added together cancel out each other.

The other colors on the pallet are variations on these primary hues obtained by mixing the primaries and the various hues of natural or synthetic pigments.

One must follow a fixed protocol when mixing colors, from the primaries or other hues, to obtain the desired hue (color). Each protocol is consistent and repeatable. For shadows, we add the complement to the hue under consideration. For contrast of a given primary or hue, we juxtapose (place adjacent) the compliment. In other words, for yellow to look its brightest we juxtapose the compliment purple which is the combination of red and blue. This provides the most contrast based on how the eye perceives color. If we want orange to look it's brightest we place it beside blue. If we want to make a shadow, we add the compliment to the color. This grays or neutralizes it.

It is interesting to note that this is the case with pigments as used in painting but, with 'real light', when we mix all the colors of the spectrum (in phase), we get white light and not darkness. This is because pigments absorb the parts of the spectrum that they do not reflect. They appear as the color they reflect. If the pigment is orange, it absorbs blue light from the spectrum and reflects yellow and red. If we add blue it now absorbs the red and yellow as well and so will now appear gray as it has already absorbed the blue with the original pigment. All three primary colors are being absorbed. Hence the color now appears gray.

This is one of the basic principles one has to know when mixing colors for painting. These relationships never change and we depend on them being fixed otherwise we would not be able to mix colors and consistently achieve the same results.

This only involves the mixing of colors, which is a critical part of any painting process. Then we move on to composition, the art of placing the colors on the paper or canvas. This requires us to have a subject or if abstract, a clear idea of what we want to convey to the viewer. These are the fixed references. We use our intelligence to fix the references and proceed from there.

As we develop a painting, we must use these 'consistent' principles of color mixing and subject matter and refer to them continually as we progress. These are our references and will always influence the finished product.

As mentioned above, it is interesting to note that light and pigment display opposite properties. If we mix the components of light, the colors of the spectrum, they enhance each other. However, if we mix the primary pigments red, blue and yellow, they neutralize each other and we get grey (or black).

When painting, to obtain the desired color we use the primaries as a reference as they are unique in the fact that we cannot mix any other colors to obtain the purity or likeness to a primary color. However, we can use the primaries to obtain other colors.

White light is direct light from a source that includes all the colors of the spectrum in phase with each other; or similar light that is reflected. If a surface appears white, it has reflected all the light falling on it randomly, without absorbing any of the colors in the spectrum. If the surface is mirrored, it reflects all the light at the same angle as the angle of incidence. This is the same angle at which it contacted the surface. The surface will then appear like a mirror. Chrome plated surfaces have this mirrored property. A black surface absorbs all or most of the light.

The characteristics of the surface determines how the light is affected based on which colors are reflected or absorbed. The surface itself has no color, only the light that it reflects has color. This is why in a dark room everything is black. No light is reflected. Beauty is in the light as this is the only medium that our eyes perceive.

Color Interpretation

Different colors are interpreted differently by the brain. Reds, oranges and browns evoke warmth whereas blues and greens are cool and evoke peace and calm. This may be the reason why the sky is blue and the trees are green. We need these colors in abundance around us to help maintain calm in our lives. This is the natural response in the way the brain interprets these colors.

Art is personal although, to some extent, cultural in our reaction to any particular work of art. Through art, certain emotions are evoked in us such as peace, turmoil, love, hate, etc. These emotions are triggered by the scenery or work of art which is being viewed.

The reference for color is light and this is fixed. This is why I now believe that we all see colors the same way. The yellow one person sees is the same yellow another person sees. In some cases, the brain may interpret it differently because of distortion due to a defect in the interpretation of the signal. That is an abnormality such as color blindness. However, I believe most of us see the same yellow, red or blue hues.

We therefore conclude that light is the reference for the color and shape of an object. All the frequencies are fixed and never change. Light is the fixed reference.

Light connects the universe. It travels long distances at great speed and makes it possible to see other planets and stars. For stars, this is the light generated and for planets, the light reflected.

Music

Music has a common scale of notes where the reference is middle C. It is mechanical vibration or sound and will be described in a later chapter. Instruments such as the piano and guitar are tuned using middle C as the reference. It is important that instruments be in harmony especially in a large orchestra or band. Otherwise, we immediately notice the discord. Our brain is looking for harmony and it becomes very disturbing and annoying when we hear discord. We are inherently ordered and will readily recognize any form of disorder as this is unpleasant to us.

Using middle C as a reference (261.6 HZ), we can compose an infinite number of tunes and will never run out of combinations. We have developed a musical scale using octaves (8 notes) as incremental bands, before repeating, both above and below the reference. Octaves are frequency bands which the ear interprets as eight separate notes.

Human hearing is within a specific frequency range from about 20 cycles per second, at the low end, to about 20,000 cycles per second, at the high end. This is for those of us who have excellent hearing. Hearing deteriorates with age and from consistent exposure to loud noises. Typically, hearing loss starts at the higher frequencies.

Our sense of hearing makes it possible for us to hear each other when we speak, to hear music, and to warn us of impending danger.

Our two ears, located on opposite sides of the head, allow us to locate the direction from which the sound originates. We not only hear sounds in stereo but we also have stereo vision. Stereo vision gives us the ability to judge distances. Do you think this all evolved randomly?

Sports

Every sport has to have rules. These rules define the sport and differentiates one from the other. The object is to achieve some goal whether as an individual or as a team. The goals are preset, with a scoring system to decide the winner. Typically, the individual or team with the highest score wins. However, in some sports, as in golf, the person with the lowest score wins.

The rules are known by all the participants and in some cases there is a referee or umpire who enforces the rules of the game. There are usually equal numbers of participants on the opposing sides. Also, the sport may be further refined by a process of elimination so that, in the end, the teams with the best record compete to determine the ultimate winner.

These rules of the game are references and they have to be fixed otherwise the game cannot be accurately defined and the final result will be in question. Each sport may be considered an 'ordered system' and so the essential components of an ordered system apply. The rules of the game are essential as they define the sport.

PHYSICS

The laws of physics define our physical or material world. I will now show you that these laws never change and work together to keep our universe ordered. These never changing laws indicate a fixed reference that never changes with time. Never-changing is one of the attributes of God. We see this fixed reference as it manifests itself in the following elements of our world.

Time

What do you understand to be the meaning of time? Is it the momentary display on the clock? Is it a finite period in which events take place? Is it the infinite period spanning billions of years? Is it a state of mind? I would say it is all of these things.

Time is the dimension in which the sequence of events is fixed. Some say time is relative, meaning it appears to go slowly or quickly, depending on our emotional state. If we are enjoying an experience, time appears to go quickly and if we are under stress, it can appear to go slowly, or even quickly if we are trying to meet a deadline. It all depends on the situation.

We use time as a reference for the sequence of events in our daily lives. If we did not have a memory, time would mean nothing to us. Time is recognized by comparing at least two events in our memory, one of which is fixed. It may then be said that, to each individual, time is a function of the mind giving us the ability to record events in the sequence in which they occur. However, time is real in that even if we are not aware, it is still progressing. Time progresses in one direction to maintain accurate sequencing.

At any given time, there are an infinite number of events occurring in our universe. In these instantaneous increments of time, which we refer to as the 'present', we are connected to each other. Once this instant has passed, the opportunity is lost until possibly some time in the future. What makes it even more difficult is that we can only be in one place at any given time. In other words, make the most of any opportunity as it may not present itself again in the future.

Our first memory is fixed at the beginning of what appears to be when life (time) began for us. When we are asleep or unconscious, as far as we are concerned, time stops. One may then say time is relative to each individual.

In our daily lives, time is our reference. Our current calendar is referenced from the approximate date when Jesus was born. Jesus must have made an indelible impression on the minds and lives of those who knew him, for us to now use the date of his birth as the fixed reference for time.

Time began at the absolute zero reference. In order for time to have been initiated it had to have had a fixed reference and still does. The reference cannot change because order has to be maintained. Time keeps all events in ordered sequence. The absolute reference, itself, is outside of the sequence of time because it initiated the sequence. Time is a product of this reference. Time is the dwelling place of the sequence of events that unfolds during our lives. The creator of time does not exist in this sequence. The creator is outside of time.

Time makes it possible for an intelligent mind to develop a plan, set a fixed reference, and initiate the sequence of actions necessary to complete that plan. It is the medium in which we are given access to event sequencing. We use time as a reference for the sequence of events in our daily lives. Again, we see that if we did not have a memory, time would mean nothing to us.

The only time we have available to us is the present. It is instantaneous, it is very short, yet everything we accomplish is in the present. The good thing is that the present is continuous and our accomplishments are cumulative. Yet the present is detached, in that, even though they are so close, we do not have physical access to the past or the future. If you want to maximize your accomplishments, you should focus on the present.

Age is a function of time. If time is not a factor then there is no aging. This concept is compatible with eternity. Time is relevant only to our realm.

The fact that every four years we add a day to the calendar is an indication of the approximation of our seconds, minutes and hours to the natural cycle. What we are doing is trying to maintain synchrony with nature's cycle. Nature is the reference and we have to synchronize with this fixed reference in order to maintain accuracy.

Time is designed to go in one direction because sequencing is critical. Order must be maintained.

Distance and Space

We live in a three dimensional world and so, in order to communicate location and relative separation, we need to define the concept of distance and direction.

In physics, this is referred to as a vector, having both magnitude and direction. We developed units of measurement such as the inch, foot, meter, mile etc. These are the standards or references for distance (magnitude). We have also developed 'degrees' to denote angle or direction from a fixed reference point.

In order for us to be consistent and accurate in relating or communicating distance and direction, we either have to use the same standard and use the proper conversion when changing from one standard to the other. It is critical that we have a standard or reference before we can proceed to using it in any practical application.

All we are trying to do is simulate and apply it to the three dimensional world in which we live, using a system to which all can relate. Using these standards, we have developed coordinates that are used to accurately define any point on earth or in the universe.

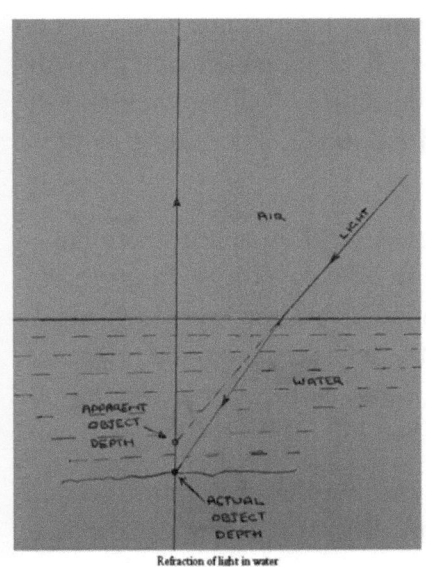
Refraction of light in water

We also use this system of measurement to define three dimensional objects and so it has universal application. It not only defines the size but also the shape of all things in the universe.

For infinite distances, we use light years which makes much more sense based on the size of the universe. For those who may not know, a light year is the 'distance' light travels in one year. At a speed of 186,000 miles per second, we can see why the speed of light is used in calculating the distance of stars or planets billions or trillions of miles from earth.

Light is a part of the spectrum of electromagnetic radiation, all of which travel at the same speed. They travel in straight lines in a vacuum. If there is no change in the medium in which they are traveling, they will continue in a straight line. In space, they travel in straight lines but variations in atmospheres do bend light to varying extents.

When light goes from air into water, refraction takes place, the light beam changes direction. This is why water appears shallower than it really is, when viewed from above. (See diagram on page 79 that demonstrates the effect of refraction)

Navigation

Now that the measurement standards have been developed for defining distance and direction, I will look at navigation.

Using the same standards of measurement used for distance and direction, with some modifications, they can be incorporated into navigation.

For navigating a three dimensional world, three axes must be used, the X, Y and Z, to represent the three dimensions. Now that there are three axes there has to be a set reference, a fixed reference.

For this purpose, we have used the equator (latitude) and Greenwich, England (longitude), as the X and Y axis (respectively), with the Z axis extending 90 degrees upward from the point of intersection of Greenwich and the equator (See Fig. A below). Using this point as a reference, any point on the surface of the earth can be defined as well as extended out from the surface, on the Z axis, into space. At the intersection of Greenwich and the equator the coordinates are (0,0,0), the fixed reference. The third 0 is the Z axis starting point.

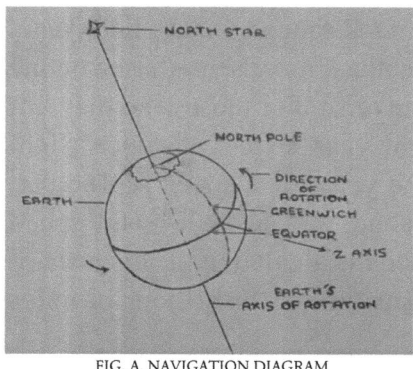

FIG. A. NAVIGATION DIAGRAM

To define a point in space, there has to be at least two fixed points on earth and then triangulate to fix the third in space. In every case there has to be a fixed reference. All points are defined relative to the reference. (See Fig. A)

Before the current navigation system was developed, using strategically placed satellites in fixed orbits around the earth, the North Star was used as a reference. You see, the North Star is located in our galaxy at a point where it is on the same axis as that on which the earth rotates. It is located on the side of our North Pole, hence the name North Star.

If one were to draw a straight line from the North Star to earth, passing through the North Pole, it would be on the same axis as that on which the earth rotates. It was, therefore, the perfect reference for ships navigating at nights, since as far as the earth's rotation is concerned, the North Star is fixed. Satellites are now used in fixed orbits around the earth because we are able to strategically place them in orbit. By triangulation, any coordinate can be located on earth. If you know the coordinates, then your GPS can be used to take you to that coordinate anywhere on our planet.

Mechanical Energy/Kinetic Energy

One of the first equations one learns in physics is that the mechanical force on an object is defined as the mass multiplied by the acceleration of that object.

$$\text{Force} = \text{mass} \times \text{acceleration}$$

$$\text{Also, Acceleration} = \text{velocity}/\text{time}$$

$$\text{Power} = \text{force} \times \text{velocity}$$

Now, all matter has mass so every object has mass. If sufficient force is applied to an object, it will start moving and accelerate if the force is strong enough to overcome friction. In other words, if the mass is constant, as for any given object, the acceleration will be proportional to the force on the object. Acceleration is the rate of change of velocity.

This relationship was developed from experimental data. The above equations can be used in the estimation of the maximum speed of a car in relation to the power of the engine. It is universal in that it is applicable to any object being moved by a force of any origin.

These equations are used to simulate motion on earth. Simulating natural occurrences was the basis for their development so that we can predict how mechanical forces act on objects.

For objects moving under the force of gravity, the same principle is true. Gravity is a constant force acting on an object. It is generated by the mass of the earth resulting in a type of magnetic attraction on objects within its range.

This force has been found to be constant resulting in an acceleration of 32 feet/second/second, on objects close to its surface.

Technically, the force on an object only produces kinetic energy if the object starts to move. In this case we say that work is being done on the body which gives it kinetic energy. (Kinetic energy is energy due to the movement of an object)

Potential Energy

Potential energy is the energy in an object by virtue of its position. It is stored energy. Gravity is one force that gives an object potential energy. This stored energy can be released by letting go the object. Potential energy may be stored under controlled conditions, as in a clock spring or wind up toy. Under gravity, if you release an object that you are holding, it will fall to the ground. You can also jump from a high place and gravity will take you to the ground. You should always remember that, in order to gain potential energy, you have to do the work by climbing to an elevated place or lifting the object off the ground. To gain energy, you have to expend an equal amount of energy. You cannot get something for nothing. When you release the body the potential energy is now transformed into kinetic energy (by virtue of its motion). The different types of energy are therefore, interchangeable.

Electrical Energy

In circuit theory, electricity is defined in terms of current and voltage in a relationship that is proportional to the amount of resistance in the conductor in which the current is flowing. If the resistance remains constant, as in a copper wire of fixed length and diameter, current flowing in the wire will increase with voltage in a linear manner.

$$V = I \times R$$

$$V = \text{Voltage}$$

$$I = \text{Current}$$

$$R = \text{Resistance}$$

The resistance, R, is constant in any given conductor. The voltage or potential difference is used to drive the current, which is a flow of electrons.

Since electrons are negatively charged, they will flow towards the positive pole, as in your car battery or any battery for that matter. Electrons are always negatively charged. The term conductor means that the molecules of the material have free electrons that will flow through the conductor when a potential difference(voltage) is established. Current is defined as the flow of electrons. Electricity can be obtained from a chemical reaction as in a battery or from a conductor moving in a magnetic field, as in a generator.

The equations developed from these relationships never change. The relationship is true even in the most complex electrical circuit. If it were not so, we would not be able to pass the first step in building on our knowledge of electricity.

Having mastered the basics of electricity, scientists progressed to electronics involving vacuum tubes and now semiconductors. Here, they are able to control electrical currents and voltages and channel them in a desired direction. They can also accurately amplify electrical currents and voltage with a superimposed sound wave, as in the case of an audio amplifier.

In electronics, in addition to the basic resistance in a circuit, there is capacitance, inductance and impedance as capacitors and inductors are introduced into the circuit in order to modify its characteristics, based on the application. Impedance in an electronic circuit is analogous to resistance in a purely electrical circuit. These components have unique characteristics that make the circuit an 'electronic circuit', with the vacuum tubes or semiconductors, as opposed to a purely electrical circuit. Here, the electrons in the circuit are being manipulated to produce a desired output, based on the given input.

We now use integrated circuits to store information for artificial intelligence (computers), which we can access in microseconds in the proper circuitry.

Electricity is the perfect medium for information transfer in circuits involving artificial intelligence, just as electrical signals transfer information in our brain, using a chemical process. Information travels quickly, at about 1/100 the speed of light. Again, we depend on the fact that the laws governing electron flow or current flow are constant and hence predictable. In all these calculations we have a reference that is necessary in any ordered system. The relationship between the three components referred to (voltage, current and resistance) must remain constant.

When the brain is examined, to determine if it is functioning normally, we look at the electrical current activity to determine if it is within normal parameters. This is one of the parameters used to analyze brain function.

Circuit stability is critical in any given electrical or electronic circuit. This means that the relationship between the components in the circuit must remain constant. Otherwise, we would not be able to predict the outcome or, in this case, output. Here again there are tolerances on the values of the components in a particular circuit design. The closer the tolerance of the components to the design specifications, the more accurate will be the predicted output.

For these characteristics to remain constant, there must be some form of fixed reference. If we duplicate this design in another independent circuit, then we can expect to accurately duplicate the output. In electrical and electronic circuits, the line voltage is the reference. This is the main supply voltage to the circuit and must be kept constant. Variations in line voltage will affect the relationships in all other parts of the circuit. It is assumed that the characteristics of all components in the circuit remain within tolerance.

Only if physical laws remain constant can such predictability be guaranteed. They always do. Thus, once gained, we can build on this knowledge with confidence.

Light

Light is the part of the electromagnetic spectrum which is visible to the eye. Only a specific set of frequencies is in this range. These are from red, at the longest, to violet at the shortest wavelength, that are visible to the eye. The combined visible spectrum appears to us as white light. However, when broken down, there are seven separate frequencies between red and violet as seen through a prism or in a rainbow. It is this separation and infinite combination of these colors that allow us an unlimited pallet when we view an object.

White light is made up of fixed frequencies within the visible spectrum that manifest themselves as violet, indigo, blue, green, yellow, orange and red.

Objects that we see have the color or colors of the light that they reflect. This ensures that there is one reference (light) as opposed to each object contributing to its own, unique color, independent of this reference.

Color manifests itself only in the specific properties of the surface of an object. The red rose is designed to reflect red light and absorb the rest of the spectrum. The various shades of red are developed by the reflection of some of the adjacent frequencies such as orange and yellow. If white is included in the petal, this means that part of the petal reflects all light randomly and this is interpreted by the eye as white light.

Color is therefore a property of light and not of the object seen. The object itself has no color. Light is therefore the reference. There would be no color except for the existence of light.

We possess an innate appreciation for color to which we may react differently based on our individual taste. Typically, blue and green are considered cool and, red and orange warm. We therefore react subconsciously, in this manner, when presented with these colors as represented in our environment or in a work of art.

Color, as represented by hues and shadows, molds shapes which communicate the unique characteristics of what we see.

Light transmits all the information we need to detect and interpret what we see. Light itself does not add or subtract from what is communicated to us but is consistent and true in its properties. In this way, it accurately depicts the environment around us. It is used as the reference or standard in the electromagnetic spectrum which we routinely use as the medium with our sense of sight.

You have probably heard of infrared and ultraviolet radiation. These are as named, the part of the electromagnetic spectrum before red light and the part beyond violet, respectively. Infrared is a good heating source and is often used to cure paints more quickly and for tanning. Ultraviolet (black light) also has its uses as certain chemicals glow when exposed to ultraviolet and can be readily identified.

It is also useful as a disinfectant. Because of its higher frequency, it is more penetrating and can cause damage to the skin, if overexposed.

Light frequencies within the visible spectrum never change and so we have a fixed reference. If there is no light we are visually cut off from our surroundings.

X- Rays

X-rays, as well as gamma rays generated by radioactive isotopes, are also a part of the same electromagnetic spectrum of which light is a part, but these are of shorter wavelength.

X rays are of even shorter wavelength than ultraviolet and therefore have greater penetrating effect allowing it to pass through the human body. The density of the human body varies depending on the type of tissue. The more dense the tissue the more X-rays are absorbed. This is how we can develop a picture of the inside of the body by exposing a sensitive film to the x-radiation after it has passed through the body. There will be variations of gray on the exposed film depending on the density of the tissue through which the x-rays passes. The developed x-ray film shows the bone structure of a lighter shade than the fleshy areas because it absorbs more of the x-rays. It is, therefore, like a film negative with lighter areas representing denser tissue and dark representing the less dense tissue.

For shielding an x-ray room and also for isolating radioisotopes during storage, lead lined enclosures are used because lead is very dense and will significantly reduce the amount of radiation escaping.

All these types of radiation are in the same category (electromagnetic) but have different properties. Our eyes cannot see microwaves, radio waves, infrared, ultraviolet, x-rays or gamma rays and are only sensitive to the visible light frequencies. Vision is limited to the visible spectrum which has all the information we need to enjoy a very satisfying visual experience.

Do you think that the process of evolution limited our sight to light frequencies in this entire spectrum just by chance. Or was this by design, by an intelligent creator?

Microwaves are electromagnetic radiation with frequencies longer than infrared and are also damaging to human tissue. To protect us, our microwave ovens must be shielded while operating. We must also be shielded from inadvertent exposure to x-rays, while these rays are being generated, as we can only have limited exposure without long term damage to the human body. Yet the visible spectrum is relatively harmless. The frequencies we need in order to see are those that are the least harmful. These are located in the range between radio waves and x-rays. We need to be exposed to light so it would seem that by design, we would be protected from overexposure to light.

Remember, the frequencies above and below light frequencies are very harmful to us. Why not visible light?

Sound

Sound, like light, may be mathematically represented by a wave form. It is mechanical in nature in that it induces mechanical vibration in the medium that it travels. It is detected by our ears as sound although some low frequencies are felt as vibration. Sound waves are audible to the human ear between the approximate frequencies of 20 cycles per second and 20,000 cycles per second. Sound travels at about 760 miles per hour in air and so is much slower than light. This is why when we watch fireworks, we see the explosion and there is a delay before we hear the sound of the explosion.

Throughout this range of frequencies, each one is fixed by a reference which is the number of cycles per second. The volume of sound that we hear is a function of the amplitude of the vibration. This is the distance the waveform cycles from the reference in the wave form diagram. The reference has to be fixed for us to make sense of what is happening. It is typically represented by a straight line. (Shown in the figure below.)

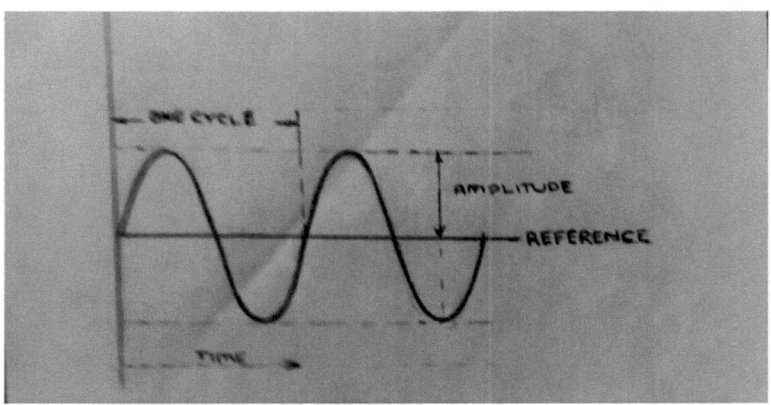

THE MATHEMATICAL REPRESENTATION OF A WAVEFORM

Resonance

Resonance is associated with the transfer of sound (mechanical) energy to an object that vibrates in sympathy with the specific frequency called its resonant frequency. This varies from material to material based on its molecular structure. When the resonant frequency is reached, the material begins to vibrate uncontrollably and could even be destroyed. What is happening is that, at the frequency of resonance, the material will start to vibrate with increasing amplitude even if the sound energy exciting it is constant. This frequency is constant for a particular material of a specific length and thickness when, ideally, it is restricted at the ends. In other words, the material can be made to vibrate at its resonant frequency if the same conditions are repeated. The resonant frequency of a given material sample is constant and it will vibrate uncontrollably at that frequency.

This principle has various applications in real life such as in a horn, siren or even a switch. The human body has a natural resonance of about seven cycles per second. When subjected to this frequency, at an optimum magnitude, you will become disoriented and get sick to the stomach. This has been used for crowd control but is considered inhumane.

Sound makes it possible to verbally communicate with each other, to help us navigate our environment, as well as for pleasure in the form of music. Because each frequency has a fixed reference (the wavelength), we will always get the same result if the same conditions are repeated. It is consistent and predictable.

CHEMISTRY

Properties

Matter manifests itself in three distinct states- sold, liquid and gas. The properties of a chemical indicate the unique characteristics of that particular molecule or compound. Within fixed environmental conditions, these remain constant. We can depend on these properties to be consistent within each state. This is the reason why we are able to obtain consistent results in the reactions of the various chemicals. Without this consistency we would not be able to duplicate our discoveries in chemistry.

Periodic Table

We have found that, in nature, atoms/molecules manifest themselves in ascending levels of complexity. The Periodic Table begins with the hydrogen atom, consisting of one proton and one electron. Then, by adding another proton and electron we get helium. The series then progresses discretely with the addition of a proton and an electron to form new elements. These all occur naturally.

It is important that for every proton added there is a corresponding electron so that the molecule remains neutral and not positively or negatively charged. If a proton is added without a corresponding electron we now have an 'ion' which is basically unstable and it will try to find an electron to become neutral and stable.

Another particle found in the nucleus of an atom is a neutron. This is neutrally charged and is thought to be a type of glue that holds the 'all positive' nucleus together.

Valency

Atoms and molecules combine in discrete amounts to form compounds based on their valencies. Valencies are fixed and so an atom/molecule will consistently react with other atoms/molecules, based on these discrete combinations. Under the proper conditions, we can make chemical compounds with repeated success by understanding the conditions necessary for these reactions. The valency is a property that results in neutralization of the two reacting atoms/molecules which is in line with nature always trying to maintain stability. Valencies typically have values represented by integers such as 1, 2 and 3.

But, even in the proper valency proportions, the reaction will not take place unless the proper conditions are met. These are fixed conditions that must be met in every case. These conditions include temperature, pressure and environment. Environment could mean the introduction of a catalyst.

When we cook, we use fixed conditions and proportions to prepare food recipes even though we are only mixing compounds and may not be changing them chemically.

All atoms/molecules react in a manner based on their valencies, to form the compounds that there are today. We use these known principles to develop new compounds and are able to consistently replicate the chemical reactions.

This is so because the chemical reactions follow fixed laws that never change. We depend on this consistency and our ability to understand and replicate the necessary conditions for making the compounds we have formulated and continue to formulate.

In sequential order are all the known elements. At the lower level in the periodic table, the nucleus of these elements are stable but, as the nucleus increases in size, instability increases and they eventually become unstable or radioactive. This means they will spontaneously break down.

Atoms and molecules, in specific combinations, produce the various compounds that make up the earth and of which we are made. In living organisms, carbon is the base, with this group labeled organic compounds.

Starting from the elements of which living things are made and the inert elements around us, there is a consistency that is repeated throughout nature. We can therefore conclude that matter is made up of the same basic elements that exist in sequential order and form all compounds. This indicates that the basic building blocks of the universe are consistent, sequential in nature and ordered. There is, therefore, order in even which appears to be random at first glance.

Atoms and molecules combine to form the compounds that are the raw materials we use in manufacturing. In chemistry, we learn to modify or change these compounds through chemical reaction to form new compounds with new properties which we find useful in providing us a better quality of life-or so we believe.

These particles of matter and their ability to combine, all abide by specific laws that must be followed for these changes or reactions to occur. Under the same conditions, we can repeat these reactions to produce the same end products again and again.

Radioactivity

At the upper scale of the periodic table, the nucleus of the atom/molecule begins to become unstable, hence the molecule becomes unstable, at the threshold where radioactive isotopes begin to consistently form. Some of these are highly unstable elements that generate harmful levels of radioactive energy (gamma radiation) in their natural state. Although radioactive isotopes have been used in medicine to treat cancer, they are inherently deadly to living tissue and must be used with great care.

The fission or breakdown of some isotopes releases intense heat which can be used for heating water to generate steam used to run turbines, to generate electricity and to power engines.

There are naturally occurring and artificially manufactured isotopes and have specific uses depending on their individual properties. Again, only certain elements are naturally radioactive and are consistent in their individual properties.

Laws Of Nature

It is not important that you understand the mathematics of these laws as this exercise is only to demonstrate that they never change. For those who do not find this 'appealing', the only takeaway should be the fact that all laws governing material things are consistent and predictable. We cannot change these laws.

It is the inherent consistency of nature's laws which ensures that anything incorporating matter must conform to them and follow an ordered process, governed by these laws, to create and develop any ordered system. Randomness is not compatible with such a process. The intent is to create order which can only be the product of an intelligent designer. Therefore, evolution, as we define it, cannot be the manifestation of such an ordered process, since, as defined, the process of evolution is not ordered but random.

If life processes were random, there would not be the interconnection and interdependence as we currently observe. We are all connected to each other and everything in the universe based on the fact that we all can interact with each other and our environment.

PHYSICS

Gravity

We are confident that when we drop an object it will fall to the ground. If we examine this further, we find that the object will accelerate, falling at a rate of thirty-two feet per second, per second (32 ft./sec./sec.). This is constant under gravity. The size, weight and density of the object are some of the variables. Yet, all objects will accelerate at the same rate if there are no other variables affecting them, such as wind resistance or some other external influence. The density, size and weight of the object do not affect this calculation.

A feather will float down slowly because of wind resistance and a rock will fall quickly. However, if we were to remove wind resistance by placing both objects in a vacuum and release them simultaneously, the feather and the rock would fall at the same rate of acceleration and hit the ground at the same time.

This experiment shows that the force of gravity is constant within the environment in which we live and has the same effect on all matter. However, in outer space, the gravitational force decreases until it becomes so small that objects float.

It has been found that the attraction due to gravity varies inversely as the square of the distance from the earth. When we are close to earth, the force of gravity may be considered constant and at a maximum but decreases exponentially as we move away from the earth's surface and into space.

There are many attributes of our universe that follow the inverse square law so we can use this principle in the formulation of the relevant mathematical equations.

Examples of these are the intensity of a magnetic field as we move away from the magnet or electrical source generating the magnetic field. This is also applicable to electromagnetic radiation such as light and x-rays as we move away from the source. This means that the intensity falls off exponentially, in proportion to the inverse square of the distance from the source.

The equation below demonstrates this principle. In mathematics, inverse means 1/ (the value under consideration). In other words, 1 divided by the value. In this case, it is the distance 'd' squared.

1/ (d x d) where 'd' is the distance from the source

As 'd' increases, the magnitude of the force begins to exponentially decrease. If the distance is 0 feet, the value will be the maximum it will ever be. If we measure this value 10 feet from the source and the force is F, when 20 feet from the source, which is twice the distance, the value will be 1/4 x F or 1/4 of what it was at 10 feet. Likewise, at 30 feet the value will be 1/9 x F. What we are doing is squaring the distance value thus resulting in an exponential decrease.

As far as the gravitational force of the earth is concerned, the equation is, the gravitational constant G x the mass of the earth(M) x the mass of the object(m), divided by the distance between the center of the earth and the center of the object(r), squared. The gravitational constant G does not change, nor does the mass of the earth.

$$\text{Force} = GmM/r \times r$$

Since the radius of the earth is so large compared with any object on its surface r x r will not start to change significantly until the distance of the object from the surface of the earth is in the order of miles.

r = the distance between the centers of earth and the object. This would then be the radius of the earth + the distance of the center of the object from the earth's surface.

This equation was developed from experimental work done in the field of physics and found to be consistent. This can also be demonstrated mathematically since the force radiates in all directions and the intensity falls off as the square of the distance from the source. Again, this shows that nature displays consistency and predictability.

The inverse square law demonstrates that, close to the source, the effect of the force is at a maximum, but as distance from the source increases, at a certain point, the effect starts to rapidly fall off. This principle keeps the force, with its maximum effect, where it is designed to have influence and with relatively no effect at long distances from the origin.

In cases where the force is designed to have significant effect at long distances, the force is of an exponentially large magnitude. The sun, being a massive object, has a large gravitational force, enough to keep the planets in our solar system in orbit around it. However, it has no effect on the planets in other solar systems. Earth has an effect only on objects within its gravitational field. Objects close to earth will be attracted and kept on its surface.

The moon is in orbit around the earth and maintains this orbit because there is a perfect balance between the gravitational pull of the earth and the equal and opposite centrifugal force of the moon in orbit.

In nature, where there is a law, there are no exceptions. These laws have been developed from close observation under controlled conditions in experiments by scientists. We are therefore able to build on this data and grow in our knowledge of our planet and the universe.

The reason for this exercise is to show you that the universe follows strict laws that do not change and that there are no exceptions to these laws. This indicates to me that the universe is an ordered system and could not have started from a random event but rather one which was planned and then carefully executed.

The laws, being constant, indicate that they have fixed references. If you observe nature, you will see that things only remain constant when they have fixed references. Otherwise, there is instability and chaos. The cycle of our day is 24 hours. This does not change because the earth takes 24 hours to make one rotation. Because this remains constant, we can use this as a reference for time. Earth also takes one year to orbit the sun and this also remains constant.

Here are some other examples of other laws that govern matter:

Matter Cannot Be Created or Destroyed

Matter is defined as anything that occupies space. When it is said that matter cannot be created or destroyed this means that the amount of matter that has always existed in the universe, since creation, is the same and will never change. It can, however, change from one form to another as in solid, liquid, gas and nuclear fission or fusion. The case of a nuclear reaction is where energy is released as the nucleus breaks down or nuclei combine, but the gross quantity of matter and energy remains constant in all cases.

As far as 'life' is concerned, there was one creation, but life is self-sustaining or self-perpetuating through the process of reproduction, manifesting itself in the form of living matter.

PHYSICS

Thermodynamics-Refrigeration

Physicists have discovered that a liquid must absorb heat from its environment in order to change from a liquid to a gas. This is called the latent heat of vaporization. During this transformation, the temperature of the liquid does not change. All the heat energy is used to perform the transformation.

It is also known that, if we compress certain gases, they become liquid at normal temperature. These are the typical refrigerants. Now, if the pressure is released, by forcing the liquid refrigerant into a larger space or volume, it will revert to gas as the pressure suddenly decreases (it boils). In so doing, it must absorb heat to make this transformation. The environment in which this transformation takes place is cooled, as a result of the heat absorbed by the refrigerant for the transformation. This would be the space in your refrigerator, or building interior in the case of the air conditioning unit. The evaporating refrigerant is passed through metal coils over which air is blown by a fan. The air is then cooled by the evaporating refrigerant in the coils and is used to cool the space.

The law of nature governing the evaporation of a liquid must be upheld and so the necessary heat to transform the liquid to a gas must be obtained from somewhere, in this case, the environment.

Using this principle, we make air conditioners and refrigerators or any appliance that is designed for cooling. The law is fixed and so we know we will get the same result every time.

Refrigerants have inherent properties that lend themselves to such applications.

The above characteristics are constant for any given system so we are able to duplicate any given system as long as we follow these principles. Again, this displays a consistency in the laws that govern matter.

Action and Reaction are Equal and Opposite

This is one of Sir. Isaac Newton's laws. It means that if we apply a force to a body it reacts with an equal and opposite force.

We use this basic concept to represent the dynamic systems that we routinely replicate in our inventions. An example of this is a jet engine or rocket engine. In the jet engine, air is pulled in, compressed, expanded and then ejected at the rear. The initial suction(action) of the air results in a forward force(reaction) on the airplane which is further enhanced by the thrust exerted by the compressed, heated air being ejected from the rear. In the case of the rocket engine, the burning fuel being ejected(action) creates a reactive force that propels the rocket upward.

These unchanging laws appear to indicate that they originated from an ordered, as opposed to a random source.

Electromagnetism

Electromagnetism implies a relationship between electricity and magnetism. Even though they manifest themselves differently and represent different forms of energy, they are interchangeable. An electrical generator demonstrates this principle.

If an electrical conductor is placed in a magnetic field and moved within the field, an electric current will be generated in the conductor. Conversely, if an electric current is run through a conductor, a magnetic field is generated. A generator produces an electric current by rotating a conductor (coil) in a magnetic field. The conductor is coiled to maximize the effect in that, as for all practical purposes, there are now multiple conductors. The effect is additive.

X-rays, a type of electromagnetic radiation, are generated by high voltage electricity in a vacuum tube, accelerating the electrons, with enough energy to generate x-rays when they collide with a fixed metal target.

Like light, these are also photons and a part of the electromagnetic spectrum. Light is the part of the spectrum with which we are most familiar. X-rays have a fixed frequency range (shorter than light) and are defined by their frequency range. The electromagnetic spectrum starts from radio waves at the longest to gamma rays at the shortest wavelength.

CHEMISTRY

The Ideal Gas Law/Boyles law

The ideal gas law enables us to develop a mathematical relationship between temperature, volume and pressure in an isolated gaseous system. This principle is used to develop the equation $P \times V/T = K$, is a constant. The symbols represent pressure, volume and temperature. In the equation below, the suffix (1) represents the initial conditions and the suffix (2), the final conditions.

Since $P \times V/T = K$, where K is constant, then it follows that, $P_1 \times V_1/T_1 = P_2 \times V_2/T_2$

If we keep one of these variables constant, we can define the relationship of the other two in the equations listed below:

with V kept constant: $P_1/T_1 = P_2/T_2$, since $V_1 = V_2$, then $P_2 = P_1 \times T_2/T_1$

with P kept constant: $V_1/T_1 = V_2/T_2$, since $P_1 = P_2$, then $V_2 = V_1 \times T_2/T_1$

With T kept constant: $P_1 \times V_1 = P_2 \times V_2$, since $T_1 = T_2$, then $P_2 = P_1 \times V_1/V_2$

We can make any one the subject of the formula and so if we know the value of three of the variables, we can calculate the fourth.

If we increase the temperature in an isolated system (V is kept constant), in order for this relationship to be true, the pressure must also proportionally increase. This has been found to be experimentally true and verifies the unique relationship between these elements in an isolated system.

As indicated above, the suffix (1) represents the initial values of the elements in the equation and the suffix (2) represents the final values. This means that within a given isolated system, if we change one of the parameters, for example the temperature, with the volume remaining constant, the pressure will proportionally change. If the temperature is increased, then we would find that the pressure would proportionally increase.

Decreasing the temperature would decrease the pressure, proportionally. The relationship in an isolated system must remain constant. One of the variables is typically fixed, in this case volume, to make the calculation simpler.

$$V_1 = V_2$$

This is the basic principle used to design a pressure cooker. The lid is secured to the cooker so that the volume remains constant. When we place the pressure cooker on the stove and turn the burner on, the temperature of the contents in the cooker begins to increase. Since the volume is constant, the pressure will proportionally increase and will continue to increase as the temperature increases. The cooker is equipped with a safety relief valve to prevent it from exploding from over-pressure. P1 would be the initial pressure which would be atmospheric. T1 would be the ambient temperature. T2 would be the final temperature, that reached inside the cooker from the burner on the stove. Since the volume is constant and also known, the value of P2 can be calculated as this would be the only unknown.

This principle is also used in determining the potential power of an internal combustion engine. The volume of the cylinder remains constant. The volume above the piston does, however, vary from a minimum to a maximum, as the piston reciprocates. The volume used in the calculation would be the minimum volume at the top of the stroke of the piston at the time of ignition (this volume is constant).

With the volume at a minimum, at the top of the stroke, the fuel is ignited. This minimum volume is constant and a function of the size of the cylinder. The ignition results in a rapid increase in pressure on the piston, from the sudden increase in temperature, resulting in rapid expansion of the air/fuel mixture inside. The piston is forced down, this force is transferred to the crank shaft causing it to rotate. The cycle is repeated, resulting in the crankshaft continuously rotating.

If the temperature is increased in a confined space, the pressure will increase, based on this law. At the time of ignition, the volume is at its minimum. The higher the temperature of the explosion, the greater the pressure or turning force on the crankshaft.

In physics, pressure is defined as force per unit area and force is mass multiplied by acceleration. The larger the cylinder the greater the potential force. This force is multiplied by the number of cylinders and made continuous with the continuous cycles of the engine ignition system. The engine therefore becomes the driving force of the vehicle.

$$P = F/area, F = P \times area \quad P = pressure, F = force$$

This relationship indicates that the larger the cylinder or area, the greater the potential force.

Consistency of Laws

We depend on the above laws to remain true. These laws are the fixed references on which all our scientific knowledge is based. A fixed reference is an indication that there is continuous, consistent control over the elements of matter. This implies order, and order implies intelligence.

In order for us to try to deduce where the universe all started, we can only use the tools and the knowledge we have available. In other words, study the dynamics of the universe and look for patterns in its characteristics that give clues as to its origin.

These laws are just an example of the consistency with which nature operates. We depend on these consistencies as references in all things scientific. If they were continually changing they could not be used as references. We could make no progress in science.

The laws of nature are designed to have a specific ordering effect on matter and so there is a relationship between them that, combined, maintains order in the universe.

Thresholds of Matter

Thresholds mark the end of one phase and the beginning of another. A dramatic change in properties as it relates to matter. Where the laws of matter are concerned, these still apply, but matter now takes on new properties that will remain consistent above the new threshold, in the new phase.

Thresholds are apparent at the freezing and boiling points of molecules. The most familiar to us is water as it changes from ice to water and from water to steam or vapor. Ice has different properties to those of water and water has different properties to those of steam or water vapor.

It is a characteristic of the molecules of all compounds and is a function of temperature, volume and pressure.

Under any given conditions, we have found that these bear a constant relationship. Variations in environmental conditions only change the threshold (temperature/pressure) at which these changes occur.

An example of this is water changing from liquid to gas at its boiling point of 100 degrees Centigrade, at a pressure of 14.6 pounds per square inch (atmospheric pressure). This is what is referred to as normal pressure, at sea level. However, if we go up into the mountains where the atmospheric pressure is less, the boiling point is less than 100 degrees Centigrade because there is less atmospheric pressure to keep the water from boiling.

In each of the three states, solid, liquid and gas, the molecules exhibit completely different properties, but within each state the properties are consistent and predictable. This all adds to the diversity of nature but shows overall consistency within the confines of the laws governing them.

Another example of this is the threshold at which a nuclear chain reaction occurs, as in an atom bomb. Below this nuclear threshold, the molecules of the radioactive chemicals are 'relatively' stable but still highly radioactive. However, once the threshold is met, a chain reaction begins, resulting in nuclear fission with an exponential release of energy until the reaction is complete. To initiate the chain reaction, there is a detonator which must be activated in order to get the radioactive atoms to this chain reaction threshold. We can control activation by controlling the detonator. The reacting radioactive atoms must also have a minimum threshold of purity for the reaction to be successful. Once the nuclear thresholds are met, a chain reaction will occur after detonation.

SUMMARY

The above are only some of the more popular examples of the basic applications of 'consistency' as it relates to ordered systems. It applies to all aspects of our conscious and subconscious lives which would not be possible without these 'TRUTHS'. This 'ORDER' is what connects us to each other and to the universe. Otherwise, we would be disconnected and unable to interact with each other and the rest of the world. In fact, we would not be aware of our existence but for the fact that there is 'ONE' fixed reference from which 'EVERYTHING' radiates and which connects us all.

Intelligence plays the most important role. It allows us to connect increments of knowledge and awareness to form patterns which we then use to come to the realization that we do exist and are a part of a greater existence. This intelligence and knowledge can then also be used by us to initiate change in our environment.

When we observe nature and look at the abundance of compatible life in one place, it would seem reasonable to conclude that it is by design. Such order cannot be randomly achieved.

Many of us believe that we are so smart, making discoveries and inventions- we should take pause and humble ourselves. We cannot create anything new, in the absolute sense. This knowledge was already in existence in another realm. The knowledge is only now being shared with you. The blessed ones are only the media through which these revaluations are transmitted. It is all already in existence in another realm even if we do not yet know it.

I will now share with you my conversion experience and try to describe GOD in the way he describes and reveals himself to us in the HOLY BIBLE. We will also look at what he has revealed about creation. I think what you will find is that creation, as we observe it, in every detail, is exactly the way he describes it. It shows a close connection between GOD and the fabric of creation, so much so that it leaves no doubt that he is the creator.

My Conversion

At an early age, I began to wonder about creation and evolution, as to which was really true. I thought there may be a creator but he/she seemed far removed from me and out of reach. How did this all come about? How did I come to be here? I had nothing to do with it and that was obvious. I could see, hear, touch, smell and taste, the senses that made me totally aware of my being and existence. This was the reason I became so interested in science and, in particular, physics which describes the physical characteristics of the universe and the laws that govern matter.

I had always believed that evolution was the way living things change from one form to a more advanced form, although there were some things that could not be explained by it.

I was now 51 years old and working for an insurance company as a Risk Control Consultant. Although this was not my passion, I was grateful to have a job that could pay the bills and provide for my family.

It was January 18,1999, a Monday morning. I went down town Phoenix, AZ, to review the operations at one of our client's properties. This was a routine survey where I would report back to the property owner my findings and recommendations associated with the visit.

It was a high rise building and it would probably take about two hours to complete. As I went around reviewing documentation and completing a physical survey, my contact brought up the subject of God. I was not interested! I listened and answered questions in a polite manner but, in fact, I was looking forward to completing the task at hand and leaving as some of his questions and comments made me feel uneasy.

Towards the end of the visit, we were in the lobby, as I had finished the survey and was about to leave. Then, he said something that caught my attention. He said, 'I think you are close' meaning close to understanding and believing in God. As far as I was concerned that was so far from the truth. Then he said, 'All you have to do is humble yourself before him and ask for forgiveness and he will forgive you and reveal himself to you'. I had no idea what that meant but I said to myself it wouldn't hurt to try what he said if there was so much to gain, based on what he had been telling me.

I then looked away from him and, in my mind, said, 'If you are there, please show me how to live my life'. I said it with sincerity but was not expecting a response.

I then went to my car in the attached garage and for some reason tuned to the Christian radio station he had mentioned. He had also given me the address of the church he attended but I had no intention of going.

Throughout the rest of the week I kept tuned to the Christian station not even realizing that I was ignoring my favorite popular station. I still thought nothing of it but, in addition to listening to this station, I began to have a new appreciation for nature. Whereas before I believed in evolution, now I was believing in creation. I did not make this change, it was made for me! It began to get scary as I now realized something was changing me, I was not doing it and had no control over it.

I did not even remember that I had prayed for guidance but, as I looked back over the week's activities, I remembered my interaction with my contact that Monday morning and the prayer that I said to a GOD I did not know.

It was then that I realized that my prayer was being answered.

A real supernatural experience came the following Sunday morning, when I was in bed about to have a late morning 'sleep in' as I always did on a Sunday morning. I began to feel very uneasy but continued to try to fall asleep for a final nap. I had a sense of something or someone communicating with me repeating 'you should go to this church'. It was not in English, my only language, but it was clear what was expected of me. However, I still resisted. The feeling became so intense that I knew I had to go to this church. I knew I had no choice and if I disobeyed I would regret it for the rest of my life.

I had not been to church regularly for several years but today I had to go because some force, now inside of me and much more powerful, wanted this of me.

I got out of bed and told my wife I was going to church. She looked at me in shock and asked 'Are you ill?'. I said, 'no, I just have to go'.

I then called the church from the invitation information my contact had given me to get the time of the service and then got ready to leave. I got into my van and started on my way to the church. I had no idea what to expect but I felt something 'real' as I drove to the church. I felt peace and joy that brought a smile to my face. I do not remember what the sermon was about that day but I did know that I would be going back every Sunday.

This was 20 years ago and I still regularly attend this church and I am now a confirmed member.

The days and weeks following my conversion, I spent trying to understand what was happening to me. I would wake up at nights thinking it was all a dream but then realize that it was all real and I was being influenced by a source that was not of me.

I would now look at a leaf and see new beauty and complexity and say in my mind 'You made this!'. I would look at a baby and say the same thing. I now had an overwhelming appreciation for creation and the fact that I was a part of this great work and had a purpose in this life.

Would you believe me if I told you that I still needed more confirmation that GOD really existed and was communicating with me. The whole experience was still unreal but every time I challenged its reality he showed me that what I was experiencing was real.

Whenever I was confronted with his response to what I was experiencing, I was so overcome that I broke down in tears.

It was now Easter and the main focus in the Christian world is the crucifixion when JESUS died on the cross for our sins. I was on my way to a Walgreens pharmacy to pick up a prescription and listening to the radio to a pastor describing the events at the cross, at the time JESUS was being crucified. When he got to the part when JESUS was dying and said 'Father forgive them for they know not what they do'. The scene was so vivid to me that I broke down in tears and said 'Why are they doing this to you'. By this time I was sitting in the car in the parking lot and so had to wait several minuets to compose myself before going into the drug store.

I have never questioned GOD since that experience. I now believe in him completely.

Since my conversion experience I have never felt alone. I am always aware of his presence in me. It is a very comforting feeling, a feeling that all should experience. It is a feeling of peace.

When I described my experience to people that I met, I initially thought they would immediately become believers. To my surprise, this was not the case. In some cases I got this 'deer in the headlights look'. In other cases I could see them trying to understand but were unable to fully appreciate what I was trying to tell them. However, there was an immediate understanding by those who were already believers.

One thing that was revealing is that, looking back, it seems as if I was being prepared for conversion. I now know that GOD first humbles you and there is no doubt that he did that to me pre-conversion. There were crises in my life that I had no idea how to deal with and how to overcome them.

Conversion Final Thoughts

After someone is converted, one of the things that begins to happen is that an internal battle begins between your old self and the new. A transformation process begins. You instinctually want to maintain control of your own life. But, then you begin to see your flaws more clearly and that you have such a long way to go to be that perfect human being.

You become much more aware of yourself, your flaws and how you now need to act as a new human being, in GOD. This means character changes that you now have to consciously make in your new life. You are given a new heart. Your conscience is renewed. You want to be a good person and sometimes fall short, but you know the road you now need to travel. You get back on that narrow and difficult road and keep on moving ahead.

When you consider the alternative, you are reminded that this is something that you have to do even unto death. You have no other choice. Who else can you lean on and be confident that you can depend on in any situation- 'ONLY GOD'.

INTRODUCTION TO THE GOD HEAD

We have looked at examples of the necessary components it takes to create and maintain an ordered system. The key components are intelligence and a fixed reference from which to build the ordered system. We also need knowledge but, for us, this grows with time as we observe, investigate and research the function of the planet we call home.

Knowledge is now being gained at an exponential rate. We are getting better at everything we do, learning how to control our environment to enjoy a better quality of life.

These key, critical components of order must have existed before we came into existence as we had nothing to do with the order already in the universe. The only explanation is that intelligence existed before the universe came into being and is responsible for initiating and developing what we are and see today. I say this because the universe is an ordered system and so its origin must also have been ordered.

So intelligence existed. Since we were not there, there is no way we can prove how life and the universe were initiated. However, we were given clues, written in what we refer to as the 'HOLY BIBLE'. Now, we can compare what we see around us to what is said in the BIBLE about GOD and creation, to see if they are consistent.

GOD tells us that one of HIS attributes is that HE does not change. HE also says that HE created the universe and everything in it. The latter would take infinite intelligence. So, we now have an intelligent source and a fixed reference, the critical components needed to create an ordered system. The ordered system is the universe and life as we know it.

He also tells us that he made us in his own image. He gave us intelligence and in so doing the ability to create. All we need is to have an idea, set a reference and start creating. We see that we do have these attributes, the ability to create, using the material available to us.

When we examine the evidence by observing what we see today and the order of it all, there is no other rational explanation. His fingerprint is all over the universe. The never changing reference and the infinite intelligence are reflected in all ordered systems that we see today. Everything we make and every positive act we perform have these basic components.

We do not have the ability to find GOD and HE only reveals to us what HE desires. But, what is true is that HE has revealed to us all we need to know in the HOLY BIBLE.

GOD gave us a jump start on the creative aspect of our attributes by giving us all the raw materials we need. All those atoms and molecules set out in an ordered periodic table. HE also made the laws to which everything material must conform and maintain constant control.

Now, we will take a closer look at how GOD describes HIMSELF and how this is reflected in everything we see in the universe.

God The Father

Do you believe in GOD? Do you believe there is some source 'up there' that directs what happens in this world or do you think it just happens randomly? Is it that same GOD that created the universe and everything in it, including life?

There is probably some reason for one to have any of these beliefs. However, if you think about it, you should come to some logical conclusion as to what you believe is really 'TRUE'.

Let us think objectively and see if we can review this together. Keeping in mind what we have already discussed on the consistencies in nature and the universe, we will now look for the similarities between what GOD says about himself and what we observe in the things around us.

God Does Not Change

One important attribute of GOD is that HE does not change. This is how HE describes himself in the Bible as revealed to us through the prophets with whom he communicated. This is also the most important characteristic of a fixed reference, a critical component of any ordered system.

Bible references- (GOD does not change-Malachi 3:6, Hebrews 13:8 Jesus)

We have seen, in previous chapters, how critical it is to have a fixed reference when defining an ordered system (creation). GOD chose to reveal this attribute of HIMSELF to us because HE thought it was important for us to know this fact about HIM. Nature's (material) laws are constant and this indicates a fixed reference as their origin.

God Is A God OF Order

We see order in the universe and the planet on which we live. The reason for any disorder we see around us has been explained by what GOD refers to as SIN.

Order, in both the spiritual and material sense, means that all the laws governing both realms must be obeyed. Man was perfectly made and would have continued to be perfect if he had obeyed all of GOD'S spiritual laws. GOD gave man one command, that he should not eat from 'The Tree of Knowledge of Good and Evil'. MAN was even warned that, if he did, he would suffer certain death (eternal separation from GOD). MAN disobeyed this command and so had to suffer the consequence of being separated from GOD.

But even after this, GOD still had mercy on us and gave us a way to redeem ourselves. HE gave us HIS 'SPIRITUAL LAWS', in the form of 'THE COMMANDMENTS', to help us in this effort. But, because of SIN, we find it difficult, if not impossible, to keep these commandments. This is the reason, as we are, none of us is acceptable to HIM. We all break these commandments every day (SIN).

Bible quote- (The Ten Commandments - Exodus 31:18, No one of us is without SIN Romans 3:10) (All have sinned - Romans 3:23)

It is interesting to note that we can break SPIRITUAL LAWS, because we were given free choice (free will), but we cannot break MATERIAL LAWS. SPIRITUAL LAWS are based on ethical references, which, because of SIN, we are no longer connected to the TRUE reference (GOD) and therefore unable to keep from SINNING.

There are fixed laws that govern the universe (MATERIAL and SPIRITUAL). GOD is a GOD of laws and will tolerate no dissent. A perfect system can have no flaws. HIS SPIRITUAL LAWS must not be broken and anyone who breaks them can no longer be a part of HIS spiritual order. You see, SIN is systemic and has affected everyone and everything since it was first committed.

God Created The Universe And Everything In It

GOD told us that HE created the universe and everything in it. When we look at the universe, it is ordered, life is ordered. Wouldn't it be rational to conclude that such order was a creation of a GOD who does not change, a GOD of order and of infinite intelligence? HIS intelligence is reflected in the complexity of creation, including life itself. We know that we are intelligent beings and we had nothing to do with the creation of our intelligence. Intelligence, therefore, must have come from our CREATOR. We have seen HIS fingerprint all over creation (it is ordered) with the interconnection of all systems and subsystems to a single fixed reference (GOD).

Bible reference - (God created the earth and everything on earth - Genesis 1)

God Makes All The Plans And Decisions Regarding Heaven And Earth

GOD planned creation and is in the process of carrying out HIS plan to the last detail. So much so that there is a set sequence in which events have been planned to occur and each sequenced event has to be fulfilled before subsequent ones can be initiated.

One important example of this is that the HOLY SPIRIT would only have been sent to us after JESUS had died for our SINS and returned to the FATHER. JESUS, himself told this to his disciples. These events unfolded exactly as planned (Bible quote - I will send the Holy Spirit after I leave - John 16:7). There are also several references in the BIBLE where JESUS fulfilled the prophecies in the old testament so that HIS actions would be as was written by the prophets.

When we read the Bible, we will soon understand that GOD, the FATHER, alone plans and makes decisions about everything in HEAVEN and on EARTH. Here, we now begin to see more similarities or the fingerprint of GOD in our universe. The universe came from a single source. HE tells us HE is the only GOD and there is no other. We also know that any ordered system not only has a fixed reference but, where there are subsystems, they are all controlled by the same, single source. In the human body, it is the brain that controls all systems. For heaven and the universe it is 'GOD the FATHER' who represents the 'BRAIN'. Even GOD the SON and GOD the HOLY SPIRIT do the will of the FATHER. GOD the FATHER is the planner and the decision maker.

Bible reference - (JESUS promises to spend the HOLY SPIRIT after HE returns to the FATHER - John 15:26)

Bible reference - ("I am your GOD and there is no other GOD but me")

Bible reference - (Jesus does the will of the FATHER - John 4:34, 5:19-21, 5:30, 6:38)

We now begin to see the recurring theme or similarities between what we see in the universe and what GOD has told us about HIMSELF. Wouldn't it be reasonable to conclude that HE is telling the TRUTH?

Remember, HE told us that he created the universe. If that is the case then HE must have infinite intelligence. (Genesis 1)

Based on the above reasoning, GOD qualifies in every respect as being the creator of the universe. HE is infinitely intelligent, HE does not change (the fixed reference), HE does all the planning and is in control of everything.

God Made Us In His Own Image

The analogy between GOD and man is that GOD made man in HIS own image. Human beings are intelligent, with creative abilities. We plan our lives and can choose what we desire. We interact with and change our environment to suit our needs. Of all living things on earth, we are the most intelligent. Man was given authority over the things of the earth.

Bible reference-GOD made man in his own image (Genesis 1:26)

God Is Eternal

God says 'I am the ALPHA AND THE OMEGA'. This mens HE is the 'BEGINNING' and the 'END'. In other words, HE is eternal. Any ordered system that has a beginning must have an external initiator. To initiate an existence such as ours, there had to be an external initiator. This is the only way to explain our existence since we had a beginning and we had no input into the process. The only source that can be the absolute initiator is something eternal. It just exists, it does not need an initiator. It is ETERNAL.

Bible quote (I am the Alpha and the Omega - Revelation 21:6, 22:13)

God Loves Us

The most important thing to note is that GOD says HE loves us. But we should also remember that HE is a 'JUST' GOD. He loves us but HE also has some expectations of us. HE gave us commandments to guide us as to HIS expectations of us. All expectations must be fulfilled. This is how SPIRITUAL order is maintained.

Bible quote - (GOD Loves us- 1 John 4:16, John 3:)

The first and most important commandment addresses love. The love we should have for GOD and the love we should have for one another. Now, do you think that this is good or bad advice as to how we should behave? Do you not think it is better to show love to one another rather than have apathy or hate? Love is order and hate is disorder. Apathy has no particular bias just like randomness. The fact that GOD loves us is an indication that GOD is on the side of order. HE is order. HE defines order.

With the above reasoning in mind, it would seem to indicate that God is the true source of all knowledge and intelligence. If the knowledge we possess comes from any other source it cannot be trusted. It cannot be TRUE. There is only one TRUTH. GOD always tells the TRUTH. GOD is TRUTH. This is part of HIS character and one of HIS attributes. Whenever JESUS was about to say something profound HE always prefaced it by saying, 'Truly, truly I say to you'. The words that followed would be the absolute, universal 'TRUTH'.

Bible quote (Truly, truly I say to you - John 6:47, John 5:24-25)

Bible reference - (GOD is a GOD of TRUTH - Isaiah 65:16)

SIN is what we commit when we disobey GOD'S will or the guidelines that HE gave us (the commandments). SIN, as HE describes it, 'separates us from HIM'. HE gave us the commandments so we would know what HE expects of us. However, we are unable to keep these commandments because we are now all flawed. SIN makes us flawed. We cannot help but commit SIN since we do not have the power to stop ourselves from SINNING. This is because we have lost our true reference. GOD is the true, absolute reference. Without his guidance, we set our own reference, which may not be in harmony with the absolute reference.

GOD is the only one who can give us power over SIN. At the time of your conversion, HE sends the HOLY SPIRIT to dwell within you and guide you. But even then you still SIN. The reason you still SIN is that you still possess part of your old self. Now, there is a continual battle between the old and the new self. Sometimes the old self wins, but ultimately the HOLY SPIRIT will prevail. This is what happens when you are saved and become BORN AGAIN, of the SPIRIT.

GOD is the one that calls you to HIMSELF and it is HIS will, as part of HIS plan. Some believe that HE calls us and also gives us the will to submit to HIM. If HE does not call us and give us the will, we will not choose HIM because we prefer darkness rather than light. Being in GOD shows up all our flaws. This feeling is very discomforting and some prefer not to know the extent of their flaws. We must therefore be willing, even with our flaws, to submit to HIM.

'Light' shows us who we really are and, being so corrupt, we do not want to see ourselves for what we really are relative to 'ABSOLUTE TRUTH'. You will come to this conclusion when you begin to understand one of the attributes of GOD, which is TRUTH. You will only know this after HE reveals himself to you. HE is nothing like you could ever imagine. HE is so much more that it overwhelms you. A typical response is to break down sobbing.

We have already been provided all the information we need to come to the conclusion that we all need GOD'S assistance and without HIS help we are lost and will remain lost. The way to HIM is outlined in HIS Holy Word, the HOLY BIBLE. What we need to do is, repent of our sins, submit to HIM and HE will guide us back to the path of 'TRUTH'.

I mentioned above that we have free choice. That is the truth. However, if GOD needs you to go in a certain direction you cannot resist. It may even appear to you that it is your choice but you would be mistaken. HE is the one that gives you your unique qualities and HE has absolute authority over you and is able to mould you as HE wishes. HE is GOD and with HIM all things are possible. Whatever needs to be done to execute HIS plan, will be done.

HE has the power to harden or soften your heart so you should pray to HIM to soften your heart so that you will hear HIM and listen to HIM when HE speaks to you.

When HE first called me, I was very unsure of how to react because I did not understand what was happening to me. It was not rational. It was supernatural and so defied logic as we know it. However, deep down I knew it was the right thing to do. I felt at peace with what I was about to do. I felt an inner peace and joy that was not of me. It was of someone in whom I could trust and have complete faith. I was now in the presence of a powerful, supernatural force and knew at that instant that following HIM would be the most important decision of my life. Ever since that moment, I have never felt alone and always have that inner peace.

Having had my conversion experience, I thought it would fade with time. However, the opposite is true. It was just the beginning of a continual process of getting to know GOD and understanding HIS character. It is a transformational process. Your conversion process may not be the same as mine as GOD is diverse in HIS ways and your conversion experience is designed for you only. It is an intimate and very private experience.

To help me in this process of transformation, I have continued attending the church to which he sent me and I still do. If it were left to me alone, I know I would not have gone to that church and continued to do so, since then.

Sometimes, when driving to church, I would ask myself, why am I doing this? But I knew the answer. I really wanted to know more about the GOD that read my mind and answered my prayer and still continues to guide me.

I am always aware of HIS presence and ask HIM questions about life and my purpose here. HE guides me daily and although I may have many trials in this life, HE has promised to be with me through them all. HE has never disappointed me.

Bible quote - (Matthew 28:20- 'Behold I will be with you always, even to the end of the age')

One important thing that I learned is, if you have not been converted, you do not know HIM and you do not hear HIM when HE speaks, because the HOLY SPIRIT is not in you. You have to repent of your sins and ask HIM to reveal himself to you and guide you. Only then will you begin to know and understand the one true GOD. Don't try to imagine who GOD is; let HIM show you HIS true self.

Bible reference - (John 10:27,28 'My children know my voice and they follow me')

GOD does not change. We discussed this previously but I will now explain further. Everything in creation changes relative to everything else, but HE does not change and will never change. HE is the absolute reference, HE is perfect, HE does not need to change. This means that you can depend on HIM to keep his promises. HE makes a covenant with you, at the time of your conversation, at which time HE starts the process of transforming you into a new person, the one HE meant you to be. When you see HIM operate in your life and you look at HIS documented words on how HE communicated in THE BIBLE, you will discover that HE is the same GOD of THE BIBLE.

'HE is GOD and there is no other'. HE makes all the decisions regarding heaven and the earth. That does not mean that if you ask HIM for something specific HE will not give it to you, if it is for your good. This is also part of HIS plan. Being your FATHER, HE knows how to give good gifts to HIS children. This is one of the reasons why HE encourages you to pray to HIM. HE wants you to communicate with HIM. Once you are adopted by HIM, HE becomes your SPIRITUAL FATHER.

Bible quote - Good gifts (Matthew 7:11)

Do you understand the honor it bestows on you to be a child of GOD? The privileges it affords you? This is the ultimate experience for any human being and no other gift supersedes it. It is the most important thing that could happen in your life. I know this because I have experienced it and continue to benefit from the ongoing experience.

We know that GOD THE FATHER is the ultimate decision maker regarding all things of HEAVEN and EARTH. We see it in everything around us because it is of one mind. In addition, HIS SON JESUS, told us so. JESUS says that HE only does the FATHER'S will and HE and the FATHER are ONE. HE knows the Father's will. The HOLY SPIRIT also does the FATHER'S will.

The fingerprint of GOD is an integral part of the universe. HIS intelligence, HIS character, HIS singularity. There can only be a single, intelligent reference at the source of the universe otherwise we would be in total disharmony. Life would not exist as it does, with the human body in harmony with itself and the environment. The universe also exhibits this harmonious trait. It is all coordinated and unified indicating one controlling force. Any disruption of this harmony is a result of SIN.

Our body has several subsystems, all operating in total harmony, which also indicates that a single, intelligent source designed it. Two or more separate, independent entities cannot be in harmony with each other unless they are designed to do so. Two separate controlling forces would have conflicting goals unless fully coordinated. The universe and everything in it act as one which confirms that one unifying force or intelligence designed and controls it all. The fingerprint of GOD is the fabric of the universe.

If there were two or more independent GODS, there would be two or more controlling forces. This could not work. There can be only one final decision maker in the design, creation, and functioning of the universe and everything in it. This ultimate controlling force coordinates and choreographs all events and has the final say in everything. In reality and for all practical purposes, there needs to be a single controller just like HE tells us. GOD THE FATHER, GOD THE SON, and GOD THE HOLY SPIRIT have separate responsibilities and there is no conflict.

All the knowledge and all the intelligence that will ever be already exists in GOD'S world. It is all that there is and all that there needs to be. This is the prefect world that JESUS calls PARADISE. It is perfection defined. It gave rise to our universe but our universe is now flawed as a result of our doing, we SINNED! We disobeyed GOD'S spiritual laws the consequence of which is death. This is total separation from HIM, the true reference. But there is Hope! HE has given us a way to redeem ourselves!

The Trinity

ENTITIES. There are 'GOD the FATHER', 'GOD the SON' and 'GOD the HOLY SPIRIT'. If this had not been revealed to us we could not have deduced it by logical reasoning. This is a supernatural concept and so not within our reasoning ability.

One very important attribute of this relationship is that, even though THEY are separate ENTITIES, THEY work together as ONE. THEY each have separate responsibilities all aimed toward one goal, creating and maintaining a PERFECT EXISTENCE.

THEY existed before time, in fact, THEY created time. We are not able to perceive such an existence as our perception is limited in time. THEY exist outside of time.

Physicists have determined that time is relevant only to our universe. This is consistent with what GOD has told us about HIMSELF, being ETERNAL.

THE TRINITY is a perfect union. There is no conflict. The THREE MEMBERS execute THEIR responsibilities flawlessly.

GOD THE FATHER is the decision maker whereas THE SON and THE HOLY SPIRIT execute the FAFHER'S will. Once you are converted, the HOLY SPIRIT is sent to guide you in the ways of the FATHER so that you are transformed into the image of his SON. This transformation process may be slow, so don't get discouraged that sometimes it is as if no change is occurring. Once you have been converted you can be assured that the transformation process will continue unto completion. You become one of the chosen few and will remain HIS eternally.

Knowing God

What I am trying to convey to you is not about religion but about developing a direct relationship with your 'CREATOR'. Even if you do not believe such a BEING exists and would have the desire to communicate with you, as in my case, 'Just do it'. Ask HIM to reveal himself to you and guide you throughout your life. You have so much to gain from asking. Is it not reasonable to deduce that, if there is a GOD, this ENTITY must be able to communicate with you? Having reached that conclusion the next step should be an attempt to communicate with HIM.

Since you do not know HIM or where to locate HIM, you must ask HIM to reveal HIMSELF to you. The way is described in HIS holy book, THE HOLY BIBLE.

Trust me, you don't have to say it out loud. I did not! Yet HE read my mind and responded.

GOD speaks the universal language that we all understand but of which we are unaware. HE can contact you any time HE wishes. HE has given us his 'WORD' (THE HOLY BIBLE) and wants us to willingly come to HIM.

GOD initiates the relationship and this is part of HIS plan. The way is described in HIS book, THE HOLY BIBLE. If HE did not influence us to turn to HIM we would go in the opposite direction and continue in 'darkness'. However, once you submit and HE chooses you, you cannot change your mind. When you fully understand what is happening to you, you do not want to resist, you willingly submit. You then begin to wonder how you lived your life before, without HIM. You then humble yourself and submit to HIS guidance. After the initiation of a relationship with GOD, you will come to find it to be the most satisfying relationship you will ever have in your lifetime.

When you are called to be in a relationship with GOD, you can only enter HIS presence if you are pure. Since none of us is pure, a sacrifice of absolute purity had to be made. This is one of the requirements of a 'JUST GOD'. JESUS, became the LAMB, the perfect sacrifice to save us. JESUS, one of the TRINITY, had the purity to save us.

While you are still on this earth, the HOLY SPIRIT, the COMFORTER, is with us to help lead us to all TRUTH. You will, however, have the opportunity to be with GOD when you pass from this world and are fully JUSTIFIED.

When you read GOD'S word, you begin to develop an understanding of how HE communicates with us. Some of the ways HE communicated with people thousands of years ago are some of the same ways HE communicates with us today. HE communicates with us directly, as well as indirectly, through other people. In my case, HE told different people the same message or different parts of the same message so that when it got to me, it was communicated completely, and with redundancy. This ensured there were no errors in communication. When all put together, you understand the full message. In each case the Person interceding is the HOLY SPIRIT.

It is not possible for nonbelievers to understand this process because it is a supernatural process. I know this because I was once a nonbeliever. When I first read the Bible, it was just a story to which I did not relate. After my conversion, its whole meaning changed. Now, when I read the BIBLE, my FATHER is speaking to me through his word.

One of the most comforting things of being one of GOD'S chosen children is the 'peace' that HE gives you even in the most difficult circumstances. It is truly a 'peace' that passes all understanding'. It is a supernatural 'PEACE'.

GOD always keeps HIS promises. You can therefore depend on HIM to be with you through all your trials.

GOD made man in HIS own image. This includes our emotions and our ability to create. Remember, HE created an ordered system and this is an indication of HIS character.

To help us understand who GOD is, we can examine ourselves, our approach to creating things and our emotions. The difference is that HE is perfect and all of his character traits are designed to maintain order.

GOD tells us, and it is also my belief, that everything comes from and is of HIM. Without HIM we are nothing. We would not exist. Something happened to break (we sinned) our direct connection to HIM and as a result, we are lost.

Bible quote (John 1:3, KJV - 'All things were made by HIM; and without HIM, there was nothing made, that was made')

You cannot find GOD because you don't know who HE is or where to find HIM. Can one even begin to look for someone if you do not know who you are looking for or even where to start? GOD has to reveal HIMSELF to you. Otherwise, HE is completely out of reach. Remember, alone, without GOD, we are lost and without the 'TRUE' reference. If you are yourself lost, how can you find someone whom you do not know?

I do not believe that, if you ask, HE will not respond to your request to reveal himself to you. HE may even reveal HIMSELF to you before you ask if it is a significant part of HIS plan! When you read about Paul's conversion, in the bible, he was not given a choice. He was persecuting GOD'S people and was struck down, made blind and converted in an instant. GOD had a specific purpose for him in HIS plan, to become one of HIS servants and be one of HIS advocates. Paul was uniquely qualified and so was selected without question.

GOD may also answer your prayer when you ask or some time later. HIS response is based on HIS plan for you. GOD knows your heart and this is how you are judged. If you genuinely repent of your sins and ask GOD to forgive you and guide you, it is because HE has already selected you for conversion and the process has been initiated.

GOD is merciful and therefore willing to save us. We also can be merciful. But, because of our SINS our mercy is flawed. HE is a JUST and perfect GOD, HE knows our hearts and so will provide the perfect response when we pray to HIM. We are far from perfect and so should not even try to predict what HIS response will be. What I do know is that HIS response will be 'JUST'. If you are sorry for your SINS and want to make a genuine effort to do good (repent) HE will respond favorably to your request.

Bible quote (1 John 1:9 - 'If we confess our sins, HE is faithful and just to forgive us our sins and cleanse us from all unrighteousness')

Just like JESUS spoke to us in parables or symbolic terms, when the HOLY SPIRIT speaks to us it can also be in symbolic terms. Like the vine and the branches, if we are not connected to JESUS we will perish. Everyone can understand that. When GOD said to man "Be fruitful and multiply", fruitful means bear a lot of fruit or have many children.

Bible quote - (John 15 - 'I am the vine and you are the branches')

I have included an abstract representation of creation which was custom painted by a true believer, Jack Sependa, at my request. I will also interpret the painting since it may not be immediately apparent to most readers. (See Painting below)

GOD calls HIMSELF "THE ALPHA AND THE OMEGA". GOD is the center of the universe. HE is surrounded by white light. God spoke creation into existence as shown by the exploding colors representing creation. Everything is of GOD. Light connects the universe and it is no wonder JESUS calls himself 'THE LIGHT OF THE WORLD'.

This painting was done in one day. Jack has a unique approach to getting inspiration before he starts each painting. He puts a symbol of the cross on the blank canvas and then prays to GOD for inspiration. The adjacent painting was inspired by the HOLY SPIRIT.

The HOLY SPIRIT also revealed a prophesy to Jack, which he relayed to me in real time as we sat for breakfast at a Hotel in Santa Fe. I knew it was real

ABSTRACT REPRESENTATION OF GOD

because, as we were talking, he suddenly stopped and tears came to his eyes and his words were, "John, the HOLY SPIRIT is giving me a message for you". It is this message that is outlined in this book. I will tell you the full story some time in the near future.

God's Fingerprint

Everything we do, say or think is ordered if it is in harmony with the universal order. This principle applies to all aspects of the universe. If we extrapolate, we can conclude that the universe, being ordered, originated from the same intelligent source. We all, therefore, originated from the same fixed reference from which everything originates.

'Absolute Intelligence' and 'Never Changing' are two of the attributes of GOD. It is therefore much more logical to conclude that GOD created the universe rather than a 'random' occurrence such as the evolutionist's 'Big Bang'. 'Random' is foreign to the order which we see and to life as we know it.

GOD stays in contact and in complete control of the universe and everything in it even if you may not believe this to be true. HE intervenes when necessary otherwise we would be completely out of control. None of us has the knowledge or ability to maintain order in such a complex system because there can only be a single source of control. We cannot each be a reference as this would lead to confusion. There can only be one ABSOLUTE SPIRITUAL and MATERIAL reference. One fixed reference. One perfect reference. Looking at these requirements we can see that none of us qualifies as we are all imperfect.

GOD'S attributes are reflected in the universe. He is our rock, our reference, HE tells us so and HE does 'NOT CHANGE'. This is critical for creation and in any attempt to construct or fabricate an ordered system using the materials available to us, matter, as we know it.

Based on what we have already discussed, is it not reasonable to conclude that this universe and everything in it must be of intelligent design?

Only a source with intelligence can create an ordered system. Intelligence defines order. This is the only logical conclusion.

Spiritual Laws

'MAN'S' current spiritual nature is such that we do not have the ability to live peacefully with each other. Disharmony begins with family conflicts and spreads to communities, cities, countries and the entire world. 'MAN' cannot break 'MATERIAL' laws, but can and does break 'SPIRITUAL' laws.

To try to maintain order in our world, we develop and implement laws relative to our conduct and a means to enforce them. These laws are based on standard ethical practices that all are expected to obey. These are not 'MATERIAL' but 'SPIRITUAL' laws based on BIBLICAL principles. They are behavior based. Because of SIN, if we did not have laws, society would end up in a state of anarchy.

There will always be some of us who are unable to comply with the laws of the land and have to be disciplined. This is an indication that we are flawed.

In order to maintain absolute order, no disorder can be allowed. The laws of GOD are much more strict than the laws we have developed. None of us, on our own, is capable of obeying all of GOD'S laws. HE created a perfectly ordered system but, through what HE calls SIN, we introduced disorder, and became separated from HIM and HIS guidance.

The first created human being, ADAM, made the choice of eating from 'The Tree of Knowledge of Good and Evil' so he became aware of good and evil. In so doing, he introduced disorder as he now had the ability to choose evil, which he did and we all continue to SIN.

God made us in HIS image so we can make our own choices except the one which HE withheld from us. HE gave us direction as to the path we should take, we disobeyed and are now suffering the consequences. We lost our absolute reference and as a result, lost our way.

We have discussed how a fixed reference is critical to an ordered system. Once we were disconnected, we substituted our own references which is not universal and absolute. We have not only ourselves made our own references, but we sometimes keep changing the reference. We can only have harmony among ourselves if we are following the 'ABSOLUTE SPIRITUAL REFERENCE' which is fixed and never changes. When we each have our own reference, it is impossible for us to live together in harmony (GODLY). We become a disordered system. We introduce negativity into our behavior, hence disorder. We are still intelligent beings so we can create order, but we now choose to create disorder in our society, resulting in disharmony.

Our FATHER does not tolerate disharmony as this is not a part of his nature. HE is a 'MERCIFUL' GOD and so HE has given us a way to redeem ourselves.

Since HE is also a GOD of 'JUSTICE', it was not possible for HIM to forgive us without a sacrifice. Remember, the consequence of SIN is death. This is one of GOD'S SPIRITUAL LAWS. Since we now have a degree of disorder within us, none of us is capable of making this correction on our own. Once a degree of disorder comes onto a SPIRITUAL system, the only way it can be corrected is by making a 'SPIRITUAL SACRIFICE'. This satisfies 'SPIRITUAL JUSTICE' which must be upheld. HE therefore sacrificed a part of HIMSELF, HIS son JESUS, as none of us was pure enough to be or to provide this sacrifice. HE did this because 'HE LOVES' us.

Satan

SATAN is a fallen ARC ANGEL who was cast out of heaven, to earth, because of his rebellion against GOD. He is a negative (Evil) force that works to create disorder in our spiritual realm. He is against GOD and his people. He attacks us by trying to influence the way we think and reason, and so tries to manipulates us into doing his will. He tempted and influenced MAN to SIN against GOD and once this happened he then had power over MAN.

SIN is the result of this negative force in our lives. It destroys anything good and ordered and so must be stopped. Its aim is to destroy anything GODLY because GOD represents all goodness and order. If we are believers in JESUS, SATAN only has power over us when we give it to him through SIN. By ourselves, we do not have the power to overcome such an evil force so we have to rely on GOD to help us.

When JESUS died for us on the cross, JESUS became our SIN (see Bible quote below) and believers have been given power over SIN and SATAN. Jesus destroyed the works of SATAN and took his power away from him. HIS people now have power over SIN, but can still sin because they are not yet perfect and still possess part of their old selves. They are still tempted and sometimes make the wrong choice. They will not be perfect until they are again fully joined with GOD.

Bible quote - (2 Corinthians 5:21 - 'For HE made HIM who knew no SIN to be sin for us, that we might become the righteousness of GOD in HIM')

The Bible tells us that the thief (SATAN) came to steal, kill and destroy but that JESUS came that we would have life and have it more abundantly.

Man

The TRINITY has always existed. The FATHER designed and planned creation together with his son JESUS and the HOLY SPIRIT. They have always and will always work in harmony. MAN was the subject of creation. Everything else was created to accommodate us.

MAN was made in GOD'S image. This means that MAN was given similar attributes to GOD'S. Some of the attributes given to man are creativity, love, empathy, mercy and even hate. The difference is that GOD uses HIS attributes in accordance with HIS holy laws in order to maintain harmony in heaven and on earth. HIS attributes are absolute in all respects.

MAN was given free choice but was also given a limit as to what he was allowed to do.

However, man sinned and was disconnected from GOD. In so doing, we lost our moral reference and became morally and spiritually corrupt. SIN is rebellion against GOD'S SPIRITUAL laws, thus creating moral and SPIRITUAL disorder. God cannot tolerate SIN because it disrupts SPIRITUAL order. Those that SIN are rejected from HIS ordered world. SIN is punishable by death. But, even in our SIN, GOD still loved us and sacrificed HIS SON to save us from eternal death. Eternal death is eternal separation from HIM. JESUS willingly gave HIMSELF as the sacrifice to save us from eternal death.

GOD is also a GOD of 'TRUTH and JUSTICE' and made laws designed to uphold spiritual order that must be obeyed. If we disobey them, we will most certainly suffer the consequences. This is where we come face to face with the fact that GOD has absolute authority over all things and commands our awe, love and respect.

Order cannot survive with disorder active in its midst. The only way to eradicate disorder is to attack it with a perfectly ordered system. JESUS was that perfectly ordered system.

In any ordered system, there can only be one ultimate leader. GOD is that ultimate leader.

When man first sinned, he immediately inherited the death sentence. All of us being direct descendants must also suffer the same fate, as the effect is systemic.

Fortunately, GOD is merciful and has given us the opportunity to be accepted back into HIS fold, if we confess our sins and commit to HIS guidance in our lives. We must have faith that he loves and wants the best for us.

With all this in mind, we should realize that HE did not have to create us. He did not have to give us the opportunity to live, know that we 'ARE', be able to interact with our environment and to give and receive love.

HE was perfect without us, but wanted to share HIS 'WONDER' with us. This is the greatest gift you could ever imagine.

Some say that GOD'S appreciation for loving and sharing comes from the fact that HE has always shared such a relationship with HIS SON and the HOLY SPIRIT. THEY are three separate entities but act as one. EACH ONE has a separate responsibility but act as an integrated unit.

We are eternal beings but in order to spend eternity with GOD, we have to repent of our SINS and commit to following his guidance. Otherwise, we will be eternally separated from HIM. I don't think it takes much debate as to which option we should choose.

GOD has shared intelligence and knowledge with us. But GOD has the power to conceal knowledge and wisdom from us, if HE so chooses, and HE sometimes does. No matter how intelligent you think you are, you can never discover certain truths unless HE chooses to reveal them to you. HE will conceal such knowledge from you until HE is ready to reveal it to you. This is because HE is and will remain in control. HE has a plan and all revelation will be done in accordance with HIS plan.

Each of us has a purpose in life. The ultimate goal is for goodness and 'TRUTH' to prevail. However, bad things do happen but in the end, everything, even the bad, now happen for the ultimate good.

We are each designed to perform different tasks over a lifetime and are given special gifts to perform these tasks effectively and efficiently. GOD has a plan for each of us and has given us the necessary gifts to carry out HIS plan. HE coordinates our actions to fulfill HIS overall plan for the world. HE remains in total control. We cannot change this plan as we do not have the power to alter it.

Since we are lost, we create our own order which may not be in harmony with the universal order. When we chose our own path, by detaching ourselves from our source, we became lost. Once lost, we have no true reference and it is impossible to find our way back. Only GOD, who knows the path, can direct us back.

If we examine what it is like to be lost, unless someone who knows the way directs us back to the true path, we cannot find our way. To find the true path, we either have to be lead back by someone who knows the way or find a reference, on that path, which we can use to guide us back. GOD is the only one who can lead us back to the true SPIRITUAL REFERENCE. It is impossible to find the reference as we do not know what to look for on whichever path we may travel. Even if we saw the true reference we would not recognize it. GOD must reveal it to us.

God made us in his own image and gave us free choice but limited our knowledge and intelligence. How could he have given us all his power knowing we could not be trusted once we were given free choice and access to the knowledge of good and evil.

Since we had nothing to do with our being here, we have no reason to be proud of our intelligence and ability to accumulate knowledge. Whatever we are able to accomplish are gifts freely given to us.

Jesus

Jesus is the one and only SON of GOD. When man sinned, the only way he could be saved was if a 'SPIRITUAL' sacrifice was made to satisfy GOD'S requirement for JUSTICE. There was nothing in creation that could meet these requirements because we are all tainted with the curse of 'SIN'. Only the prefect sacrifice was good enough and JESUS willingly became that sacrifice. He still loved us even though we were now SINNERS and no longer loved HIM.

GOD never stopped loving us even when we did not love HIM. HE was willing to sacrifice HIS SON to redeem us and JESUS was willing to be the SACRIFICIAL LAMB. This is the true definition of unconditional love.

When I read the Bible, I always consider JESUS'S quotes to be profound and with much more depth than initially meets the eye. We must always remember that HE is part of the TRINITY and so sometimes speaks in absolute terms. HE speaks universal TRUTH. HE was there in the beginning and so participated in everything that was made. When HE makes a statement such as 'l am the (spiritual) VINE and you are the branches, without ME you cannot live'; This is absolute TRUTH, in every sense. The branches of a tree must always be attached to the vine. The vine or trunk supples the nutrients of life. The same applies to the human body. The body parts must all be attached otherwise they die. Earth supplies water through rain, rivers and streams; otherwise, life could not exist on our planet. The elements of life must be constantly flowing in the body in order to support life, any disconnected part will die.

Jesus is our only hope for eternal life as HE died for us so we can now be reconnected with GOD through HIM. The closeness of the relationship between the GODHEAD and us is seen in the way JESUS refers to us once we are saved. JESUS tells us HE is now our 'BROTHER' and GOD is our 'FATHER'. This is an indication of who we are in the family relationship with the 'FATHER' and the 'SON', once we are adopted (SAVED) by the FATHER. But, to gain this honor, we have to be BORN AGAIN not of flesh but of the 'SPIRIT'.

GOD tells us that the first step to redemption is to accept HIS SON JESUS CHRIST as our savior, repent of our sins and commit to following HIM. GOD then reveals HIMSELF to us directly (through the HOLY SPIRIT) and through HIS 'HOLY' book, 'THE BIBLE'. However, in order to properly interpret the BIBLE we need the HOLY SPIRIT to guide us in understanding it. It is impossible to understand the full meaning of the BIBLE without the help of the HOLY SPIRIT. But GOD'S word, as manifest in the BIBLE, is how we are initially made aware of HIM and are converted.

After you are converted, you become one of GOD'S chosen people. HE cares for you and nourishes you like a mother would a child. HE sends you the HOLY SPIRIT, the HELPER, COMFORTER, SPIRIT OF TRUTH to assist you in knowing and understanding HIM. This is what is referred to as being 'BORN AGAIN'. You are now born of the 'SPIRIT'.

Once this happens in your life, this means that you have been selected to be a SPIRITUAL CHILD of GOD and GOD THE FATHER is now your 'FATHER'. From this point onward, you are forever saved and no one can take you from JESUS'S hands. GOD selects you and gives HIS SON, JESUS, authority over you. JESUS is now given the responsibility to protect and care for you and you are guided by the HOLY SPIRIT. Your life is changed forever.

Bible quote - 'No one will snatch them out of my hand' (John 10: 28-30)

A transformational process now begins in your life which will continue until you are fully made into the image of JESUS. While you are still here, the process will not be complete, but will be when the FATHER calls you to be with him and you are fully JUSTIFIED.

If we consider this objectively, we see that GOD sacrificed HIMSELF to save us since GOD and JESUS are one. JESUS humiliated HIMSELF, being cursed by us and put to death on a cross. During this period 'JESUS' was separated from the 'FATHER', giving up HIS power as a GOD for 'SINFUL' man.

JESUS lived with us from birth to mature adulthood, experiencing our world first hand and overcoming all its temptations. HE is the only one that has been able to do this in human form. HE was tempted much more than we are but as HE so aptly said 'I overcame the world'. HIS promise to us of eternal life depended on the fact that HE overcame the world.

Bible quote (Jesus overcame the world- John 16:33)

Overcoming the world also means that HE overcame DEATH. This is important to understand because of HIS promise of eternal life. The only way HE can make such a promise to us is if HE has power over death, which is the only thing that separates us from eternal life. Having died for our SINS, JESUS was given complete authority over us. Even though HE is our brother, HE is also our KING and commands our respect.

Jesus rose from the dead on the third day after being crucified, just as HE prophesied. HE now sits on the right hand of the 'FATHER' until HE comes back for HIS chosen people.

Sin

SIN creates disorder. It is systemic. It not only affects our thought process but also matter, the material world. SIN results in our separation from the ABSOLUTE (spiritual) reference. GOD is the absolute reference and had no choice but to separate us from HIMSELF as a result of our breaking HIS spiritual law. By SINNING, we caused a disruption in the spiritual realm to which we were directly connected through HIM. This is why we had to be separated from HIM, as SIN has no place in that perfect world.

The order given to man was clear. 'Do not eat from 'The Tree of Knowledge of Good and Evil.' But man did eat from that tree and as a result was cursed, lost the absolute reference (GOD) and substituted it with his own limited reference which is in disharmony with universal order.

God is completely and totally ordered and made us in HIS image. However, we chose to introduce disorder (SIN) by disobeying HIS command. Once this disorder was initiated, it spread like a cancer affecting everything and everyone with which it came in contact. It is a destructive force. It is non-productive and has no place in an ordered world. SIN'S only function is to create disorder and so it is unacceptable in an ordered world.

As indicated above, SIN has affected not only our minds, in the way we think, but also matter as it exists (in ordered systems) in our universe. You see, GOD also cursed the earth, and SIN is systemic. What was initially ordered and perfect is now tainted with disorder.

HE is a 'JUST GOD' so HE had to sacrifice part of HIMSELF to redeem us and he made a covenant with us to again accept us as HIS own. Since we could not redeem ourselves, the sacrifice was HIS SON JESUS.

Love

There are four types of love described in the Greek language.

These are: Storge - Empathetic love

Phila - The love of a friend

Eros - Romantic love

Agape - Unconditional love

The love I will address is 'Agape' or Unconditional Love, which is the love that GOD has for us. This LOVE is the most powerful and binding force in the spiritual world. It is a spiritual law. It is a commitment to maintain a covenant relationship with another which does not change (it is unconditional). Love, the emotion, is secondary as this is subject to change based on the emotional state of the parties involved.

UNCONDITIONAL LOVE incorporates the SPIRITUAL reference that binds us all together in perfect harmony.

Treat another individual as you would have yourself be treated. We should love one another in spite of challenges we may encounter in the relationship and not only because of the benefits we will get from it. This love is self sacrificing and will make sacrifices to ensure that order is maintained in the relationship.

(1 Corinthians Chapter 13 defines unconditional love)

LOVE is an order producing force. This is the reason why our GOD has told us that 'Of the three, FAITH, HOPE and LOVE, the greatest of these is LOVE'.

Bible quote - (Greatest of these is love, 1 Corinthians 13:13)

Our secular definition of love is selfish. It is self serving and so not the same as the 'LOVE' GOD has for us. When we say we love someone, we typically mean that the individual satisfies a selfish need in us. Without them we are unfulfilled. As we examine this concept, we see that our definition of love is not giving but receiving or taking. In other words, expecting something in return for the 'love' we give.

GOD'S LOVE gives not requiring anything in return. Similar to the love a mother feels for her child. It is instinctual and ordered. For order to be maintained, we must overcome outside influences and use only the 'TRUE', absolute reference as a guide. LOVE is an ordered system. One should love not because of the benefit to us but because of the benefit to others. The benefit to you will follow. It is the natural order of things. Remember, GOD still loved us even after we SINNED.

Love is a conscious decision that we make that sometimes sacrifices our own happiness. This is necessary to ensure that order is maintained in the universal system. This is critical in our fallen world. Each part or member of the universal system must sometimes make sacrifices for the whole.

In our fallen world, this is essential for order to be upheld. It therefore seems logical that, to save us and help us return to order, GOD had to sacrifice HIMSELF in the form of HIS SON. This is what 'UNCONDITIONAL LOVE' does. It is self sacrificing.

No human being could ever have come up with such a definition of love as we typically think selfishly. If we look at the 'Big Picture' of us being the ultimate ordered system in our world, it becomes clear that this LOVE is critical to maintaining order. It binds us together, making us unified. Its cohesive force is necessary for our survival. If there is no love, there is no unity. If there is no unity, we become self destructive and will become even more disordered and lost. We will have become totally detached from our universal reference, on our own, and unable to get back on course.

In order to stay on course, we need to understand the definition of UNCONDITIONAL LOVE and practice it. This is not something we can do on our own because most of us are already lost and look at the world from a selfish perspective. We may think this selfish perspective is natural but it is not conducive to the universal order. We need the guidance of the absolute reference, but first we must recognize that we are lost and need help.

The only ones that will be redeemed are those who recognize the need for help and ask for and receive guidance. Even after conversion, we still need constant guidance and support to keep on track. It is a daily battle and we must depend on continual guidance from our ABSOLUTE REFERENCE, our HEAVENLY FATHER.

When GOD created us HE created a need in us that only HE can fulfill. Things of the world will never satisfy this need.

Trust

Trust is a concept that is so delicate it takes only one mistake or act of mistrust for it to be completely lost. Trust develops from one being consistently trustworthy and so it is earned. Someone we trust is someone we can depend on to look objectively at the situation at hand and judge impartially. Such a person can be depended upon to tell the truth and be fair in every situation. Once an individual is given the honor of being trustworthy, he or she is used as a standard and expected to continue acting in character. Any change in this individual that questions this attribute can result in trust being lost. (Always remember, GOD does not change and so he can always be trusted)

Trust can sometimes take years to build but can be lost with a single negative act. Consistency is critical for trust to be maintained. This is another example of how important it is for us to live by unchanging, upright standards and be a 'light' to all.

If our reference is not absolute, there is always the chance for error. The only absolute reference is GOD. GOD never changes. We can absolutely trust HIM. For us to trust someone, it is critical that the person remains the way we perceive them. We develop trust over years of association with someone by analyzing their actions and responses in various situations. Trust is very fragile and so we should always be aware of how we are perceived and uphold the highest ethical standards.

God's Forgiveness

GOD promises to forgive us if we repent of our sins, humble ourselves before HIM and commit to keeping HIS commandments. This is necessary in order to recreate the order that was disrupted. There is a consequence for breaking spiritual laws and a sacrifice necessary to redeem us. Forgiveness is only possible if a suitable sacrifice is made. This is GOD'S JUSTICE and is one of HIS attributes.

There had to be a sacrifice for HIM to again accept us as one of HIS own. None of us is worthy to be that sacrifice as we are all impure. HE therefore sacrificed a part of HIMSELF for us by sending HIS SON to die for our sins (GOD and JESUS are one). There is no sacrifice we could have given HIM that would have been sufficient because we are all unworthy. Only the perfect sacrifice would be acceptable to a JUST GOD. We do not qualify and nothing we posses qualifies.

Only a GOD of LOVE, MERCY and JUSTICE would consider doing this for us. HIS SON, also part of the Godhead gave HIMSELF to be that sacrifice. This is a very powerful and humbling acknowledgement.

Our Forgiveness

When we forgive someone of an offense against us, we are not only forgiving the person but are also freeing ourselves from dislike, hate or contempt for that person. Holding resentment against someone that has offended us hurts us more than it hurts the person who committed the offense.

This is one of the reasons why forgiveness is so important as it provides us relief from the burden of dislike, hate, and contempt.

Prayer

Prayer is direct communication with GOD. We know that GOD has a plan for us and a plan for creation but HE still encourages us to pray to HIM. In prayer, we can ask for blessings even regarding specific things. HE will grant us such requests if they are things that are 'GOOD' for us.

Even though GOD has a plan for us, HE will listen and respond to our prayer requests. HE gives gifts to HIS children and, being our father, HE knows better than we do how to give good gifts to HIS children.

Bible quote (Good gifts - Matthew 7:11)

Our prayers should always include requests for wisdom so that we will better understand what he desires of us. But most of all, we should recognize and address GOD'S sovereign power over us. Even JESUS said, 'Not MY will but YOUR will be done'

Bible quote 'not my will (Luke 22:42)

The more we pray to OUR FATHER, the faster and deeper our relationship with HIM will grow. When we pray we should believe or have faith that our prayers will be answered. GOD will always answer our prayers but the answer may sometimes not be what we expect. Remember, HE only gives us things that are good for us or for the universal good. HIS response may be delayed so you must be patient. HE does everything in HIS own time in accordance with HIS plan. Sometimes HE really blesses us and gives us even more than we expect.

Always remember that prayer is direct communication with GOD and the best way of getting to better know and understand HIM. Also, remember that you are in communication with 'GOD' and must remain humble. You are HIS child and HE loves you. In this context, you can talk to HIM about anything and should. HE will always look out for you and give you good advice.

Faith

Bible reference – Hebrews 11:1' Faith is the substance of things hoped for, the evidence of things not seen.'

Faith is believing in something you cannot see or even rationally explain. However, you believe because of GOD'S supernatural influence.

True faith is not initiated by us but by GOD. We know that it is supernatural because it does not fade but gets stronger with time as our relationship with GOD grows. It is not something that can be rationally explained. Yet we know it is real and we can put trust and have confidence in its source.

FAITH cannot be fully understood by someone who is not a believer. Typically, non believers argue that unless you can see it and touch it, it is not there, it does not exist, it is all in the mind.

Having true faith is an indication that one is in GOD and the HOLY SPIRIT is in you. A measure of FAITH is one of the gifts given by GOD when you become a believer and you are adopted as one of HIS chosen children.

With this FAITH, you have no doubt that GOD exists and that HE created the universe and everything in it. When you read the Bible, you believe everything HE says about himself, what HE says about the world and what HE says about you. You will find that HE tells you things about yourself that you find later to be true. One can only conclude that, if HE knows more about you than you know about yourself, then HE must have created you. Your faith then grows even deeper when these truths are revealed to you.

Hope

Hope is the desire that things wished for will come true. If we pray for something, we hope it will come true. But, it is up to GOD as to whether or not HE will grant you your desire. You should always remember that HE will only grant you things that are good for you or contribute to the universal good. We may not always ask for things that are good for us so don't be disappointed if you do not get the response you desire.

Even when bad things happen, there is good that comes out of that experience or event. We should remember that we are not in a position to see the big picture, but HE is. Things will always go according to HIS divine plan and we should not question it.

Parents, understand that your children are 'from' you but not 'of' you. You do not own them. They are all GOD'S children, as are you. Their destiny is up to HIM and you need to respect HIS sovereign will.

The fate of your child is not for you to decide. It is for GOD only to decide. I know this is difficult to accept but this is the truth. You may believe you know what is best for your children and you do not want to see them harmed or suffer but, ultimately, it is GOD'S will that will prevail. Accept that His will provides the best outcome even if you may not understand it now.

With the above in mind, we can still hope that what 'we' consider all good things will be given to us. We should always remember that we live in an imperfect world and will experience suffering. But, if you are one of GOD'S chosen children, HE will always be with you through any difficulties or trials this life presents.

Final Message

There is a fixed reference that connects all of us and everything in the universe. This fixed reference is an intelligent source. The intelligent source is God.

The only reason I have been able to write this book is because I have a fixed reference, the true reference, my FATHER GOD. My focus was always on HIM to guide me in every thought and word. Whenever I had a problem with expressing a thought, I would ask HIM to tell me what to write to convey the intended message. In fact, HE has guided my thoughts throughout this process, from beginning to end. My hope is that this book has reflected HIS divine guidance and offers some more clarity as to your purpose in this world. Keep in mind that I am human, but I think I have accurately recorded the basic principles I am trying to convey to you, the reader.

Throughout my life I have been in training to write this book. Looking back, I now see the complete picture as all aspects of my life are relevant to the included message.

The idea for this book was given to me by my 'FATHER GOD'. HE sent me the message that I would have to do nothing and 'IT' would all come to me. That was so true as all I have written have been my life experiences and training as an engineer.

HE showed me how to look at the 'big picture' and put it all together. I now see that throughout my life I was being prepared for this project. I take no credit for the message, I am only the messenger.

GOD knows that we are flawed, yet HE loves us anyway. HE knows that our tendency is to go the other way, yet HE loves us anyway. HE accepts us the way we are because HE knows we cannot change ourselves. HE knows HE is the only one that can direct us back into harmony with the ordered system that HE created. By ourselves, we are lost, we have no 'TRUE', fixed reference.

We should not be self righteous believing that we know what is best for us. Believe me, any one of us is capable of committing any sin or find ourselves in difficult circumstances, so we should not look at others and say, 'How could they do such a thing'?. Hence the quote 'There but for the grace of GOD go I'. GOD is the only one that keeps us from being the worst we can be.

We have too little information and knowledge to be able to make some life changing decisions by ourselves. We need HIS help. We are like a small blip in the annals of time. We are insignificant in the overall order of things. HE is the only one that has the power to give us any significance so we have to look to HIM for guidance. Who better to look to for guidance than your HEAVENLY FATHER.

'GOD chooses ordinary people to do extraordinary things'.

DEFINITIONS AND CONCEPTS

Here are some 'definitions and concepts' to keep in mind as you analyze the principles in this book.

Reference

The reference defines the origin as well as each subsequent point in an ordered system. This is a critical component in any ordered system. The reference must always be active to maintain influence or control over the whole. It must be set by an intelligent source. This is a universal truth.

The absolute reference governs the laws of matter as well as the way we, as humans, behave. It is the source of the fabric of both the mental (spiritual) and physical manifestations of the universe.

When awake, we are aware of where we are at all times. If we are not, then we are in trouble; we are lost. We are always aware of our location because we have stored references from our experiences to which we have multiple access. We navigate using references both to locate where we are and in the way we process information. This is how critical it is to always have a reference. It is important that we understand the significance of a fixed reference as this is an essential component in maintaining order.

If we want to start on a journey, to go from one place to another, we must take the first step. Then we continue in a direction which we know will take us to our destination. Through all this, we need to keep in mind where we are going and use references along the way. This is a natural process. If, however, we lose our reference, we become lost. We will never be able to continue on our journey if we don't find a reference on the original path so that we can regain our bearing. This is a simple illustration but it applies to all ordered systems. We have no idea where we are until we find a reference in the ordered system that we are navigating.

If we become lost, we have to ask someone for directions. We have to communicate very clearly where we want to go and the person we ask must know the way. We can then be guided back to the original path to get to our final destination.

Without a reference, we are unable to process information. Our brain must search for and find a reference in order for us to think and interact with the environment. Intelligence allows us to perform the process of thinking and to gain and use knowledge.

Life is all about references. Without references, we would be unaware of our existence. In the absolute sense, without a reference we would not exist. GOD THE FATHER, GOD THE SON AND GOD THE HOLY SPIRIT have always existed. Since we did not always exist, for us, there must have been a beginning, a creation that started in the mind of GOD. This is our initial reference and is fixed in time. Our experiences then begin in a sequential order, fixed in time. We are then born on this earth and our experiences then begin in a sequenced order fixed in time. Memory allows us to tap into that sequence in an ordered manner.

Our brain is the medium where all of this takes place. It is designed to record this sequence and for us to access this information at will. Our brain allows us to breathe, walk, talk, hear, smell, feel, think and solve problems. The brain is also the connection between us and the material world in which we live.

Our 'BORN AGAIN' recreated spirit is the gateway to the 'SPIRITUAL REALM' where all things are possible. It connects to the 'SPIRITUAL REALM' where ideas originate to initiate our creative ability. Much of our knowledge and wisdom is communicated to us in this manner.

Our absolute reference is spiritual and is eternal. Using the absolute reference as the point from which everything originates, there is nothing new under the sun. Our universe has always been part of GOD'S plan and at some finite point in time, HE spoke it into physical existence.

Here are some everyday examples of fixed references being used to regain order and perspective in an ordered system.

Computer Reboot - program goes back to its reference point to regain accuracy.

Negative Feedback—Part of the output fed back to an input stage to reduce distortion. This cancels unwanted signals introduced by the circuit. Differences between input and output are cancelled except for proportional input amplification. The input signal is used as the reference.

Absolute Zero Temp- thermodynamics. Temperature at which all atomic movement ceases. This is the 0 energy reference.

0 - Zero - mathematics —The reference

In any experimental test, there has to be a control or reference that is the standard to determine if the experiment has succeeded or failed. Even our name and signature are types of references.

The universe has to have a singular reference otherwise there would be total chaos. It all has to be coordinated under one unifying force as this is the only way it can operate in harmony.

We have shown this by looking at isolated ordered systems and noting that each has to have a fixed, singular reference for the whole to act as one. If there are several subsystems, as in the human body, then they all have to be under one central command in order to operate in harmony. The control center must be intelligent. This feature is critical in any ordered system. In the human body, the control center is the brain.

Reference is an important component of any ordered system. However, what is also critical is that the 'TRUE' reference be used. If the TRUE reference is not the foundation of the ordered system, then it will not be in harmony with universal 'TRUTH'. This would mean that we have developed our own truth which will lead nowhere but to eventual failure.

If you don't have a reference you are lost. If you don't have the 'TRUE SPIRITUAL' reference, you are truly lost. There is only one 'TRUTH' and one 'TRUE SPIRITUAL' reference.

In 'SPIRITUAL' applications, there is only one 'TRUE' reference. In material applications, we can fix our reference depending on what we are trying to accomplish. The important thing is that once a material reference is set it can never be changed for that system. We only have the option of deciding where we want the reference to be.

When it comes to 'SPIRITUAL' references, we are not given that option. We cannot fix or change 'SPIRITUAL' references. These are fixed by GOD. This is the reason why once we become disconnected from our 'SPIRITUAL" reference, we are 'SPIRITUALLY' lost and only 'HE' can lead us back to the 'TRUE' path.

Instinct is a programmed reference and is common in living organisms. It is from the source of ABSOLUTE intelligence. If there is no fixed reference, order cannot be maintained.

Once the reference is set in a manufacturing process and the sequence is initiated, any changes must be communicated back to the original fixed reference. For any given ordered system, there is an active connection between all components and subsystems. As long as the change is in harmony with the component to which it is connected, it will be in harmony with the system as a whole. The change must therefore be of an intelligent source which is able to see the big picture and understand the relationship between all components in that system.

Spiritual Realm

It is only logical that there is another realm which is at a higher level than the material realm, as the material realm has not always existed and came into existence at some finite point in time. Also, the material realm alone does not fully explain the order we observe in our universe. In other words, the order that is evident around us cannot be fully explained by the material laws. Material laws control only matter. It would require control that is significantly more far reaching than those that govern matter alone.

Life is also another ordered process that cannot be explained by material laws alone. Material laws are designed for the stability of matter and such stability requires fixed references. However, these fixed material references are not apparent as a function of 'living matter'. Matter, in general, is subject to the law of entropy which degrades matter toward randomness. The references for life oppose entropy and must have been introduced by an intelligent source, the same source that developed the absolute laws that govern matter, as matter is designed to be used as the raw material for all living matter.

Therefore, for matter to develop life, other laws would have to be at work, neutralizing entropy or any other laws that are in conflict with the laws governing the development of life. Such laws would also have to have originated from the absolute reference that defines all things both MATERIAL and SPIRITUAL. These laws are designed to combine the 'MATERIAL' and the 'SPIRITUAL' in perfect harmony, suggesting the same intelligent source.

The MATERIAL REALM had a beginning so it is temporal. This was 'The Big Bang', as some refer to it. The initiation of the 'MATERIAL' realm must have been some external source which I refer to as the 'SPIRITUAL' realm. Even though we do not see it, it exists as there is much evidence of its existence. We have had numerous interactions with this realm which cannot be ignored.

The origin of our temporal, 'MATERIAL' world cannot be explained in any other way. It must have come from another realm, an ordered realm that pre-existed our 'MATERIAL' realm. This realm must be eternal, with its source (GOD) absolute in knowledge and intelligence. Only such a source could create a universe. This is also the source of limitless knowledge, some with which we have been blessed. We also have the ability to recognize and use this knowledge, our intelligence.

In the 'SPIRITUAL' world, absolute order is imperative. It is the realm where virgin thoughts and ideas originate. We see changes in our world that occur from day to day indicating growth in knowledge such as the vast technological growth we are currently experiencing. We, as human beings, may not have had this knowledge before, but it always existed. We are only tapping into the source. It is being revealed to us as part of GOD'S natural order of things. It is not new. It is only new to us.

We may think that it is of us but this is not the case. It always existed as TRUTH. It is eternal. It is absolute. Absolute knowledge and intelligence have always existed in the SPIRITUAL realm and is gradually revealed to us in the MATERIAL realm.

To be in harmony with the 'SPIRITUAL' realm, we need to follow 'SPIRITUAL' laws. These have been communicated to us as ethical guidelines (the commandments) in the Bible which we find impossible to fully obey without the help of JESUS and THE HOLY SPIRIT. This is because we were given free choice, we sinned, and have lost our true reference. These commandments direct us to focus on complete harmony in the way we live and interact with each other.

Eternity

Eternity 'IS'. Eternity has no beginning and no end. GOD refers to HIMSELF as 'The Alpha and the Omega'. He is eternal. This is a very difficult concept for us to grasp as we all live in a temporal world. In fact, it is impossible for us to fully understand it. We can, however, look at what factors or components must have had to be eternal for the universe to be possible.

Intelligence, as it relates to GOD, must have been eternal because time cannot explain the concept of intelligence which is not a function of time, and is of GOD. Similarly, absolute knowledge is also not a function of time. It must have always existed. The accumulation of knowledge in our world, however, as is revealed from GOD, is a function of time. The rate of knowledge gain was initially gradual but now appears to be growing exponentially.

GOD'S absolute intelligence always was and always will 'BE'. We inherited intelligence from our GOD. We only get more intelligent when our ability to process information improves as a result of changes in our brain function. This is part of the growth and maturity process. It is a function of the material world.

Intelligence is an absolute concept but the ability to accumulate and use knowledge varies between species. Man has been given a great gift, the ability to accumulate and use knowledge to a greater degree than any other specie. Our ability to gain and use knowledge is limited by brain function and this is the main variable. All the information that is available already exists and has always existed, we only discover it.

GOD has absolute intelligence which means HE is not limited by a brain in HIS ability to use knowledge. HIS knowledge is also absolute. With unlimited intelligence comes unlimited knowledge. Absolute knowledge and intelligence have always existed in GOD. They are eternal. The only change is that now we have been given the opportunity of these being shared with us from GOD at His desired fate to release.

Purpose

What do you think is your purpose in life? One of the first clues is where you were born. Also, your nationality! Whether you are male or female! What gifts were given to you! What situation you find yourself in at any given time in your life!

If this universe was created by an intelligent source, that source must have had a plan and a purpose for everything in it. It is only logical to come to that conclusion based on how an intelligent mind reasons. If we look at ourselves, we see that we do plan our lives and make an effort to have a desired outcome for our undertakings. We are made in the image of our intelligent creator, GOD.

Material

This is the realm in which we live. The realm in which our universe exists. It is made up of space and matter. We interact with it through our five senses. We can see, touch, feel, smell and hear. In this realm, the laws that govern are fixed and we cannot change them or break them.

Singular

Singular, in this context, means from a single source or the same source. Creation manifests certain patterns and similarities that can only be explained by a singular origin. All ordered systems are in harmony which is indicative of a singular source.

Matter

Matter is anything that occupies space. Even light occupies space as light rays are photons which are very tiny particles. Some may argue that light is not a particulate but a wave form, but it still occupies space. It is a form of energy.

Matter was created and creation is complete. We have not been given access to the ability to create something from nothing. We have, however, been given the raw materials to use as we see fit.

In physics, it is said that 'matter cannot be created or destroyed'. Scientists have demonstrated this theory. We cannot create matter and we cannot destroy matter. We can only change it from one form to another, as we discover in the study of thermodynamics and nuclear science. This is done at thresholds that are fixed, under given conditions. Knowing these thresholds we are able to initiate changes to get the desired results.

Using this knowledge we are able to make a nuclear device by careful preparation of the raw materials and controlling the conditions so that the thresholds are met. Our aim is to obtain a chain reaction, but the path to success is very narrow and must be closely followed.

Matter is either ordered or disordered. I will consider randomness as a form of disorder. However, disorder is more than randomness. Whereas randomness, by definition, has no bias, disorder is biased toward the negative.

We live in a material world and we know that matter cannot create itself or come from nothing. The law of physics that says 'matter cannot be created or destroyed' implies that, if you have a void, matter cannot suddenly appear in that void. Any matter that exists in our world must have always been in existence in one form or another, since its creation.

We may therefore conclude that if there is a void and matter suddenly appears in it, then it must have come from somewhere else and, if created, it must be of a source outside that void.

Matter cannot create itself. If it could, we would see evidence of such creation in this realm. The ability to create must therefore be of another realm. Creation meaning something from nothing.

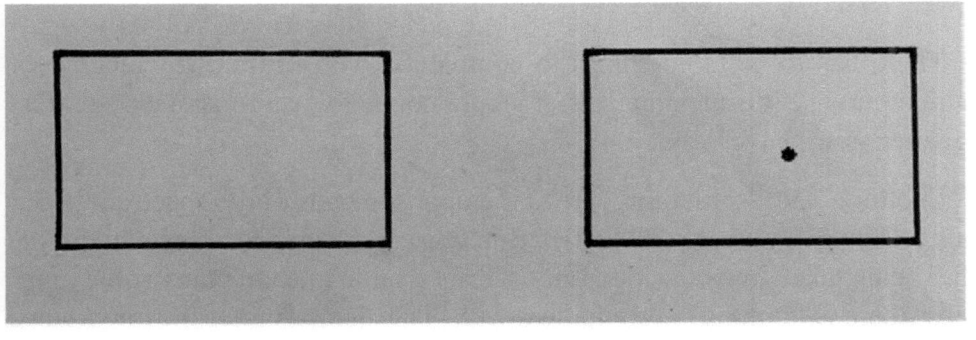

FIG. 1　　　　　　　　　　　　　　FIG. 2

Let us look at the diagrams above to try to explain this concept.

In the first diagram, Fig. 1, there is nothing, which represents a void. There is no solid, liquid, gas or energy present. No matter present. The second diagram, Fig. 2, has a 'dot' representing some form of matter. It does not matter what it is but, if it suddenly appears it must have come from somewhere. Now, if before there was only a void, then the dot must have come from somewhere else. Its source must have been something outside that void. In other words, an external source. This is the only way it could have appeared in what was once a void.

We can look at our universe in the same way but on a much larger scale. Its source must have been outside our universe.

Properties

Properties are characteristics that are unique to a particular material, process or system. They are sometimes used to define the material or process because of their unique characteristics.

If we look at 'salt', as an example, some of the properties would be that it is white, it is crystalline/granular, it is a solid at normal temperature and its taste is 'salty'.

Something

If we look at matter and the concept of something versus nothing, we have to start with a reference. There can only be 'something' if there is 'nothing' or vice versa.

There has to be 'something' to compare it with in its own category or realm otherwise it is meaningless. It would not be relevant. (something refers to matter as we have defined it)

'Nothing' and 'something' are at the opposite ends of the spectrum in our realm. They are mutually exclusive. But, one cannot be perceived without the other being taken into consideration. We are unable to appreciate 'something' unless there is 'nothing' to compare it with. This demonstrates the importance of reference. We may look at space as representing nothing and matter representing something. Similarly, we would not understand light unless there is darkness.

If we say, 'in the beginning, there was nothing", then we have defined nothing, which will become our reference. But, how can we perceive nothing unless there is something.

If there was 'nothing' in the beginning, a new realm must have been created when 'something' was created. The 'something' being matter as we know it today. The concept 'nothing' was defined the instant 'something' was created or vice versa. To define 'something' the concept of 'nothing' had to be invented. This could only have been perceived by a source outside the material realm as there was nothing in our realm to perceive it.

In the beginning, there must have been some other realm in order for 'nothing' to have been perceived. Remember, 'nothing' only became relevant when 'something' was created.

The preexisting realm, I believe, is the 'SPIRITUAL' realm. The realm in which we now live is a new realm, the 'MATERIAL' realm. That initiating source had to have been outside the realm of 'something/nothing' as we know it today.

Now, if we look at this concept in the absolute, SOMETHING must have existed in the realm from which 'something' in our realm was created. This 'SOMETHING' is a different 'something' as we define it in our realm, the material realm. It must have been eternal since, if it had a beginning, there would have to have been something external to initiate it. This leads us to conclude that this 'SOMETHING' must have always existed- the SPIRITUAL realm. This is the only explanation.

In conclusion, the 'SOMETHING' from which our universe was created must have always existed since only such a source could be the absolute initiator. We have no understanding of what eternal means but our interpretation is that it has no beginning and no end. The "Alpha and the Omega". (This is how GOD describes himself)

It is impossible for us to perceive 'eternal' as we are constrained by our experiences which are of a temporal world. But 'SOMETHING' must have existed prior to the creation of our universe otherwise our universe could not have come into being. If we look at the product of this prior existence, our universe, an ordered system, then there must have been INTELLIGENCE to define order in that realm. That order was then communicated to the 'material' realm to manifest itself as the order we see in our universe. This is as far back as I can rationally go as that realm is beyond our sense of reasoning.

Nothing

Nothing, as we define it in our world, is a 'void' where no matter exists. Like in outer 'space' where there is no matter. Here, space is also loosely defined as, even where there may be no visible matter, the 'space' is occupied by small particles separated by large areas of void. What we are actually doing is comparing it with what we know as 'something'. But 'something' cannot come from a void. To initiate 'something' in a void there must have been an outside source, independent of the initial void. 'Nothing' became relevant or valid when 'something' was created. In other words, in our world, both 'something' and 'nothing' are dependent on each other for validation. One cannot exist without the other.

We can only imagine what our brain allows us to perceive. Since we are a part of the 'something' it is not possible for us to objectively examine this (something/nothing) concept and come to any meaningful conclusion. We would need to be outside the system to think objectively. Unfortunately we will never have that opportunity in this world.

Positive

In our universe, we see evidence of attraction and repulsion relative to the very building blocks of matter. An atom consists essentially of protons, neutrons and electrons. Protons are positively charged and electrons negatively charged. Neutrons have no charge and are neutral, as the name implies. Neutrons are believed to bind the all positive nucleus of the atom/molecule together despite the all positive charges on the protons repelling each other. However, as the nucleus gets bigger, it becomes more unstable and eventually so unstable that it breaks down at a certain threshold, as we see in nuclear isotopes.

The positive and negative charges on the protons and electrons, respectively, are the forces that bind the electrons to the all positive nucleus (protons) and also make it possible for new compounds to be formed in chemical reactions.

The movement of free electrons in a conductor manifests itself as an electrical current. Electrons move towards the positive pole since opposite charges attract.

We also use the term positive to describe something good and negative to describe something bad.

Negative

Negative is only a way of expressing the converse of positive. We have found that negatively charged electrons are attracted to positively charged protons which means that, relative to the electron, the proton has an opposite or positive charge. The reference may be either the proton or electron but once we set this reference it has to be fixed for order to be maintained. Electrons are in one category and protons in the opposite category.

Protons, neutrons and electrons are the main building blocks of matter and combine in fixed proportions to form different molecules and compounds. It should be noted that the combination of atoms (protons neutrons and electrons), exist sequentially in the periodic table. Compounds are formed from atoms combining, but only relative to their valencies. Both the sequencing of the periodic table and valency manifest order and consistency.

Good

Good is anything in harmony with the universal order, both material and spiritual.

Bad

Bad is anything in disharmony with the universal order, both material and spiritual.

Truth

There is only one TRUTH. TRUTH is FACT as it relates to our very existence and interaction with each other and the universe. There is no alternate truth; there are no alternate facts. Man is a singular system with multiple subsystems, all connected. For us to live in harmony, with the ability to communicate with each other, there can only be a single source controlling it all. That source defines and executes 'TRUTH'.

In our judicial system, we seek to find the truth before making a judgement. We may not always find it, but we should always do everything to uncover the truth. This is instinctual because we are products of order. However, if our reference is not (absolute) TRUE (THE TRUTH) we can never discover 'TRUTH'. The only 'TRUTH' originates from our source (GOD). Any variation from this is false. We must therefore only seek 'TRUTH' from our source.

But, how do we find the source of TRUTH? We do have access to this source (GOD) but unfortunately most of us do not know the source and are not able to identify HIM. We need to faithfully ask the source that the TRUTH be revealed to us. This is what faith is all about. GOD is the source of 'TRUTH'.

This is where our creative abilities originate. Is it not logical that, if we get our intelligence from our source, the source would have more intelligence than we do? If so, wouldn't the source have made a way to communicate with us and vice versa?

As I mentioned before, there is only one TRUTH and only one TRUE source. This is how order is created, from TRUTH. Absolute TRUTH. Absolute or universal TRUTH defines all perfectly ordered systems.

Today we live in a world where 'Fake' news is a misleading influence on society. There is also nothing such as 'alternate facts' to which we sometimes hear people refer. The TRUTH is absolute and there are no variations on the TRUTH. Facts are facts and there are no 'alternate facts'. We are sometimes unable to get to the truth but the failure is on our part. For any given situation the truth never changes.

Fabric

This is a metaphor indicating that the ordered systems of the universe are interwoven. The fabric extends from a fixed point into infinity in all directions.

Intelligence

Intelligence is the most important component for the creation of an ordered system. It is the initial requirement for developing an ordered system. Intelligence is the ability to create order. Intelligence defines order.

In any ordered system, the designer must have some concept of the finished product and its intended function. This takes intelligence. Also built into the design of a system is the ability to adapt to changing environmental conditions, to ensure that the system will continue to operate as designed.

Humans use a similar principle when we design an ordered system. We anticipate changing conditions and compensate for those in the design. In order to do this, we build into the system the ability to recognize the perceived environmental changes and then to initiate the steps to compensate, so that the system will continue to operate as designed, in a stable state.

The design of living organisms indicates that adaptive changes were anticipated by the designer and built into the initial design of the system. Any changes that occur, whether adaptive or 'evolutionary', were done by the intelligent designer.

In our case, when we design something, we sometimes do not anticipate these environmental changes and so need to modify the initial design based on failure experience. Failure may also be the result of flaws in our initial design. We must then modify the design to correct the design flaw. All this takes intelligence which is the most important requirement for the design and development of an ordered system.

If we assume that our universe is an ordered system, then it must have been created by an intelligent designer.

Intelligence allows us to understand the concept of sequencing and timing thus giving us the ability to recognize disorder and to create order.

Intelligence applies in every discipline and every aspect of our lives.

We have to comply with the laws and rules of the ordered system in which we find ourselves, otherwise we would be out of harmony and unable to integrate. The laws governing matter are fixed and never change. All we can do is discover and familiarize ourselves with them and use them to our advantage. This is the process of gaining knowledge. Intelligence gives us the ability to understand these laws which we then use when we create and invent, using matter which is subject to them.

The intelligent designer must develop a plan and execute it from beginning to end. To do so, a reference must first be set and this reference must be fixed. Execution of the plan is done using continual feedback to ensure that the plan is being executed in the right sequence, with monitoring for errors at every stage. The intelligent designer must be a part of the process in order to know when the plan is complete and whether any other changes or modifications are necessary.

Intelligence, by definition, is order creating, it is logical, it is rational, it is systematic. It is an ordered system. Only order can create order. Only intelligence can create order. Intelligence is able to recognize disorder and also has the ability to change it into order. But, there must first be a fixed reference with subsequent sequential, positive steps and continual feedback to ensure that progress stays on track. Feedback is built into the system.

This is what we refer to as checks and balances or quality control. Intelligence gives us this ability.

We use the thought process that we have inherited from our creator. We think rationally and logically to solve problems and to act. We develop a plan and then execute it in a logical sequence. We inherited this method as the natural way to approach systems design. Wouldn't it then be logical to conclude that this is how nature works. You see, we are a product of nature. Then, evolution, which is random, does not fit this profile. The process of evolution would have to have intelligence but there is no evidence that there is feedback or planning, only chance change and adaptation resulting in random or natural selection.

Based on the way we think and reason, there is a plan with a preplanned outcome. This is far from random.

If one thing is changed in an ordered system it affects everything in which it is in immediate contact and sometimes other areas as well. In order to compensate for this change, modifications will need to be made in the affected parts of the system. It takes intelligence to recognize and make these changes as they will sometimes be significant. In the case of a living system, the changes will also need to be communicated to subsequent reproductions (in the DNA). This process is very complex and unlikely to be random in nature. It would take a source with complete knowledge of the system design, awareness of the change and ability to modify the design to compensate to the degree necessary.

We see intelligence in our world but there is no indication that any element on earth contributed to the creation of this intelligence. Matter has to be instructed on how to be ordered. I think it is reasonable to conclude that the intelligence we have inherited is not of this earth as there is no evidence of such intelligence in the basic building blocks of nature. If it is not of this earth, then its source must be of something external.

Using our intelligence, we create and invent things that never before existed on this planet. It must therefore have been initiated by the same external source which directs us to study our environment where we will find the raw material that will make things that are imagined become real. It tells us that all we have to do is search and we will find it. If we can imagine it, we can make it real. This is creativity in its virgin form.

INTELLIGENCE does not evolve. INTELLIGENCE exists. It 'IS'. It is absolute. GOD is the source of all intelligence. However, as far as it relates to us, our individual abilities do vary. Each of us is a medium with the ability to be intelligent, hence the ability to gain and use knowledge. Our species has the greatest ability to be intelligent based on its design. It was made in the image of GOD.

Within our species, intelligence is distributed in the form of a normal distribution. Most people are of average intelligence, then there are those at the extremes. The normal distribution represents an ordered system, with the distribution of the sample or group being about equal on both sides of the median (reference).

This is the reason why we have named this statistical measure of distribution 'normal' because this is how specific features of groups manifest themselves in nature. None in any group is exactly the same and specific features in that group can be represented by a normal distribution.

(See diagram of normal distribution on page 63)

If you look at the curve of a normal distribution, you see that it is a combination of an exponential increase and an exponential decrease on either side of the reference (median) or statistical norm. Neither side reaches infinity as in the typical exponential graph. They flatten and meet at the maximum then decrease back down to zero.

Intelligence can only be derived from something already intelligent. GOD, the source of all intelligence is eternal. Absolute intelligence does not evolve. It already exists in the infinite.

Artificial Intelligence

We are now in the age of artificial intelligence. Our reference is the human brain or mind and how it controls the body and solves problems. We are trying to simulate how the human brain works with its subtle abilities to display logical and rational reasoning in problem solving and make the best decision in any given situation.

To be comparable with our brain, artificial intelligence must be able to switch command and control from one situation to another, as it is confronted with changing issues, making split second decisions where needed. This is a difficult prospect. Currently, artificial intelligence can be designed to do a single task very well but does not have the flexibility of switching to another task, using a process of logical reasoning, for which it is not designed. It therefore lacks the flexibility of the human brain in these circumstances.

One of the reasons for this shortcoming in artificial intelligence is that our brain gets input from five sources, our five senses. The brain is able to prioritize these inputs and make split second decisions as to what it needs to react to as well as how to react. What we should do first, followed by the next sequential step and so on, until the response is complete. There are also responses at the subconscious level from other inputs. All this is done by one small mass of cells that has complete control over our body functions.

We can identify danger by vision, sound, scent, touch and taste. We are continually monitoring all these inputs and are always ready to react to any situation presented to us. We immediately know how to prioritize in any situation with which we are presented. Sometimes our reactions are subconscious in cases where rational thought process is not fast enough.

This is a challenge for artificial intelligence. But this is not the only obstacle. We have complex emotions such as love and hate and empathy. These are very difficult to quantify and hence to program into artificial intelligence.

Order

How we recognize order is fairly evident. We look for consistent patterns, cycles, harmony, theme, balance, aesthetics, timing and sequencing of events.

We are programmed with the ability to recognize order using the above clues as a guide. References define who we are and provide evidence as to our source. They tell us that our source is ordered. They reveal the character and essence of our source.

Order is the opposite of randomness or disorder. Order includes anything we do that involves intelligence and knowledge and utilizes a logical sequential process based on a fixed reference. This would be applicable to the design and fabrication of something as simple as an arrow head to the complex manufacturing of an automobile or a spacecraft. The only difference is that the arrow head is significantly less complex than the automobile and the automobile significantly less complex than the spacecraft. They each require different levels of intelligence and knowledge.

We must start from somewhere, but first, we have to have intelligence so that we are able to obtain, retain and use the applicable knowledge. Intelligence is therefore a critical requirement. It is the first and most important requirement in the development of an ordered system.

We were given intelligence with the ability to problem solve. We should always remember we all came into this world and we contributed nothing to the process. We were given intelligence, so intelligence must have existed before us.

We have everything we need to survive and grow. We are able to connect with our environment through all our senses. This is as complete a set of senses as we could desire. So, did these senses develop randomly as suggested by evolution or was it an ordered development? Could their development have been stimulated by the environment? That seems irrational as the process of evolution would not know how or where to begin.

Order is apparent both in material and spiritual systems. In fact, the material order is modeled after the spiritual from which it originated. Material order is a reflection of spiritual order. They were both developed with strict laws that govern the behavior of the systems that are subject to them.

An ordered system is interconnected. The interconnection is active. It has to be! Every element in an ordered system is connected to the other either directly or indirectly. This is how order is maintained. This is critical in both the 'MATERIAL' and the 'SPIRITUAL' worlds.

The reason why there is no absolute SPIRITUAL order in our world is because we are disconnected from GOD, our 'SPIRITUAL' source, our 'SPIRITUAL' reference.

GOD had to disconnect us because the 'SPIRITUAL' world has to be perfectly ordered and uncontaminated. Once we sinned we had to be disconnected.

Order and intelligence are synonymous. Order is what intelligence is all about. It is true that some intelligent sources set out to create disorder, but this is intentional. Such disorder is with a negative bias of which the source may be fully aware and consciously pursuing.

Sequence And Timing

To create order, in our world, requires sequencing and timing. To monitor sequencing and timing, there has to be an intelligent source in both the design and execution processes.

Disorder

Disorder may be similarly defined as order but with a negative bias, and also of an intelligent source. Disorder implies it is with the specific intent to disrupt and negatively impact an ordered system. Disorder may also be described as the effect of a random force on an ordered system. Disorder is the result of a negative influence on an ordered system.

Our source, GOD in heaven, is where perfect order exists. Disorder entered into our world because of original SIN.

Randomness

Randomness implies disorder but with no particular bias. There is no bias because there is no fixed reference. There is no single force influencing the final outcome that significantly outweighs the others involved.

Randomness is inert unless acted on by an intelligent source. The intelligent source extracts order from randomness by giving it a fixed reference. It now becomes part of an ordered system and under its control. It now has direction and purpose.

We recognize that something is random because we cannot comprehend it. Randomness is foreign to an ordered mind.

Knowledge

We gain knowledge by observing, understanding and recording the patterns and laws we see in nature and the universe. We cannot change the laws that govern them. These are fixed. However, by understanding them we are able to use them to invent or create ordered systems that we use to enhance our quality of life. Our brain has the ability to identify and store these patterns or bits of information in the form of knowledge, for later use.

The order we see in nature and the universe existed before we came into being. These consistencies that we refer to as laws suggest that there was an intelligence that created them.

All the knowledge that we now have and will discover in the future, has always existed. All we are doing is discovering and learning about our realm. All scientific discoveries are only being revealed to us now but these truths have always existed and have always been known.

Entropy

Ordered matter always gradually declines into disorder or randomness and loss of energy if not interrupted by an instruction for order to be initiated or maintained. This gradual decline is known as entropy. The instruction for order has to be of an intelligent source, a typical example being the manufacturing process. Entropy will continue until a stable state is reached and, after this, random change will continue to occur. Entropy is the second law of thermodynamics.

Error

Error is disorder. Error multiplies! If an error is made at the initial phase of an ordered process, it is multiplied in subsequent stages and therefore amplified as the process progresses. This is why it is critical for quality control procedures to be in place, at every stage, from start to finish in any manufacturing process.

In a manufacturing process, at each stage, the quality of the individual components must be checked, before assembly begins, to ensure they are within the design tolerance. If a component is found to be out of tolerance, it is rejected. All components in the assembly must be within the specified tolerance otherwise the finished product will not be acceptable.

In other words, it takes very close monitoring at all the production stages to make an acceptable finished product. Error creates disorder and disorder multiplies, if left unchecked.

One reason why error multiplies is that the reference is always changing. For example, if we are making a batch of similar components, after the first error, if the original reference is not used, error will begin to accumulate. As manufacturing progresses, if a new reference is continually used, as opposed to the original reference, errors will continue to increase.

Using the original reference in all cases will ensure minimum error. It is also important that the reference be a true reference and be accurate.

If a story is told by one person and subsequently by others, each time the story is told there will be variations on the true narrative. These variations will increase as the number of people telling the story increases. By the time it gets to the 10th person, it will not be the same story. The reason is that the reference is always changing and so error increases each time the story is told. The references become the different people telling the story.

Aesthetics/beauty

Aesthetics play an important part in our culture. We appear to have an innate sense to appreciate beauty that sometimes transcends cultural boundaries. We inherently like things we consider beautiful and dislike what we consider ugly. Here again, there must be some universal reference with which we compare in order to come to a conclusion as to what we consider ugly or beautiful. Some of the things that probably guide our sense of judgement are symmetry (balance) and color, that appeal to our emotions in some positive or negative way based on our individual taste.

We associate beauty with order and ugliness with disorder. Beauty has a positive emotional effect on us. It affects our state of mind and hopefully evokes something 'good'.

CONCLUSION

The examples I've given in this book are only a small sample of the ordered systems that we encounter in any discipline that we may analyze. This indicates that all ordered systems must have these basic components (intelligence and a fixed reference), and they define such systems. To test this theory, you should select any ordered system that you encounter and analyze it to see if you can verify the theory. You will find that, on analysis, any ordered system that you choose will have these basic components, and there is no exception.

I wrote this book to convey a message that answers some of the questions we all have about our existence. The included definitions and concepts should help give you a better understanding of the principles that govern our material and spiritual worlds. Both manifest one common element, and this is order. This order must be defined by intelligence, as this is the only attribute by which order or disorder can be recognized.

In our world, we are always trying to create order and is the desired end product. This is reflected in all aspects of our lives and is something we are always striving to achieve. As indicated in the definition of intelligence, this is the most important component of any ordered system. Intelligence defines order and so must be the source of any ordered system.

Since the universe manifests itself as an ordered system, it follows that it must be of an intelligent source. It is also clear that creation is from a singular source, which scientists refer to as the Big Bang. Since the Big Bang produced an ordered system (our universe), this must have been the product of an ordered source, and by definition, that source must have been intelligent. This is because order cannot be produced from disorder or randomness unless directed by intelligence.

Finally, we can conclude that the source that produced the universe must have been intelligent because it produced an ordered system. This concept is at the core of what I am trying to convey to you, the reader. Based on how God describes Himself, He is the only one that fits this profile and thus, puts everything in perspective.

AUTHOR'S BIO

I was born in Oracabessa, a small town on the north coast of Jamaica. For those who are familiar with the popular vacation spots on the island, it is thirteen miles east of Ocho Rios.

I have always had several hobbies and would switch from one to the other depending on my mood at any point in time. My hobbies included drawing, from an early age, which led to oil painting. I also made model cars, boats and planes as a teenager. I then became interested in electronics and made amplifiers and preamplifier, from circuits printed in Popular Electronics, to enhance the fidelity of the music of the time, as music had now become one of my interests. Woodworking was my later interest and I made furniture and speaker enclosures.

I enjoy using my hands and would sometimes skip meals from being totally absorbed in a project. This gave me an understanding of the properties of materials and the knowledge required to plan, initiate and complete a project.

This experience influenced me in my choice of career to become an engineer. It seemed natural as I was always interested in how things work and tried to gain the level of knowledge required to successfully complete my projects. This later led me to seriously consider what initiated the universe and the forces needed to maintain order. It is apparent that the universe is ordered, but how did it get that order?

This book is designed to give the reader a rational explanation as to the origin of the universe, from a scientific viewpoint, based on its physical manifestations with which all of us are familiar. We will see that the same principle is also true for spiritual order.

REFERENCES

The Holy Bible.

Google to verify the equation of the formula for the gravitational force on a body and the theories associated with evolution.

Printed by Libri Plureos GmbH in Hamburg, Germany

Construction de la figure
du nouvel ennemi

Ninou Garabaghi

Construction de la figure du nouvel ennemi

Essai

© L'Harmattan, 2018
5-7, rue de l'Ecole-Polytechnique, 75005 Paris

http://www.editions-harmattan.fr

ISBN : 978-2-343-13559-5
EAN : 9782343135595

Introduction

Bien nommer les choses peut aider à résoudre les problèmes qui se présentent à nous. « Développement »[1], le mot d'ordre promu au sein du système des Nations Unies durant la période de la guerre froide a fait place pour quelque temps à l'insipide notion de « bonne gouvernance »[2] suite à la victoire du capitalisme dérégulé[3]. A faible pouvoir mobilisateur, cette notion a été scotomisée par la montée en force et en visibilité du concept d'« islamisme ». Un des trous noirs de la géopolitique du XXIe siècle, ce concept s'est imprimé dans les consciences comme l'expression d'une réalité incontestable. Les inconséquences de cet état de fait soulèvent la question de savoir comment un concept et/ou un mot d'ordre se crée et se diffuse au sein des sociétés. Si l'on se limite à la catégorie « concept », on constate que, nonobstant les processus de démocratisation des différentes sphères de la vie des sociétés, c'est l'élite, sous-entendu les puissants, qui s'arroge ce droit. D'après

---

[1] Axées sur l'objectif d'indépendance dans l'interdépendance, les décennies du développement des Nations Unies devaient œuvrer à la réduction des rapports de domination structurels Nord-Sud.
[2] C'est la mauvaise gouvernance des gouvernements des pays du Sud qui fut considérée comme la cause essentielle de l'échec des politiques de développement.
[3] Victoire des institutions de Bretton Woods rendue possible par la crise de la dette du Tiers Monde en 1982. Cette crise fit perdre à ces pays toute capacité de négociation et les astreignit à se soumettre aux programmes d'ajustement structurel du FMI et de la Banque mondiale. C'est ainsi que la quatrième décennie du développement qui fut qualifiée de « décennie perdue » mit un terme aux décennies pour le développement des Nations Unies.

Nietzsche[1], « Ce droit de maître en vertu de quoi on donne des noms va si loin que l'on peut considérer l'origine même du langage comme un acte d'autorité émanant de ceux qui dominent ». Si l'histoire s'accélère, on ne peut pour autant ignorer que le terme d'islamisme, jusqu'en 1980 utilisé pour désigner une religion, a depuis acquis le statut de concept politique ce qui, loin de toute idée de complot, a contribué à mal nommer les choses ajoutant, selon la prophétie de Camus, du malheur au malheur des hommes.

D'après un des architectes du Conseil Français du Culte Musulman (CFCM)[2], pour une large fraction de l'opinion, « l'islamisme, c'est l'islam, et l'islam est le problème ». Le pire est que la pratique de la confusion des genres ne se limite pas à la France, il correspond, selon cet expert de l'Islam de France, à une tendance que l'on retrouve aux Pays-Bas, en Belgique, au Royaume-Uni et ailleurs. Problème ici, solution ailleurs ; comme nous serons amenés à le constater, ce ne sont pas seulement des musulmans qui considèrent l'islam ou l'islamisme comme l'unique solution aux problèmes du monde musulman, certains orientalistes[3] et islamologues[4] partagent ce point de vue. Solution pour les uns, problème pour d'autres : serions-nous condamnés à une vision antagoniste qui conduit à légitimer la posture conflictuelle de l'idée du « choc des civilisations » imaginée par Bernard Lewis, idée reprise et théorisée par Samuel

---

[1] Frédéric Nietzsche, *La Généalogie de la Morale*, (Traduit par Henri Albert), Paris, Mercure de France, 1900, p.20. Nous avons préféré nous référer à la traduction de Henri Albert pour la simple raison qu'elle exprime mieux notre propos.
[2] Bernard Godard, cf. Laurent de Saint Périer, « Les nouveaux musulmans » – interview de Bernard Godard (responsable du Bureau central des cultes au ministère intérieur), *Jeune Afrique* N° 2823, 15-21 févr.2015.
[3] Exemple, Bernard Lewis.
[4] Exemple, François Burgat.

Huntington et mise en pratique aux fins de la prise de décision stratégique par l'équipe de George W. Bush ? Pour mieux circonscrire les origines de cet état de fait, il est intéressant de chercher à savoir qui a décidé de dévoyer le terme d'islamisme. Pourquoi a-t-on accepté le dévoiement de ce terme ? Qu'est-ce que ce mot, qui a acquis le statut de concept, définit exactement ? Est-ce que ce concept a permis de solutionner des problèmes ou a-t-il au contraire généré des problèmes ? Ce concept a-t-il contribué à une plus grande intelligibilité de la réalité ?

Comme il ressort du chapitre qui suit, le terme « islamisme » n'a pas toujours eu un sens négatif, c'est après l'avènement d'un régime théocratique suite à la révolution iranienne de 1979 qu'on a commencé à parler d'islamisme radical et c'est après l'odieux attentat du World Trade Center perpétré par al-Qaïda le 11 septembre 2001 que le qualificatif de radical a été abandonné et que le terme d'islamisme a été dévoyé de sorte que « islamisme » est devenu synonyme « d'islamisme radical ». Mais l'histoire du terrorisme islamique ne s'arrête pas là, Oussama Ben Laden, chef d'al-Qaïda, meurt le 2 mai 2011. Trois ans à peine après sa mort, le 29 juin 2014, Abou Bakr al-Baghdadi s'autoproclame Chef de l'Etat islamique en Irak et au Levant (EIIL, DAECH en arabe), sous le nom de "calife Ibrahim". C'est fin 2014-début 2015 que le terme de « djihadisme » sera consacré par l'entrée sur la scène géopolitique mondiale de Daech qui, non content de lui-même, va s'auto-reconnaître en tant qu'entité étatique. Posture résolument offensive et recours à une redoutable propagande multilingue et multimédia, ce nouveau protagoniste fascine et donne l'illusion de puissance par ses mises en scène d'actes barbares d'une violence inouïe.

Si hier, vers la fin du XXe siècle, c'est l'islamisme qui posait problème ; aujourd'hui, au XXIe siècle bien entamé, c'est le djihadisme qui prend la relève et devient le concept clef majeur destiné à désigner la nouvelle figure du terrorisme. Islamisme, djihadisme, terrorisme islamiste[1] deviennent des termes quasi interchangeables dans la bouche et sous la plume de nombre d'acteurs politiques et médiatiques. Et comme, à la longue, le subconscient, qui est l'archiviste de nos habitudes de penser, rappelle que le terme « islamisme » est synonyme d'islam c'est l'islam qui *in fine* devient le problème. Ce qui n'empêche pas lesdits politiciens de chercher à utiliser à titre de solution ce qui est censé faire problème. S'il est clair que l'islam a été et continue à être *de facto* instrumentalisé à des fins idéologiques et politiques, la question consiste maintenant à savoir pourquoi nous en sommes arrivés là et qui sont les véritables perdants et gagnants.

Après avoir analysé l'évolution des différentes définitions du terme « islamisme », nous procéderons à l'examen des tenants et aboutissants de ce processus de dévoiement du sens originel de ce terme. Pour finir, nous serons amenés à voir comment, en quête de clés et de grilles de lecture permettant une nouvelle perception des rapports interculturels et interreligieux dans un monde globalisé, l'Occident a confectionné au cours des trente dernières années tout un arsenal terminologique qui a influé sur la vision et la marche du monde ; la dynamique des changements échappant à son contrôle, prisonnier du paradigme qu'il s'est forgé, le tout a fini par se retourner contre lui. Refusant de changer d'orbite et de logiciel, l'Occident poursuit obstinément sa course folle nonobstant les dévastateurs effets de boomerang. La question se pose alors de savoir comment nous pourrions nous hisser hors de ce bourbier.

---

[1] Il est à noter que l'on parle de « terrorisme islamiste » et non de terrorisme islamique ; Olivier Roy dirait « cherchez l'erreur ».

CHAPITRE 1

Processus de dévoiement du terme « islamisme » : la question de la légitimité du nouveau concept d'« islamisme »

Pour concilier autant que faire se peut « rigueur » et « réalisme », nous allons essayer de saisir l'évolution du sens du mot « islamisme » dans le temps, en tenant compte des différents points de vue et/ou courants de pensée. Pour ce faire, nous commencerons ici par analyser l'évolution du sens du mot « islamisme » dans différents dictionnaires de langues française et anglaise. Bien que nous nous situions délibérément dans une perspective globale du fait de la mondialisation, nous ne pouvons pour autant éviter par moments de donner l'impression d'avoir un regard franco-centré sur la question de la construction du concept d'« islamisme » et celle de l'islamophobie pour deux raisons.

La première tient au fait que cet essai est rédigé en français, certes, mais dans un contexte où la laïcité est parfois malencontreusement perçue par une partie non négligeable de l'élite française comme un quitus à une posture d'« anti-multiculturalisme », alors même que le multiculturalisme ne se présente pas d'emblée comme une alternative politique en tant que choix de société mais comme une réalité factuelle et que la loi sur la laïcité n'est pas une loi anti-diversité culturelle mais, au contraire, une loi qui a pour fonction de garantir les conditions propres à permettre l'exercice de la liberté individuelle de conscience et de conviction pour tous.

La seconde raison tient au fait que le concept d'« islamisme » est une construction française en ce sens que, s'il est vrai que ce sont des religieux et des politiques qui ont instrumentalisé l'islam et d'autres religions à des fins géopolitiques et/ou politiciennes, ce sont des chercheurs français qui ont à l'origine décidé de dévoyer le terme « islamisme » (cf. chapitre 5). Comme nous serons amenés à le constater, d'apparence anodine, nous sommes avec la question de la construction du concept « islamisme » en fait en présence d'un écheveau de questions complexes et imbriquées qui touchent tout à la fois les sphères de la science, de la religion et de la politique[1], la première ayant pour objet la réalité, la seconde la vérité et la troisième le juste, le meilleur et le légitime[2]. La question du champ des responsabilités des différents acteurs sera longuement analysée plus loin, nous allons pour l'heure commencer par examiner l'évolution du sens du mot « islamisme » dans différents dictionnaires de langues française et anglaise.

Comme nous serons amenés à le constater dictionnaire à l'appui, jusqu'à la décennie 90, le mot « islamisme » était quasi synonyme d'islam. "Quasi", car, comme nous allons le voir, d'après les définitions transcrites ci-dessous, à l'origine, islamisme avait un sens plus restreint que l'islam, puisque le mot « islam » était et continue à être défini

---

[1] La politique ici doit être comprise au sens de politique pure. Pour plus de précision, cf. Bertrand de Jouvenel, *De la politique pure*, Paris, Calmann-Lévy, 1977 (la dernière édition augmentée d'une nouvelle préface datée d'août 1976 qui s'avère ô combien d'actualité).
[2] Comme on le constate, le problème *in fine* consiste à savoir : Sur quelle base décide-t-on de ce qui est juste, meilleur et légitime ? Quelle est l'instance qui tranche sur ce qui, à un moment donné, est juste, bien et nécessaire ? Une décision démocratique est-elle exempte de tout sexisme ? Une décision prise sur la base des connaissances scientifiques de ce jour est-elle exempte de tout risque d'erreur ?

comme « religion et civilisation des musulmans » tandis que le mot « islamisme » était dans la plupart des dictionnaires limité à la « religion musulmane ». Il importe ici d'apporter quelques précisions quant à l'usage de la majuscule et/ou de la minuscule dans l'écriture du mot « islam ». D'après Tayeb Chouiref, invité de l'émission Questions d'islam de Ghaleb Bencheikh[1] intitulée « Considérations sémantiques à propos du vocable islam », le terme islam sans majuscule est conventionnellement réservé pour désigner la religion et Islam avec majuscule se réfère à la civilisation islamique. Nous nous conformerons donc à cet usage mais ce qu'il importe de souligner ici, est que, contrairement à ce qui nous a été argué pour justifier le dévoiement du terme « islamisme »[2], l'existence de deux mots pour désigner la religion et la civilisation n'était pas de trop puisque « islamisme » pouvait être assigné à la religion et « islam » à la civilisation islamique ou vice versa sans qu'on ait besoin de faire appel au distinguo majuscule/minuscule.

Venons-en maintenant aux définitions du mot « islamisme » dans les dictionnaires. Dans le Petit Robert publié en 1970, le mot « islamisme » est défini comme suit :
- « Islamisme. n.m. Religion musulmane V. Mahométisme (vx) ».

L'édition de 2014 du Petit Robert fournit deux définitions pour le mot « islamisme » comme suit :
- 1697 de *islam* ■Religion musulmane ➤islam, vx mahométisme. « *L'artiste avait abjuré*

---

[1] Docteur ès sciences et physicien, Ghaleb Bencheikh est également islamologue, théologien et philosophe ; il est depuis le 8 mai 2016 responsable de l'émission Questions d'Islam sur France Culture.
[2] Notre entretien du 12 février 2015 avec le *Chief Editor of the international Desk* de Télévision Française 1.

l'islamisme, et lui et sa femme n'avaient de musulman que le bonnet turc » Nerval.
- Mouvement politique et religieux prônant l'expansion ou le respect de l'islam.

Comme on le constate, catégorie religieuse dans l'édition de 1970, « islamisme » devient une catégorie politique dans l'édition de 2014. Comme nous allons le voir ci-après, par comparaison avec d'autres éditeurs, la définition politique fournie par l'éditeur des dictionnaires Robert (grand et petit) est relativement ambigüe. Qui plus est, la citation littéraire accompagnant la première définition comporte une connotation athéiste qui contraste d'autant plus avec la qualification de mouvement expansionniste retenue aux fins de la seconde définition. Au train où vont les choses, l'on est en droit de se demander s'il n'est pas à craindre que la citation de Nerval ne soit d'ici peu remplacée par une des multiples citations sur l'islam de Michel Houellebecq qui, comme tout un chacun le sait, ne sont pas des plus élogieuses et parfois même frisent l'indécence.

Comme il apparaît clairement, le choix des citations littéraires servant à illustrer et à justifier la définition des mots traités dans les dictionnaires n'est pas anodin, pour ne pas dire innocent. En effet, lorsqu'on se réfère au Littré, et plus spécifiquement à l'édition publiée en 1966 par Gallimard – Hachette, on constate que le mot islamisme est défini comme suit :

♦1° La religion de Mahomet. « Ce ne fut point par les armes que l'islamisme s'établit dans plus de la moitié de notre hémisphère, ce fut par l'enthousiasme, par la persuasion », Volt. *Mœurs*, 7.

♦2° L'ensemble des pays où règne cette religion, dans le même sens que chrétienté par rapport aux pays chrétiens. – **E.** *Islam*

En date de 1966, la définition de l'islamisme fournie par Le Littré est exempte de tout jugement d'ordre politique. Cette définition est identique à la définition de l'islam fournie par le dictionnaire Robert, de sorte que l'islamisme définit à la fois la religion et la civilisation des musulmans. Ce qui conforte la remarque formulée ci-dessus quant à l'utilité des deux mots. Pour ce qui est du choix de la citation littéraire, la citation de Voltaire laisse perplexe. Celle-ci paraît, en effet, comme des plus élogieuses lorsqu'on sait à quel point Voltaire a pu être critique à l'égard de toutes les religions, et plus spécifiquement des trois religions monothéistes.

Il est vrai que, si le chapitre VI de l'*Essai sur les mœurs et l'esprit des nations* de Voltaire qui est consacré à l'Arabie et à Mahomet n'est pas très tendre, le chapitre VII s'avère des plus cléments. Et les deux seules phrases de l'*Essai* qui font usage du terme islamisme se présentent comme les plus élogieuses de ses « descriptions » puisque Voltaire prétend vouloir décrire plutôt que juger[1]. Citons, à toutes fins utiles ces deux mémorables phrases qui font usage du terme « islamisme » : « Cette religion s'appela *l'Islamisme*, c'est-à-dire résignation à la volonté de Dieu ; et ce seul mot devait faire beaucoup de prosélytes. Ce ne fut point par les armes que *l'Islamisme* s'établit dans plus de la moitié de notre hémisphère, ce fut par l'enthousiasme, par

---

[1] Voltaire avec l'humour grinçant qui lui est propre se plaît à juger des hommes et de leurs mœurs tout en feignant de les décrire, c'est ainsi qu'après avoir fourni des arguments à la faveur de la pratique de la polygamie il déclare : « Ce n'est pas ici la place d'une dissertation ; notre objet est de peindre les hommes plutôt que de les juger »

la persuasion, et surtout par l'exemple des vainqueurs, qui a tant de force sur les vaincus. ». Il est vrai que le terme « islamisme » est aujourd'hui un mot fourre-tout et qu'il semble de bon ton de souligner que l'islam signifie soumission comme le titre du livre de Michel Houellebecq en témoigne. C'est pourquoi il paraît nécessaire d'ouvrir ici une parenthèse pour préciser que l'« islam » est un mot polysémique, ainsi que l'Iman Tareq Oubrou[1] le rappelle « parmi les étymologies du mot *islâm*, se trouve la notion de « paix ». D'après Soheib Bencheikh[2], ancien Grand mufti de la mosquée de Marseille, « l'islam veut dire se donner, s'offrir, un choix libre... Celui qui est soumis exclusivement à sa conception de la divinité par engagement, celui-là par définition ne peut pas être soumis... Il ne peut être soumis devant aucune force ».

Nous allons maintenant nous pencher sur les explications fournies par les dictionnaires Larousse. Dans le dictionnaire étymologique publié en 1970 par Larousse, les termes « islam » et « islamisme » sont tous deux datés de 1765, tandis que les mots « islamisation » et « islamiser » sont datés du XXe siècle. La date de 1765 pour le premier emploi du mot « islamisme » paraît inexacte puisque l'*Essai sur les mœurs et l'esprit des nations* de Voltaire, qui fait usage du mot, a été publié en 1756. Venons-en maintenant à l'essentiel et examinons l'évolution du sens du mot « islamisme » à partir des définitions qui en sont fournies dans Le petit Larousse illustré à dix ans d'intervalle.

---

[1] Tareq Oubrou (entretiens avec Michael Privot et Cédric Baylocq), *Profession Imâm*, Paris, Albin Michel, 2015, p.195.
[2] *Islam : foi et valeurs par Soheib Bencheikh*. Intervention au séminaire Psychiatrie, psychothérapie et culture(s) du vendredi 31 mars 2000 mise en ligne le 19 juillet 2012 sur le site « www.parole-sans-frontrière.org ».

Dans le petit Larousse illustré de 1983, publié en 1980, le mot islamisme est défini comme suit : « Religion musulmane ». Dans le Petit Larousse illustré de 1993, publié en 1992, le mot islamisme est enrichi d'une définition de plus comme suit :
1. Vieilli. Religion musulmane, islam.
2. Mouvement politico-religieux préconisant l'islamisation complète, radicale, du droit, des institutions, du gouvernement, dans les pays islamiques.
Comme on le constate, en 1992 le pli est pris, nous avons droit à deux définitions. La première qui définit l'islamisme comme religion musulmane est considérée comme désuète ("vieillie"). Selon la seconde définition, le mot islamisme est considéré comme une catégorie politique mais, à la différence du Petit Robert qui demeure relativement « ambigu » en tenant compte tout à la fois des mouvements radicaux et modérés, le Petit Larousse va de l'avant par une prise de position qui fait évoluer le sens du mot de sorte qu'il ne définit plus qu'un groupe spécifique au sein du « mouvement politico-religieux » : les « radicaux ». C'est ainsi que le mot islamisme devient synonyme d'« islamisme radical ».

Fait non moins important, c'est suite à l'évolution du sens du mot islamisme et à l'avènement de la catégorie politique du mot islamisme que le mot « islamiste » est introduit dans la langue française. Il convient néanmoins de préciser qu'au début de la décennie 90 la situation n'est pas encore désespérée puisque, si l'édition de 1993 fait référence au mot islamiste[1], elle ne fait en revanche aucune allusion aux mots islamophobie et/ou islamophobe.

---

[1] Il convient à cet égard de garder présent à l'esprit que le mot islamiste ne figure pas dans le Petit Robert de 1970.

Concernant le mot « islamiste », Olivier Roy[1] se plaint en l'an 2001 qu'il n'ait pas été dûment vulgarisé par les médias en ces termes : « Le mot est au mieux un adjectif, mais pas un terme opérant au niveau du grand public ». Preuve, d'après lui, des « limites de l'influence du chercheur sur les médias ». Il précise à cet égard que « le fait qu'il soit complaisamment interviewé n'implique pas que les catégories qu'il construit soient adoptées ; autant que je sache, la presse parle toujours de fondamentalistes, d'intégristes et de radicaux islamiques, plutôt que d'islamistes ».

Lorsqu'on se réfère au Petit Larousse illustré de 2012, publié en 2011, on constate que la définition du mot islamisme a quelque peu changé. La définition originelle du mot islamisme, qui est qualifié de « vieilli », est reléguée au second rang tandis que la définition du mot comme catégorie politique est hissée au premier rang et modifiée comme suit :
1. Désigne, depuis les années 1970, les courants les plus radicaux de l'islam qui veulent faire de celui-ci non plus essentiellement une religion, mais une véritable idéologie politique par l'application figée de la charia et la création d'Etats islamiques intransigeants.
2. Vieilli. Religion musulmane ; islam.

Cette nouvelle définition du Petit Larousse de 2012 pose un double problème. Il y a d'abord un problème d'anachronisme. En effet, en réalité c'est la révolution islamique iranienne qui a été à l'origine de l'avènement dans la décennie 80 du nouveau concept d'islamisme. Il importe à cet égard de garder présent à l'esprit qu'en fait, l'essor des mouvements religieux dans les années 70 est à l'origine

---

[1] Olivier Roy, « Les islamologues ont-ils inventé l'islamisme ? », revue *Esprit*, août-septembre 2001.

davantage un phénomène de résistance culturelle, il s'agit bien plus d'un mouvement mystico-religieux que d'un mouvement à proprement parler politique. C'est grâce à l'affrontement américano-soviétique que ce mouvement d'ordre essentiellement spirituel a mué de la catégorie religieuse à la catégorie politique. En effet, s'il est vrai que l'islam a, dès le début du XXe siècle, été instrumentalisé à des fins politiques par des grandes puissances, ce n'est que vers la fin des années 70 que cette religion a été instrumentalisée à des fins idéologiques. Comme nous allons le voir ci-après, c'est en pleine guerre froide que, inquiétés par l'inexorable avancée de l'URSS, les Etats-Unis d'Amérique ont senti le besoin impérieux d'instrumentaliser l'islam pour contrer l'avancée de l'Union soviétique. Et, en tout état de cause, l'islam politique n'étant pas un phénomène nouveau, sa réactualisation ne peut à elle seule justifier le dévoiement du terme « islamisme ».

Le second problème, et non des moindres, est le problème de la légitimité du concept d'« islamisme ». En France, la construction de l'Etat a précédé la formation de la nation de sorte qu'on peut dire que la « nation » est le produit de l'« Etat » et non l'inverse, comme il se devrait. En effet, si la légitimité des institutions repose sur leur caractère démocratique, il est évidemment préférable que cette légitimité soit obtenue a priori comme en Suisse et non pas a posteriori comme en France. Ce qui est valable pour les institutions l'est pour les concepts. Il est vrai que, pour les mandarins, détenteurs de la vérité, le savoir n'a rien de démocratique et échappe à la discussion. Mais, comme l'ère du « savant roi » est révolue, comme Habermas[1] le souligne à très juste titre, le savoir, en général, et dans le domaine des sciences sociales et

---

[1] Voir le numéro Août/septembre de la revue *Esprit* intitulé « Habermas, le dernier philosophe ».

humaines, plus spécifiquement, doit être le produit d'un consensus obtenu via un processus de délibération démocratique et non une invention de l'élite qui s'impose d'autorité.

Or, le nouveau concept d'islamisme est une invention propre à une élite spécifique, en l'occurrence des orientalistes islamologues français et non point propre à l'ensemble de la pensée occidentale et encore moins propre à la pensée islamique. Ce qui n'est pas sans poser des questions quant à la légitimité du concept d'un point de vue scientifique et/ou du point de vue de son utilité instrumentale dans la lutte contre le terrorisme. Le problème est que le concept a été si bien intériorisé et ancré dans l'usage des différents acteurs des sociétés occidentales qu'il peut paraître impensable que l'on puisse changer quoi que ce soit à la pratique courante. Il serait en conséquence anachronique de vouloir ouvrir un débat sur la question de la légitimité du concept d'islamisme. Mais, comme dit le proverbe « qui ne tente rien n'a rien », pourquoi ne pas tenter d'échapper au défaitisme ambiant ? Pourquoi ne pas tenter de voir s'il est encore possible, à l'image du baron de Münchhausen, de nous hisser hors du bourbier ?

Avant d'examiner *l'usage* du concept d'islamisme par les scientifiques (politologues et islamologues), il est nécessaire de prendre connaissance des définitions fournies dans quelques dictionnaires anglais d'origine britannique et américaine.

Dans le dictionnaire américain Shaffer Webster de 1828, le mot *"islamism"* est défini comme suit : *"The true faith, according to the Mohammedans ; Monhammedism"*. Comme il ressort clairement des définitions fournies dans l'édition datée de 2003, du Merriam Webster, le terme

"*islamism*" cesse d'être exclusivement utilisé pour la définition d'une religion :
1 : *the faith, doctrine, or cause of Islam* ;
2 : *a popular reform movement advocating the reordering of government and society in accordance with laws prescribed by Islam.*
Non moins intéressant, lorsqu'on se réfère aux différentes éditions du dictionnaire britannique *Oxford Dictionary of English* parues depuis 1998, on trouve une seule et même définition très succincte mais fortement chargée : *"Islamic militancy"* [1] *or "fundamentalism"*. Comme on le constate, *"islamism"* a cessé d'être un mot « neutre » utilisé pour désigner une religion pour devenir un terme péjoratif porteur d'un jugement de valeur réprobateur.

La première question qui vient à l'esprit consiste à savoir pourquoi dévoyer le terme « islamisme » lorsqu'on peut faire appel à toute une panoplie de termes et de qualificatifs d'ores et déjà disponibles tels que : « activisme », « fondamentalisme », « intégrisme », « radicalisme », etc. comme ceci a été fait et continue à être pratiqué pour les autres religions. Le fondamentalisme et l'activisme ne sont pas des phénomènes nouveaux dans le monde musulman : à la fin du XIe siècle, on a eu droit à un mouvement terroriste dirigé par Hassan ibn al-Sabbâh qui a d'ailleurs fait l'objet d'un roman en langue slovène en 1938[2] et la première organisation fondamentaliste du XXe siècle date de 1928 (la Société des Frères musulmans). La question se pose dès lors de savoir pourquoi on décidé de

---

[1] Il est utile de signaler que le mot *"militancy"* est défini comme suit : *"The use of confrontational or violent methods in support of a political or social cause"*.
[2] Roman traduit et publié en français. Cf. Vladimir Bartol, *Alamut*, Paris, Phébus, 1988.

dévoyer ce terme maintenant dans les années 80 et pourquoi en France.

Si l'on fait abstraction des questions d'ordre théorique et épistémologique, objet du chapitre 5, la réponse est apparemment simple. La France est la patrie des révolutions. Elle a accueilli l'imam Khomeiny et l'a aidé à mener à terme la révolution iranienne qui a été soutenue par le pouvoir politique, les médias et les grandes figures de la gauche intellectuelle, tous subjugués et excités par la perspective, pour les uns, des avantages économico-financiers que la France pourrait en tirer, pour les autres, par la possibilité d'expérimenter une solution alternative "non occidentale" dans un contexte caractérisé par la crise de l'idéologie marxiste. C'est ainsi que le concept d'« islam politique » forgé et travaillé tout au long de la décennie 80 sera reconsidéré et consacré par le terme « islamisme » qui fera figure de surnom accolé à son nom de baptême, suite, d'une part, au désenchantement de l'élite intellectuelle jadis fascinée et, d'autre part, à la construction de la nouvelle figure de l'ennemi par le pouvoir politique de l'époque qui a décidé de soutenir politiquement et militairement l'Irak en guerre contre la nouvelle République islamique d'Iran. Il nous faut ici garder présent à l'esprit, que nonobstant son issue, la guerre Iran-Irak a, par défaut, été une aubaine pour la théorie de l'arc de crise de Bernard Lewis.

CHAPITRE 2

Echec des politiques du développement : de la montée de l'islam spirituel et mystique à la construction de l'« islam politique »

La révolution iranienne n'a pas été une révolution monolithique comme certains esprits paresseux se plaisent à croire. Essentiellement plurielle, cette révolution a été récupérée et accaparée par un groupe d'opposants qui a trouvé un dénominateur commun en la personne de l'ayatollah Khomeiny. C'est ainsi que, cherchant à instrumentaliser le religieux, le politique a fini par être instrumentalisé par le religieux.

Construction postmoderne, l'« islam politique », présenté comme réponse à la crise de la modernité importée, ne peut être considéré comme "non occidental" c'est-à-dire comme n'ayant rien à voir avec l'Occident. Pour bien comprendre ce qui s'est passé, il importe de garder présent à l'esprit le contexte de la guerre froide des années 70. Au Nord, nous n'avons pas affaire à une mais à des crises de tout genre, et la non moins importante étant la crise de la pensée marxiste-léniniste en général et de l'idéologie communiste en particulier. On notera à cet égard que c'est en 1970 que le prix Nobel de littérature est décerné à Soljenitsyne. Il y a, par ailleurs, la montée, au sein du courant de pensée tiers-mondiste, des revendications en faveur d'un nouvel ordre mondial qui devaient déboucher sur l'amorce d'un prétendu dialogue Nord-Sud. Avec les revendications en matière de nouvel ordre économique international, ce n'est plus le Noir mais le Blanc qui se sent opprimé : le temps du sanglot de l'homme blanc est venu[1]. Certains intellectuels et hommes politiques se situent dans l'ordre du faire semblant pour calmer les ardeurs du Sud. D'autres, inquiétés par les cris d'alarme du Club de Rome, adoptant une posture du chacun pour soi, sont franchement opposés à tout dialogue. Un dernier groupe de gens de bonne volonté se situe dans une perspective de réconciliation de l'homme avec la nature et avec lui-même : nous sommes à l'époque de la Commission Brandt[2].

Au Sud, le contexte est marqué par l'échec des politiques de développement. Après l'euphorie des années 60, nous en sommes à la désillusion au regard du modèle de développement fondé sur la théorie des étapes de la croissance économique de Rostow[3] : on ne croit plus au modèle de développement inéluctable, on parle plutôt du «mythe» du développement[4]. Les modèles de modernisation «top-down», caractérisés par le rôle prépondérant de l'Etat dans l'industrialisation des pays du Tiers Monde, ont généré des problèmes et des distorsions de tout genre dans les domaines politique, économique, social et culturel. Tous les éléments relevant du domaine de la culture endogène, qui avaient été disqualifiés et évacués du champ de la conscience des économistes, réapparaissent par un effet de boomerang. En effet, considérés comme des freins au développement : «le poids des mentalités», «les interdits», etc., sont au mieux ignorés, au pire combattus.

---

[1] Pascal Bruckner, *Le sanglot de l'homme blanc*, Paris, Seuil, 1983.
[2] Rapport de la commission indépendante sur les problèmes de développement international sous la présidence de Willy Brandt, *Nord-Sud : un programme de survie*, Gallimard, 1980.
[3] W.W. Rostow, *Les étapes de la croissance économique*, Paris, Seuil, 1963.
[4] Candido Mendès (sous la dir.), *Le mythe du développement*, Paris, Seuil, 1977.

La connotation péjorative attribuée à ces éléments d'ordre culturel avait empêché de les examiner objectivement et de les comprendre pour y apporter des réponses idoines. Ces agressions culturelles allaient mettre en route un certain nombre de mécanismes de défense, de contre-acculturation.

C'est ainsi que, dans nombre de pays musulmans du Tiers Monde frappés par la volonté de résister, le refus de se soumettre au diktat d'un processus de modernisation paupérisant et excluant, la couche la plus défavorisée et/ou la plus révoltée par les excès de la modernisation des mœurs se replie sur elle-même et se réfugie dans les mouvements religieux en guise de carapace protectrice. Couche aisée désabusée par la crise de la modernité, couche moyenne révoltée par les excès de la modernisation ou couche défavorisée exclue des bienfaits de la modernisation, dans tous les cas c'est l'islam spirituel et/ou mystique qui est d'abord et surtout essentiellement à l'ordre du jour tandis que, parallèlement, des mouvements de résistance politique d'inspiration islamo-marxiste commencent à prospérer dans la clandestinité.

Ce phénomène du retour du religieux et/ou de la quête du spirituel ne se limitait évidemment pas à l'Orient. En Occident aussi, le désenchantement est de rigueur, par exemple, Aldous Huxley, dans son livre *Les Portes de la perception* (*The Doors of Perception*), qui est la principale source d'inspiration du groupe hippie dans les années 60, écrit : « Aujourd'hui, après deux guerres mondiales et trois révolutions majeures, nous savons qu'il n'y a pas de corrélation entre la technologie plus avancée et la morale plus avancée ». Dans la préface à l'ouvrage de Weisskopf intitulé *Aliénation, idéologie et répression*, Marc Guillaume précise que : « les valeurs morales sur lesquelles se fonde la société industrialisée négligent ou répriment des dimensions humaines essentielles, spirituelles et affectives ». Désaveux du capitalisme d'Etat et désenchantement du capitalisme libéral, la crise de la modernité dans les années 70 marque le retour non pas tant de la religion mais surtout du religieux ; retour du religieux comme manifestation du besoin individuel de spiritualité et du besoin collectif de marqueur identitaire dans des espaces et contextes caractérisés respectivement par le maintien de l'entité Etat-nation et la crise concomitante des idéologies politiques.

Il faut garder présent à l'esprit que nous sommes à une époque où l'idée de « projet de société » continue à faire sens. C'est dans ce contexte de guerre froide qui est une guerre des idéologies que Jimmy Carter[1] prononce son fameux discours du « malaise », au cours duquel le Président des Etats-Unis d'Amérique fait expressément et explicitement état de la crise de sens vécue par la nation américaine en déclarant à sa nation : « Je souhaite maintenant vous parler d'une menace fondamentale qui pèse sur la Démocratie de notre pays... Je ne fais pas référence à l'influence exercée par l'Amérique, une nation actuellement en paix avec le reste du monde, et dont la puissance économique et militaire est inégalée... Cette menace est à peine perceptible par des moyens ordinaires. Il s'agit d'une crise de confiance. Il s'agit d'une crise qui frappe la volonté de notre nation en son sein même, en son âme et en son esprit. Nous percevons cette crise à cause du doute croissant que l'on porte sur la signification de nos propres vies et de la perte d'un objectif unique pour notre nation. » Sens et/ou Puissance, comme on le constate, forts de leur puissance économique et militaire, les Etats-Unis ne peuvent se contenter du *hard power*, il leur faut aussi et surtout investir dans le *soft power* et répondre à la question

---

[1] Discours télévisé du 15 juillet 1979.

de sens pour l'individu et la nation américaine dans son ensemble.

Désœuvrés par la crise économique faisant suite à la flambée des prix du pétrole dans un contexte hanté par la rareté des ressources et les limites écologiques certes mais aussi et surtout par la crise de l'idéologie technocratique et positiviste et par la crise politique et épistémologique du marxisme dogmatique, à l'instar des jeunes, certains intellectuels occidentaux se tournent vers l'Orient pour y puiser quelques idées régénératrices au sein de l'islam spirituel et mystique en plein essor comme auprès d'autres religions et pratiques spirituelles « exotiques » telles que le bouddhisme, l'hindouisme, le taoïsme etc. Désillusionnée, toute une frange de la jeunesse s'affilie à des mouvements mystico-religieux ; certains jeunes adeptes du New Age vont jusqu'à sombrer dans la déchéance suite aux excès des pratiques psychédéliques.

La conversion de dernière heure de Michel Foucault témoigne de l'ambiance mystico-religieuse qui submerge tous ceux qui s'approchent de la révolution iranienne, même un communiste tel que Sartre qui semblait être resté à l'abri de tout risque de biais théologico-religieux est pris au piège du transcendantal. Purgatoire de toutes les frustrations[1] de gauche et de droite, terrain d'expérimentation des fantasmes les plus divers, l'esprit de la révolution iranienne embrase tous les cœurs et tous les esprits. Les revendications du Sud en faveur d'un nouvel ordre

---

[1] Se croyant assagi en 2012, Pascal Bruckner déclare : « Avant la révolution iranienne, j'étais un grand admirateur de la culture arabo-musulmane et j'aimais beaucoup me rendre dans les mosquées. Depuis, je suis un peu refroidi. L'islamisme nous a fait oublier la grandeur de l'islam d'autrefois », voir « Pascal Bruckner : L'Islamophobie, ça n'existe pas ! », *Causeur*, 29 octobre 2012.

économique international, d'une part, et les prétentions qualifiées de « mégalomanes » du Shah d'Iran dans le Moyen-Orient, d'autre part, le Nord ne doit pas laisser passer une occasion d'en découdre, ne serait-ce qu'un tant soit peu, avec les acteurs de « désordre ». Le moment est donc venu pour lui de laisser tomber le Shah pour tirer au mieux profit du mouvement de révolte "spontané" qui, jour après jour, prend de l'ampleur et semble de plus en plus échapper à tout contrôle. En effet, comme le Général d'armée Jean Delaunay[1], ancien chef d'état-major de l'Armée de terre française, le souligne, fin 1978 début 1979, l'Iran était « confronté à une dangereuse agitation fomentée de l'extérieur ».

Embarrassante ? Pour peu qu'on y réfléchisse, la révolution islamique peut s'avérer bénéfique pour l'Occident. Après tout, cette révolution n'a pas été aussi spontanée que ça, chacun a apporté sa pierre à l'édifice. Le Shah a bien balisé le terrain tout au long de son régime : réforme agraire, dépossession des biens du clergé iranien (il est à noter au passage que la Grèce n'a toujours pas réussi à imposer les biens de l'Eglise orthodoxe) ; politique de modernisation tous azimuts, style « top-down » sur le modèle du despotisme éclairé ; fêtes de Persépolis ; répression de tout mouvement d'opposition hormis le « pouvoir noir »[2] (les lieux de culte sont les seuls endroits où il est encore possible de s'exprimer librement), *last but not least* pour ne pas dire le plus important, le Shah entre ouvertement en conflit avec les grandes compagnies pétrolières qu'il fustige dans les médias occidentaux sans

---

[1] « Allocution du Général d'armée Jean Delaunay (CR) aux obsèques du Général d'armée Gharabaghi, dernier chef d'Etat-major des Forces Armées Impériales d'Iran », Père Lachaise, Paris, 13 octobre 2000.
[2] Expression du Shah, en référence à la couleur de la tenue vestimentaire des religieux.

pour autant lutter contre l'extension de la corruption qui gangrène le pays.

1973 : choc pétrolier, 1979 : choc islamiste ! Le Shah d'Iran aurait probablement mieux fait de laisser les compagnies pétrolières piller les ressources de son pays en paix au lieu d'aller chercher à négocier ailleurs des prix plus justes. On ne s'attaque pas impunément à plus fort que soi, aussitôt fait, aussitôt évincé. Donc, rien d'étonnant à ce que les Américains aient été pressés de voir le Shah d'Iran partir. Le Général d'armée Abbas Gharabaghi[1], dernier chef d'Etat-major général des Forces armées impériales d'Iran, témoigne à ce sujet : « J'ai insisté auprès de l'Empereur afin qu'il ne quitte pas le pays. Il me répondit que Sullivan, ambassadeur des Etats-Unis, et Huyser, commandant-adjoint de l'OTAN, s'étaient enquis de la date et de l'heure de son départ ». En exil, le Shah le confirmera dans ses mémoires en ces termes : « Je ne le (le général Huyser) vis qu'une fois pendant son séjour. Il accompagna l'ambassadeur Sullivan lors d'une des dernières entrevues que j'eus avec ce dernier. Ce qui les préoccupait, l'un et l'autre, c'était de savoir quel jour et à quelle heure je partirais »[2].

Les Américains n'étaient apparemment pas les seuls à vouloir son départ, il y avait aussi les Britanniques qui, via la propagande de la BBC en faveur de Khomeiny[3] et son

---

[1] Abbas Gharabaghi, *Vérités sur la crise iranienne*, Paris, La Pensée Universelle, 1985.
[2] Mohammed Reza Pahlavi, *Réponse à l'Histoire*, Paris, Albin Michel, 1979.
[3] Il convient de préciser à cet égard que l'arrêt de la propagande de la BBC en faveur de Khomeiny était une des trois conditions posées par le général Abbas Gharabaghi pour rester en fonction. Témoignage personnel de l'auteur du présent essai, cf. également A. Gharabaghi, *op.cit.*, p. 113.

« soutien logistique »[1] aux rassemblements des manifestants, participaient activement à la révolution, à croire qu'ils n'avaient toujours pas bien digéré d'avoir été évincés par les Américains. Le Shah rapporte à ce sujet « Après que des émeutiers eurent incendié l'ambassade d'Angleterre, un de mes généraux rencontra l'attaché militaire britannique. Celui-ci hurla : "Vous n'avez donc pas compris que la solution est une solution politique". » Et M. Georges Lambrakis, premier secrétaire de l'ambassade des Etats-Unis, de confier : « Il y aura bientôt en Iran un nouveau régime ! »

Personnage encombrant aux yeux des grandes puissances occidentales réunies à la Guadeloupe, le Shah, privé du soutien de son peuple, devait se résoudre à quitter son pays. Lorsqu'on parle aujourd'hui de mise à l'écart du peuple, il faut garder présent à l'esprit qu'à l'époque les politiques du développement opéraient dans le cadre du paradigme « top-down », la question de la participation n'est donc pas de rigueur. En Iran, comme dans beaucoup de pays du Tiers Monde, nous avons affaire à une politique de modernisation sans peuple, ce dernier n'a donc *a fortiori* pas de rôle à jouer dans le domaine de la politique étrangère qui aujourd'hui encore s'avère être, partout, y compris dans les démocraties dites libérales, la chasse gardée des gouvernements et de leurs services secrets. En effet, dans le domaine de la politique étrangère, contrairement à l'idée répandue, l'opinion publique est, autant que faire se peut, traitée comme une variable d'ajustement. L'autocratie peut rendre aveugle, ce qui est certain c'est que le pouvoir absolu du Shah l'ayant coupé du peuple, il n'a pas compris qu'il aurait dû, d'une façon ou d'une autre, associer le

---

[1] Diffusion des heures et des lieux de rassemblement, il s'agit là d'un témoignage personnel de l'auteur du présent essai.

peuple à son projet de modernisation et de pleine souveraineté sur les ressources du pays.

Les Etats-Unis d'Amérique avaient pour leur part bien miné le terrain : politique d'ingérence tous azimuts dont l'intervention de la CIA dans le renversement du gouvernement de Mossadegh qui, dans les années 50, avait cherché à acquérir la pleine souveraineté du pays sur ses ressources pétrolières ; et pour finir Jimmy Carter allait, dès sa prise de pouvoir, mettre le feu aux poudres en brandissant l'étendard de la démocratie et des droits de l'homme. Alexandre de Marenches[1], qui, à l'époque, dirigeait les services secrets français, déclare à ce sujet avoir mis en garde le Shah en ces termes : « Je l'ai prévenu que ce personnage désastreux sur le plan national et international qu'était le président Carter avait décidé de le remplacer. Le président américain était totalement ignorant des réalités proches et moyen-orientales et, entre autres, iraniennes. Dans la courte vue de ce personnage boy-scout au visage poupin qui devait tout juste savoir où se trouvait l'Iran, le Shah était un vilain dictateur, qui mettait les gens en prison et, donc, il s'agissait d'y implanter le plus tôt possible le système démocratique, façon U.S.A. » Il précise à cet égard : « J'ai mentionné un jour au Shah le nom de ceux qui, aux Etats-Unis, étaient chargés d'envisager son départ et son remplacement. J'avais même pris part à une réunion où l'une des questions abordées était : « Comment fait-on pour faire partir le Shah et par qui le remplace-t-on ? »

Reste à savoir quelles étaient les motivations réelles et les priorités du gouvernement américain de l'époque : la défense des principes démocratiques et des droits de

---

[1] Ockrent & Marenchez, *Dans le secret des princes*, Paris, Stock, 1986, chapitre 18, Le Shah et l'Ayatollah.

l'homme, la défense des cartels pétroliers américains, la guerre froide ? Probablement toutes à la fois ; s'il nous semble erroné de penser en termes de pur complot, il nous paraît tout aussi faux de raisonner en termes d'explication à base de cause unique de la chute du Shah ou de la décision des Etats-Unis de le destituer. Différents facteurs et calculs sont à la base des décisions politiques. Comme nous allons le voir plus loin, il est peu probable que Carter ait seul décidé de la destitution du Shah comme De Marenches semble le croire. Les présidents sont épaulés, freinés et influencés par leurs conseillers et l'ensemble de l'establishment, il n'en demeure pas moins que comme De Marenches le souligne, les connaissances, les compétences et la personnalité desdits présidents ont leur rôle à jouer[1]. Mais ce qui est certain pour ce qui concerne Carter c'est que, de son propre chef ou téléguidé, ses déclarations intempestives sur la démocratie et les droits de l'homme ont joué le rôle de feu vert pour le renversement du Shah que les opposants au régime du Shah, qu'ils soient au sein du pays ou à l'étranger, n'allaient pas laisser passer.

La France, qui avait reçu, hébergé et préparé en grande pompe le retour en Iran de l'Ayatollah Khomeiny, allait aussi et surtout précipiter le départ du Shah dont les prises de position au sein de l'OPEP et les critiques parfois acerbes sur l'Occident, en général, et les compagnies pétrolières, en particulier, avaient pesé sur la décision du groupe des cinq à la Guadeloupe de laisser tomber le monarque. Mais la révolution iranienne allait aussi et surtout conforter la stratégie de la ceinture verte de Zbigniew Brzezinski et valider après coup la théorie de l'arc de crise de Bernard Lewis. Comme nous allons le

---

[1] Comme les diatribes de Donald Trump nous le rappellent quotidiennement. A croire que les temps changent mais que l'histoire se répète.

constater ci-après, avec cette dernière, nous sommes en présence d'une prophétie auto-réalisatrice tandis que la première transparente s'affiche clairement et est honnêtement assumée par son concepteur. Il importe de préciser qu'il est communément admis que « Brzezinski s'est inspiré de la théorie de l'« arc de crise » de Bernard Lewis, qui devint son conseiller au département d'Etat à partir de 1977. Ensemble, ils ont imaginé la stratégie de la « Ceinture Verte » attribuée depuis à Brzezinski », ce qui a pu faire croire à tort, encore faut-il le souligner, que les deux hommes avaient exactement la même vision des choses et la même visée, ce qui est parfaitement faux.

CHAPITRE 3

De l'ingérence à l'instrumentalisation des religions à des fins géostratégiques et politiciennes : la construction et l'expansion du djihadisme

En décriant le dévoiement du terme islamisme, il ne s'agit évidemment pas de renier la réalité de l'« islam politique » et de sa radicalisation par les mouvements djihadistes pour certains et terroristes pour d'autres. L'islam a de tout temps été instrumentalisé, de l'intérieur, certes, mais aussi et surtout de l'extérieur par les Anglais d'abord et les Américains ensuite. Le fait géopolitiquement important au regard des mouvements politiques d'inspiration islamique est que, si l'islam politique est une construction locale en réaction à des facteurs estimés exogènes par ceux-ci, le djihadisme comme mouvement politique est une construction américaine qui s'est retournée contre les Etats-Unis prouvant encore une fois que la fin ne justifie pas les moyens.

Si l'« islamisme » est, comme d'aucuns l'ont précisé, le résultat de l'instrumentalisation de l'islam par des acteurs locaux, le djihadisme est le résultat de l'instrumentalisation de l'islamisme par les Etats-Unis (EU) dans le contexte de la guerre froide. Pour ce qui est de l'instrumentalisation de la religion par les EU, celle-ci ne se cantonne pas à l'islam. On ne peut en effet ignorer que Zbigniew Brzeziński, conseiller du président Carter pour les affaires de sécurité nationale, est un émigré polonais catholique. Fin 1977, il se rend à Varsovie en compagnie de Rosalynn Carter où, d'après le *New York Times*[1], ils ont le 30 décembre un entretien avec l'archevêque Stefan Wyszynski lequel archevêque allait, au conclave d'octobre 1978, participer à l'élection du premier pape polonais Jean-Paul II. Ces faits vont créer du côté soviétique quelques suspicions à son égard, suspicions qui se renforcent avec le mouvement de grève qui secoue la Pologne durant l'été 1980 et donne naissance au syndicat Solidarnosc (Solidarité) qui va jouer un rôle non négligeable dans la chute du mur de Berlin.

Pour ce qui est de l'instrumentalisation de l'islam par l'Occident, c'est avec la stratégie de la ceinture verte que commence l'histoire du terrorisme islamique mondialisé[2]. La promotion du djihadisme ne sera évidemment qu'un moyen parmi d'autres de la guerre tous azimuts que les Etats-Unis mènent contre l'URSS. Il n'est pas inutile de rappeler ici que Nixon n'a jamais eu de coup de foudre pour la Chine, c'est pour partie contre l'Union soviétique que les Etats-Unis se sont rapprochés de la Chine en 1972. Pour en revenir à la stratégie pro-islamique des Etats-Unis dans le

---

[1] David A. Andelman, "Brzezinski and Mrs. Carter hold discussion with Polish Cardinal", *The New York Times* (December 29, 1977).

[2] Il est vrai que ce sont les Britanniques qui ont les premiers instrumentalisé l'islam mais c'est avec la stratégie de la ceinture verte des Etats-Unis que cette instrumentalisation prend la forme de soutien et de promotion du «djihadisme». Pour ce qui est du terrorisme islamique, il est à noter que la secte musulmane ismaïlienne dirigée par Hassan ibn al-Sabbâh --que nous avons pris soin de mentionner précédemment et que Bernard Lewis aime à citer comme les premiers terroristes-- n'avait que des visées politiques internes au monde musulman. Nous aurons l'occasion de faire état du rôle de l'Arabie saoudite et du Qatar dans le soutien aux entités djihadistes, nous limitant pour l'heure au rôle de l'Occident, on ne peut décemment ignorer que sans l'aide, le soutien logistique et la technologie des grandes puissances, nous n'aurions probablement pas eu droit au djihadisme mondialisé.

contexte de la guerre froide, Brzeziński n'a cessé de se vanter d'avoir été l'instigateur du « Vietnam soviétique » qu'a été la guerre d'Afghanistan. D'après Robert Gates[1], ancien directeur de la CIA, « les services secrets américains ont commencé à aider les moudjahidines afghans six mois avant l'intervention soviétique ». Information que Brzeziński[2] confirme en ces termes : « Selon la version officielle de l'histoire, l'aide de la CIA aux moudjahidines a débuté courant 1980, c'est-à-dire après que l'armée soviétique eut envahi l'Afghanistan, le 24 décembre 1979. Mais la réalité gardée secrète est tout autre : c'est en effet le 3 juillet 1979 que le président Carter a signé la première directive sur l'assistance clandestine aux opposants du régime prosoviétique de Kaboul. Et ce jour-là j'ai écrit une note au président dans laquelle je lui expliquais qu'à mon avis cette aide allait entraîner une intervention militaire des Soviétiques. »

Le fait le plus intéressant est que, lorsqu'en 1998, c'est-à-dire trois ans avant l'attentat du 11 septembre, le journaliste de l'hebdomadaire Le Nouvel Observateur demande à Brzeziński s'il ne regrette pas « d'avoir favorisé l'intégrisme islamiste, d'avoir donné des armes, des conseils à de futurs terroristes », l'intéressé répond en toute honnêteté : « Qu'est-ce qui est le plus important au regard de l'histoire du monde ? Les talibans ou la chute de l'empire soviétique ? Quelques excités islamistes ou la libération de l'Europe centrale et la fin de la guerre froide ? ». Il est vrai que nul politique n'est prophète sur terre mais il est tout aussi certain que tous savent que la fin

---

[1] Robert M. Gates, *From the Shadows*, Simon and Schuster, 2007. Cité par l'hebdomadaire *Nouvel Observateur* du 15-21 janvier 1998.
[2] Interview de Zbigniew Brzezinski : « Oui, la CIA est entrée en Afghanistan avant les Russes... », in l'hebdomadaire *Le Nouvel Observateur* du 15-21 janvier 1998.

ne justifie en aucune façon les moyens comme l'Histoire dans sa réalité passée et quotidienne le leur rappelle constamment, la suite de l'histoire de l'« intégrisme islamiste » est là pour nous le confirmer.

Nous en arrivons maintenant au cas de l'Irak de Saddam Hussein. La question, qui se pose au commun des mortels concernant la seconde attaque de l'Irak présidée par la famille Bush en 2003, consiste à savoir pourquoi, après le drame du World Trade Center de 2001 qui a clairement mis en exergue l'absurdité du soutien aux mouvements islamistes anticommunistes qui se sont mués en mouvements djihadistes antioccidentaux, Bush fils s'attaque à un régime « laïque »[1]. Il est à noter à cet égard que, dans un premier temps, le gouvernement américain voulait, en guise de prétexte pouvant justifier une attaque armée, faire croire à l'existence de relations de connivence entre le régime et des cellules dormantes de l'organisation terroriste al-Qaïda, prétexte qu'il a dû abandonner à la faveur d'une autre contrevérité : le crime de détention d'armes de destruction massive.

Etant donné que, bien avant la sortie tardive en juillet 2016 du rapport Chilcot sur la guerre en Irak, l'opinion publique était partout au courant que le prétexte des armes de destruction massive avait non seulement été invalidé par les faits mais que tout un chacun savait déjà qu'il s'agissait là d'un prétexte, pour ne pas dire d'un mensonge échafaudé pour les besoins de l'attaque dite préventive, reste donc à savoir quelle a été la cause réelle de l'attaque et du renversement du régime irakien en 2003. La cause

---

[1] Si le régime de Saddam Hussein ne peut être qualifié de laïque au sens occidental du terme, il ne pouvait à l'époque être considéré comme pro-religieux. Voir, entre autres, Michel Chatelus, *Stratégies pour le Moyen-Orient*, Paris, Calmann-Lévy, 1974, p.84.

primordiale de l'invasion de l'Irak a été la même que celle qui a présidé à la première guerre du Golfe (1990-1991) plus de dix ans auparavant, à savoir le pétrole. Etant donné qu'aujourd'hui les Etats-Unis, grâce à l'exploitation du gaz de schiste, ont réduit leur dépendance vis-à-vis du pétrole du Moyen-Orient, la réalité ne nous saute pas aux yeux. Mais, comme Alan Greenspan[1], ancien président de la Réserve fédérale américaine des Etats-Unis, le souligne dans ses mémoires, il est peut-être « politiquement incorrect de le reconnaître » ("politically inconvenient fact") mais il faut admettre que « la guerre en Irak a été essentiellement une guerre pour le pétrole » ("the war was largely about oil"). Il est intéressant de noter que, d'après Alan Greespan, l'invasion de l'Irak par les Etats-Unis a évité un « chaos au niveau de l'économie mondiale » ("chaos to the global economy"), elle a permis « le fonctionnement normal du marché du pétrole en attendant que les Etats-Unis trouvent des énergies de substitution »[2].

Lorsque le dictateur Saddam Hussein a envahi le Koweït, il avait probablement raison de penser que les revenus pétroliers de ce pays pourraient être utilisées à meilleur escient par un grand pays que par un petit pays. Mais le problème est qu'il semble ne pas avoir été au courant de la stratégie britannique qui avait présidé à la création des petits Etats pétroliers, stratégie reprise et mise au goût du jour dans la théorie de l'arc de crise échafaudée par des penseurs néoconservateurs américains spécialistes de relations internationales ! Nous ne rentrerons pas dans les détails de cette théorie qui, à elle seule, pourrait faire

---

[1] Bob Woodward, "Greenspan: Ouster Of Hussein Crucial For Oil Security", *Washington Post*, September 17, 2007.
[2] "Making certain that the existing system [of oil markets] continues to work, frankly, until we find other [energy supplies], which ultimately we will.", *op.cit.*

l'objet d'un ouvrage de plusieurs volumes. Cette théorie, qui date des années cinquante, a si bien servi de source d'inspiration pour l'action politique de la première puissance mondiale d'abord et de ses alliés ensuite, qu'elle a fini par être validée après coup, de sorte qu'elle peut aujourd'hui servir de prototype pour les cours d'enseignement pratique en matière d'ingérence "humanitaire".

Pour revenir au cas du *Rais* d'Irak, la question consiste à savoir comment ce dernier a pu croire qu'il pouvait impunément envahir le Koweït, petit pays à forte population shiite. Comment, avec l'expérience de la chute du régime du Shah en guise de mémoire, Saddam Hussein, qui était au courant de la tactique de « la cause de la démocratie et des droits de l'homme » comme instrument de déstabilisation de régimes dictatoriaux en quête de souveraineté sur leurs ressources, a-t-il pu croire qu'il allait pouvoir impunément enfreindre la Charte des Nations Unies en envahissant un pays voisin ? Il aurait, paraît-il, eu un entretien avec une représentante américaine qui lui aurait fait croire qu'il pouvait mettre main basse sur le Koweït ! Même en admettant que ceci soit vrai, comment a-t-il pu être aussi naïf pour croire à une telle chose ? Non, Saddam Hussein n'était pas naïf. Le problème est qu'il n'était pas novice non plus en matière d'invasion de ses pays voisins. En effet, l'intéressé n'était pas à son premier coup d'essai, dix ans auparavant, en violation totale du droit international, il avait déjà tenté d'envahir l'Iran, mais, à ce moment-là, il bénéficiait de la bénédiction des grandes puissances qui non seulement ne s'opposaient pas à son agression mais, au contraire, lui apportaient tout leur soutien nécessaire. Il se souvient qu'il a même été jusqu'à utiliser des armes chimiques contre l'armée iranienne et la population civile sans que la fameuse communauté internationale s'en indigne pour autant, ceci alors même que son pays était signataire du protocole de Genève interdisant l'utilisation des armes chimiques !

Si des grandes puissances ont soutenu son agression contre un grand pays comme l'Iran et fermé les yeux sur ses multiples manquements au droit international, elles n'allaient pas prendre au sérieux l'invasion d'un petit pays comme le Koweït. Oui, mais Saddam Hussein ne semble pas avoir compris que c'est à lui qu'il appartenait de mettre à jour son logiciel et non aux grandes puissances. Il est certain que les grandes puissances n'ont pas toujours respecté les règles qu'elles ont elles-mêmes fixées mais c'était pour la bonne cause, elles devaient défendre leurs intérêts. Il n'a pas compris que la guerre Irak-Iran était conforme aux intérêts des grandes puissances et confortait la stratégie de l'arc de crise de la superpuissance américaine.

En envahissant l'Iran, Saddam Hussein (SH), dictateur connu pour sa cruauté, se voyait déjà futur Saladin ; il espérait ainsi pouvoir profiter des dissensions et des désordres consécutifs à la révolution pour élargir sa fenêtre maritime sur le Chatt-al-Arab (Alvand Rud en persan) et faire main basse sur les ressources pétrolières iraniennes. Le problème est que, aveuglé par sa folie des grandeurs, SH n'avait pas l'air de se rendre compte que, s'il avait bénéficié à l'époque du soutien occidental, ce n'était pas parce que les Occidentaux approuvaient ses projets mais parce que, d'abord et surtout, la République islamique d'Iran, contrairement à toute attente et en violation des principes de base de la théorie de l'arc de crise, refusait de se comporter en vassal et de se soumettre au diktat des puissances occidentales, à l'instar des monarchies pétrolières et, qui plus est, ses dirigeants n'avaient pas pu ou voulu empêcher certaines bavures des extrémistes. Car il ne faut pas perdre de vue qu'à l'époque, l'instauration de la République islamique équivalait à la victoire du courant des religieux antimarxistes contre des communistes et des religieux marxistes-léninistes, ce qui équivalait *in fine* à la victoire du camp occidental contre l'URSS.

Il faut effectivement garder présent à l'esprit que, à la grande joie et avec le soutien des Etats-Unis, ce n'est ni le communisme, ni la version marxiste-léniniste de l'islam qui allaient l'emporter en Iran mais un régime islamique antimarxiste. Ce qui, dans le cadre de la guerre froide, équivalait à une victoire du camp occidental. Car, comme on le sait, le problème des Occidentaux à l'époque n'était pas l'islam, au contraire, ce changement de régime confortait la stratégie de la ceinture verte ; ce qui dérangeait, c'était le caractère inféodé du nouveau régime islamique qu'ils avaient aidé à s'instaurer en Iran.

En ce qui concerne SH, son problème résidait dans le fait que, lorsqu'il décida en septembre 1980 d'attaquer l'Iran, il fonça droit sur son adversaire sans prendre soin de jeter un coup d'œil dans son rétroviseur pour voir s'il avait bien perçu ce qui s'était passé en Iran. Il semble qu'il n'ait pas compris pourquoi le Shah d'Iran avait été éjecté. Il n'avait même pas pris garde de regarder autour de lui pour évaluer la force réelle de son adversaire.

En effet, en 1979, le Général d'armée Abbas Gharabaghi[1], dernier chef d'état-major général des Forces armées iraniennes, après consultation et avec l'accord et le

---

[1] Le Général d'armée Abbas Gharabaghi, dernier chef d'état-major général des Forces Armées Impériales d'Iran, également ministre de l'Intérieur dans son pays avant et pendant la crise, est l'auteur de différents ouvrages dont *Vérités sur la crise iranienne*, livre écrit en persan, traduit et publié en français.

soutien de l'ensemble de son état-major (26 généraux), avait réussi non seulement à éviter une sanglante guerre civile mais, n'en déplaise à SH, à sauvegarder également tout le complexe militaire du pays intact (armées et arsenal militaire). Et ceci, malgré les purges opérées par le nouveau régime, a permis à l'Iran de faire face à l'agression de l'Irak. Qui plus est, contrairement à Chahpour Bakhtiar[1], dernier Premier ministre de Mohammed Reza Chah Pahlavi, qui a pactisé avec l'Irak et reçu des financements de l'Irak, de l'Arabie saoudite et d'autres pays arabes du Golfe persique[2] ; en exil en France au moment de l'agression de son pays par l'Irak, le Général d'armée A. Gharabaghi prit soin de lancer un appel à l'union et à la mobilisation de l'ensemble des forces armées de son pays pour combattre l'ennemi, précisant avec force et conviction qu'il était du devoir de chaque Iranien de défendre son pays.

Saddam Hussein ne semble pas avoir bien compris pourquoi les Occidentaux l'avaient soutenu dans sa guerre contre l'Iran. Les Occidentaux l'avaient soutenu parce qu'ils étaient, certes, contre le régime islamique tel qu'il se présentait à leurs yeux mais aussi et surtout parce qu'ils étaient également contre son projet expansionniste de panarabisme qui est à l'opposé de la stratégie de la « balkanisation » du Moyen-Orient. Ainsi que nous avons déjà eu l'occasion de le signaler ailleurs, la guerre Iran-Irak, qui réduira à néant des décennies d'effort de modernisation des infrastructures économiques et sociales des deux pays

---

[1] *In fine*, en pactisant avec l'Irak, C. Bakhtiar, qui se trouvait à l'étranger à cette époque, a coupé l'herbe sous le pieds des opposants non khomeynistes qui luttaient pour la prise du pouvoir en Iran.
[2] Cf. Interview télévisée du mois d'août 2011 de Manouchehr Razmara, ministre de la Santé du cabinet de C. Bakhtiar, par David Abbasi, directeur de Mehrtv, accessible sur youtube.

concernés, allait être une affaire bien juteuse pour les pays occidentaux. Elle allait leur permettre de mettre main basse sur les ressources de pétrole de ces pays voués ainsi à demeurer encore longtemps de simples pays exportateurs de matières premières autant que faire se peut à l'état brut, écartant tout espoir d'une quelconque souveraineté économique.

En effet, ces pays allaient, grâce à la guerre, devenir les plus grands importateurs d'équipement militaire, toutes les puissances industrielles chercheront à tirer profit du vaste marché qui s'ouvrait à elles, plus de 50 pays ont approvisionné en matériel militaire les deux pays et une trentaine les deux à la fois. Le président Eisenhower avait finalement tout à fait raison de mettre en garde contre le complexe militaro-industriel. Il est certain que son discours d'adieu était destiné au peuple américain et que l'intéressé ne cherchait pas nécessairement à défendre les intérêts des pays du Tiers Monde mais souhaitait plutôt attirer l'attention sur le caractère vicieux d'un système où les dépenses sociales sont sacrifiées à l'aune du complexe militaro-industriel. Il est inutile de faire étalage ici de l'ampleur des destructions subies par les deux parties, surtout du côté iranien puisque les Etats-Unis avaient prêté main forte aux Irakiens, en détruisant pas mal de cibles dont notamment les plateformes pétrolières iraniennes. Après huit années de guerre et de destruction de part et d'autre, allait commencer la phase de reconstruction des deux pays et de leur économie, c'est ainsi que les pays occidentaux allaient pouvoir encore une fois lorgner sur de nouveaux marchés.

Dévastatrice à plus d'un égard et plus particulièrement au regard des pertes en vies humaines, la guerre Iran-Irak a laissé les deux pays exsangues. Mais, contrairement à toute attente, loin d'affaiblir le régime islamique d'Iran, comme d'aucuns l'avaient espéré, l'agression irakienne a coupé court aux dissensions et aux luttes internes pour le pouvoir, créant une cohésion interne qui a permis au pays de faire face à l'ennemi. Consolidé, le pouvoir en place a pu bénéficier du temps nécessaire pour la conception et la mise en place d'institutions islamiques lui assurant ainsi les assises nécessaires à la pérennisation d'un régime théocratique, mode de gouvernement inédit dans l'histoire du pays. S'il n'y avait eu l'épisode de la prise d'otages à l'ambassade américaine, le basculement de l'Iran vers un régime islamique ne pouvait que conforter la stratégie dite de la ceinture verte des EU.

Plus tard, avec la disparition de l'URSS, la stratégie de la ceinture verte sera revue et actualisée pour être dénommée, sous le régime de Bush fils, de projet du Grand Moyen-Orient (GMO). Les stratèges tiendront dûment compte du vide créé par la perte du précieux ennemi. Fidèle à l'idéologie de « guerre des civilisations » de Bernard Lewis et Samuel Huntington, le projet du GMO, qui a pour objectif de civiliser l'arc de crise constitué par des entités islamiques (le monde musulman), va officiellement initier l'ère de la « bombocratie ». Comme il a été signalé précédemment, la cause primordiale de la destruction de l'Etat irakien a été la mainmise sur le pétrole. Une démocratie à l'occidentale et la soumission au diktat étranger ne peuvent aller de pair car la première suppose un minimum de souveraineté, il serait donc erroné de croire au « pari » sur la démocratie et de son prétendu « effet de domino ». En effet, selon d'aucuns[1], « Carter avait fait le pari, perdu, que les pays arabes rejoindraient l'Égypte dans un cercle vertueux après Camp David ; Bush, celui, tout aussi perdu, qu'un cercle vertueux aboutirait à une

---

[1] Antoine Coppolani, « L'arc de crise selon Obama et ses conseillers », *Politique étrangère* 2009/1 (Printemps), pp. 133-145.

démocratisation salutaire du Moyen-Orient après le changement de régime en Irak ». A moins d'être naïf ou manipulateur, on peut se demander comment certains peuvent continuer à croire qu'il y ait eu un quelconque « pari » à ce sujet ! Si le pari avait porté sur le chaos, là on pourrait dire oui, l'effet de domino a bien fonctionné.

Comme l'instrumentalisation de l'islam par Carter en Afghanistan a été à l'origine de la naissance du mouvement djihadiste d'al-Qaïda, la guerre d'Irak du président Bush junior a enfanté un autre monstre bien connu aujourd'hui du nom de Daech.

Pour en venir maintenant au cas de Daech, il est d'ores et déjà admis que ce groupe terroriste a été formé par les vaincus de la guerre d'Irak de Bush junior. Un bref rappel des faits s'impose. Le 29 juin 2014, le djihadiste irakien Abou Bakr al-Baghdadi qui, depuis 2010, dirige Daech (à l'origine dénommé "Etat islamique en Irak"), s'auto-proclame « Calife ». Stupéfaction ! Tout le monde se demande qui est ce personnage[1]. Après investigation, les médias révèlent la sordide vérité, Bucca, la prison irakienne des Etats-Unis, a joué le rôle d'incubateur. Selon Europe1[2], « En enfermant massivement les islamistes radicaux et gradés de l'armée de Saddam Hussein pendant la guerre d'Irak, les Etats-Unis pourraient bien avoir fait le lit de l'Etat islamique ». D'après le site d'information Médiapart : « En six ans, la prison américaine dite "Camp Bucca" du sud de l'Irak a "accueilli" au moins neuf cadres de Daech, dont son actuel dirigeant Abou Bakr al-Baghdadi ». C'est en effet dans la prison de Bucca que des hauts dirigeants de

---

[1] Voir, entre autres, Emeline Wuilbercq, « Irak : qui est Abou Bakr al-Baghdadi, le nouveau « calife » ? », in JeuneAfrique.com, du 14 juillet 2014.
[2] Pauline Hofmann, « Irak : la prison américaine qui a fait le lit de l'Etat islamique », *Europe1*, le 17 novembre 2014.

Daech ont pu nouer contact avec les officiers de haut rang de Saddam Hussein. Communauté d'intérêts oblige, les officiers de l'état-major de SH, majoritairement des baasistes laïcs, écartés du pouvoir, se sont ainsi ralliés à des salafistes djihadistes dont certains d'obédience al-Qaïdienne.

Il est à noter que les anciens officiers de SH qui ont rejoint les rangs de Daech n'ont pas tous été nécessairement en prison mais ils ont ceci en commun qu'ils ont tous été mis à pied. D'après Romain Caillet[1], historien spécialiste de la mouvance djihadiste globale, « Lorsqu'on évoque le poids de l'idéologie baasiste dans l'engagement djihadiste des anciens officiers de Saddam Hussein, il faut impérativement distinguer deux catégories. La première, celle d'anciens baasistes convaincus, et qui le sont demeurés, devenus par la suite des alliés tactiques de l'État islamique, à l'instar de Izzat ad-Duri, ancien proche de Saddam Hussein et secrétaire général du Baas. La seconde catégorie regroupe, quant à elle, des officiers dont l'appartenance au Baas était purement utilitaire : sous le régime de Saddam Hussein aucun officier ne pouvant sérieusement espérer faire carrière sans avoir sa carte du Parti ».

On comprend mieux ainsi pourquoi Daech cherche à exterminer les chiites. Il est utile de rappeler à cet égard que, après la chute du régime de SH, les Américains n'avaient d'autre choix que de s'appuyer sur les chiites qui avaient été brimés et écartés du pouvoir par SH. Les dirigeants sunnites écartés du pouvoir se trouvent pour leur part acculés à rejoindre les groupes djihadistes. De sorte que, lorsqu'en 2006, et plus exactement le 3 octobre 2006, le "Conseil consultatif des Moudjahidines en Irak",

---

[1] L'étude a été publiée le 16 juin 2015 par la fondation norvégienne Noref.

composé d'al-Qaida en Irak, d'autres groupes djihadistes et d'une trentaine de tribus sunnites, s'institue en "Etat islamique en Irak", le groupe terroriste ainsi constitué est déterminé plus que tout à éradiquer la domination chiite dans le pays. C'est ainsi que la lutte pour le pouvoir va être instrumentalisée pour prendre la forme d'une guerre confessionnelle au sein de l'islam.

Comme nous serons amenés à le constater un peu plus loin, cette guerre civile entre sunnites et chiites sera malencontreusement perçue comme une aubaine par certains dirigeants et membres "islamophobes" de l'intelligentsia occidentale mais, malheureusement pour eux, les chiites ne sont pas les seules victimes, l'animosité de Daech envers les autres religions est totale et conduit à des massacres indistincts des yézidis, chrétiens, juifs, alaouites, etc.

Il nous faut pour l'heure voir comment la poursuite de l'ingérence des grandes puissances dans le Moyen-Orient a pu contribuer au renforcement et à l'extension territoriale de Daech. Il importe d'ouvrir ici une parenthèse pour rappeler que, sans l'action proactive du pape[1] et la capacité d'écoute et l'intelligence proactive d'Obama, c'est la question de la libération de la Syrie du joug de l'Etat islamique qui aurait été aujourd'hui à l'ordre du jour des préoccupations de la "communauté internationale". On ne peut, en effet, ignorer le rôle fondamental du pape qui a réussi l'exploit d'éviter la mainmise d'al-Qaïda et de Daech sur la Syrie grâce à son intervention fatidique, qui a pris la forme d'un appel à la paix et d'une invitation à une veillée de prière pour la Syrie à la veille même d'une intervention militaire imminente des grandes puissances qui aurait inévitablement conduit à la destruction de l'Etat syrien.

---

[1] Cf. Ninou Garabaghi, « Les Organisations internationales et régionales et le progrès du genre humain : Quel avenir pour la culture de la paix et l'éthique de la non-violence ? » in revue *Géostratégiques* *N° 44*, avril 2015, p.172.

En effet, pas plus que l'histoire passée, l'histoire immédiate n'a apparemment été d'une quelconque utilité pour certains dirigeants européens qui ont pourtant coutume de se targuer d'être plus "diplomates" que leurs homologues américains. C'est ainsi qu'après la « bombocratie » d'Irak de 2003, le Moyen-Orient a eu droit aux bombardements humanitaires de la Libye en 2011, ce qui a permis la création d'une armée de réserve supplémentaire de mercenaires pour Daech qui, d'après un rapport du Congrès américain, bénéficie de sources de financements directs du Qatar, du Koweït et surtout de l'Arabie saoudite qui est le principal bailleur de fonds de ce groupe djihadiste. En effet, les musulmans de différentes nationalités qui bénéficiaient d'un emploi grâce aux investissements de Khadafi se sont trouvés au chômage et sont venus ainsi grossir l'armée de réserve constituée par les laissés pour compte prêts à être enrôlés comme mercenaires par l'organisation de l'Etat islamique en Syrie et en Irak (Daech). De plus, la Libye mise en chaos est devenue une zone de non-droit, un no man's land où non seulement Daech mais tout groupe terroriste peut se réfugier en cas de nécessité (situations de défaite, etc.). Pire encore, s'étant implanté dans plusieurs villes (Sabratha, Derna et Syrte), Daech s'offre le luxe le luxe de demander à ses supporters nord-africains de le rejoindre en Libye plutôt qu'en Syrie et en Irak[1].

Après la sortie bien tardive en juillet 2016 du rapport Chilcot sur la guerre en Irak, le rapport de la commission des Affaires étrangères du parlement britannique sur l'intervention en Libye, rendu public le 14 septembre 2016, est éloquent à cet égard. Le rapport d'enquête parlementaire relatif à l'intervention en Libye, qui n'a pas pu, à l'instar du

---

[1] Cf. paragraphe 113 du rapport *Libya - Examination of intervention and collapse and the UK's future policy options*, UK Foreign Affairs Committee, 9 September 2016.

rapport Chilcot, bénéficier d'un embargo de 13 ans, révèle le dessous des cartes.

A quelque chose malheur est bon, les rapports d'enquête du parlement britannique sur les interventions militaires occidentales en Irak et en Libye font honneur à la démocratie libérale fondée sur la séparation, l'équilibre et le contrôle des pouvoirs. Ceci dit, si le rapport sur la guerre en Irak a été dévastateur pour Tony Blair, le rapport sur l'intervention militaire en Libye n'a pas été sans conséquence pour David Cameron. Mais le pire est que ce dernier rapport est particulièrement accablant pour l'ancien président français, Nicolas Sarkozy. Nous avons déjà traité de la question des interventions militaires occidentales dans les pays jadis qualifiés de Tiers Monde et plus spécifiquement en Libye[1], nous nous concentrerons ici sur les rapports d'enquête parlementaire rendus publics après la parution de notre précédent essai traitant de la question des interventions dites humanitaires et de la culture de la paix[2]. Rappelons toutefois que, depuis la fondation des Nations Unies en 1945, seul le motif de la légitime défense est accepté par le droit international comme argument pouvant justifier une intervention militaire, intervention qualifiée par d'aucuns de « guerre juste ». Il est également utile de rappeler que la guerre d'Irak qualifiée de « guerre préventive » par le président Bush junior n'avait pas reçu l'aval du Conseil de sécurité des Nations Unies. Après ces rappels liminaires concernant des questions de légitimité en

---

[1] Pour le cas de la Libye cf. Ninou Garabaghi, « Les Organisations internationales et régionales et les révoltes arabes », revue *Géostratégique N° 32*, 3è semestre 2011.
[2] Ninou Garabaghi, « Les Organisations internationales et régionales et le progrès du genre humain : Quel avenir pour la culture de la paix et l'éthique de la non-violence ? », op.cit.

droit des interventions militaires, venons-en maintenant à l'évaluation de la réalité des faits.

La conclusion du rapport Chilcot est sans appel : « le Royaume-Uni a envahi l'Irak sans avoir épuisé les options pacifiques et sans préparation adéquate quant aux conséquences ». En effet, le mercredi 6 juillet 2016, John Chilcot, le président de la commission mise en place en 2009 pour enquêter sur la guerre d'Irak, déclare selon le quotidien Le Monde[1] que : « *Nous avons conclu que le Royaume-Uni avait décidé de se joindre à l'invasion de l'Irak avant que toutes les alternatives pacifiques pour obtenir le désarmement* [du pays] *ne soient épuisées* », estimant que « *l'action militaire n'était pas inévitable à l'époque* » et que Tony Blair s'était engagé à suivre le président américain George Bush sans questionner sa politique et le lui avait fait savoir par un mémo secret en ces termes « *Je serai avec toi quoi qu'il arrive* ».

Ce qu'il importe de souligner est que, d'après le rapport Chilcot, des dizaines de milliers d'Irakiens sont morts lors de la guerre d'Irak et des violences interconfessionnelles qui ont suivi la guerre. Quelque 45.000 soldats britanniques ont participé à cette guerre entre 2003 et 2009, dont 179 sont morts. Se basant sur des informations en provenance de différentes agences de presse, le quotidien Le Figaro[2] rapporte, le 5 juillet 2016, qu'un premier rapport officiel publié en 2004 avait conclu que Tony Blair avait exagéré devant le Parlement la menace représentée par le président Saddam Hussein. Même si son auteur, Robin Butler, a déclaré par la suite que l'ex-Premier ministre « croyait vraiment » à l'époque en ce qu'il disait. Depuis, Tony Blair s'est excusé plusieurs fois pour les vies perdues et, non

---

[1] Le Monde.fr avec AFP, le 06.07.2016.
[2] AFP, AP, Reuters Agences, *Lefigaro.fr*, le 05.07.2016.

moins important, a reconnu sa responsabilité dans l'essor de Daech en Irak et en Syrie, sans pour autant regretter le fait que Saddam Hussein ait été renversé.

Les propos de l'ancien Premier ministre Tony Blair concernant « sa responsabilité » après la sortie du rapport d'enquête ayant « scandalisé la presse », d'après le quotidien français le Monde[1] « *The Herald* dénonce un Tony Blair « *provocant* », tout comme le *Daily Telegraph*. La plupart des tabloïds britanniques ont publié des « unes » au vitriol contre l'ancien Premier ministre. « *Une arme de manipulation massive* », titre *The Sun* avec une photo de Tony Blair, en référence à l'argument très discuté avançant la présence d'armes de destruction massive en Irak, justifiant l'intervention militaire, alors que le *Daily Star* qualifie l'ex-Premier ministre de « *pire terroriste du monde* ».

Il est à noter que les excuses de Tony Blair ont été estimées partielles au regard de sa responsabilité personnelle. En effet, selon les médias, ce dernier a cherché, avant la sortie officielle du rapport Chilcot, à relativiser « "son crime" face à la responsabilité collective occidentale dans la guerre civile en Syrie » en déclarant sur CNN[2] que : « Nous sommes restés en arrière et nous, à l'Ouest, en portons la responsabilité, l'Europe plus que tout. Nous n'avons rien fait. C'est un jugement de l'histoire auquel je suis prêt à faire face ». Il faut reconnaître que Tony Blair a le mérite, malgré tout, de faire le lien entre l'invasion de l'Irak et l'essor de Daech et de reconnaître *in fine* la responsabilité de l'Occident dans l'essor du djihadisme et la guerre civile en Syrie. Ce qui ne semble pas être le cas des anciens

---

[1] Le quotidien *Le Monde* du 07.07.2016.
[2] Florentin Collomp, « Tony Blair reconnaît sa responsabilité dans l'essor de Daech », le quotidien *Le Figaro* du 25.10.2015.

dirigeants occidentaux responsables du chaos libyen qui n'ont jamais fait de *mea culpa* et encore moins reconnu le lien entre l'intervention en Libye et l'essor de Daech et d'autres groupes djihadistes au Moyen-Orient et en Afrique du Nord comme ceci a été pour partie mis en évidence dans le rapport susmentionné de la commission des Affaires étrangères du parlement britannique rendu public le mercredi 14 septembre 2016.

D'après le rapport d'enquête britannique sur l'intervention en Libye, « aveuglé par l'enthousiasme français à intervenir », le gouvernement britannique n'a pas été en mesure d'évaluer à sa juste mesure la réalité des faits, à savoir que la menace envers les civils de Benghazi avait été largement exagérée tandis que la part non négligeable des islamistes dans la rébellion avait été ignorée. Courtoisie diplomatique oblige, dans le résumé du rapport qui met en exergue les débordements, seul David Cameron est blâmé : "By the summer of 2011, the limited intervention to protect civilians had drifted into an opportunist policy of regime change. That policy was not underpinned by a strategy to support and shape post-Gaddafi Libya. The result was political and economic collapse, inter-militia and inter-tribal warfare, humanitarian and migrant crises, widespread human rights violations, the spread of Gaddafi regime weapons across the region and the growth of ISIL in North Africa. Through his decision making in the National Security Council, former Prime Minister David Cameron was ultimately responsible for the failure to develop a coherent Libya strategy". Comme on le constate d'après le bilan établi par les enquêteurs, les résultats de l'intervention précipitée sont désastreux : elle a conduit à « un effondrement politique et économique, des affrontements entre milices et tribus, des crises humanitaires et migratoires, des violations des droits de l'homme à grande échelle, la dissémination des armes du régime de Kadhafi dans toute la région et l'expansion de l'État islamique en Afrique du Nord »[1].

Si la lecture in extenso du rapport d'enquête britannique laisse perplexe et dubitatif, le chapitre consacré à la France est particulièrement déconcertant et s'avère *in fine* salutaire pour la prise de conscience des motivations réelles de nombre d'interventions dites humanitaires et de l'impact négatif qu'elles peuvent avoir sur la foi en la possibilité d'une justice internationale indépendante et impartiale, fondement du droit d'ingérence. Faisant état des propos rapportés à la secrétaire d'Etat américaine par Sidney Blumenthal, conseiller d'Hillary Clinton, alors secrétaire d'État suite à ses entretiens avec les services secrets français, le rapport d'enquête britannique énumère les motivations qui ont poussé Nicolas Sarkozy, à l'époque président de la République française, à intervenir en Libye :
- obtenir une plus grande part de la production de pétrole libyen ;
- accroître l'influence française en Afrique du Nord ;
- améliorer sa situation politique en France ;
- permettre aux armées françaises de réaffirmer leur position dans le monde ;
- contrer les plans à long terme de Kadhafi qui, selon ses conseillers, cherchait à supplanter le leadership de la France en Afrique francophone. »

Le rapport précise : « quatre de ces cinq facteurs correspondaient à l'intérêt de la France. Le cinquième représentait l'intérêt politique personnel du président Sarkozy »[2]. On comprend mieux pourquoi Nicolas Sarkozy et son ministre

---

[1] « Intervention en Libye. Un rapport étrille Sarkozy et Cameron », le quotidien *Ouest France*, le 15/09/2016.
[2] Cf. rapport du parlement britannique susmentionné, paragraphe 20.

des Affaires étrangères étaient si pressés d'intervenir en Libye[1] ; et pourquoi une opération, qui, sous la bannière du « droit d'ingérence », était destinée à aider les civils de Benghazi, s'est muée en une intervention militaire visant à renverser le régime du colonel Kadhafi. Le fait aggravant, quant à l'impact négatif de l'intervention militaire française sur l'opinion publique internationale, est qu'indépendamment du rapport d'enquête du parlement britannique, une information judiciaire concernant « un financement allégué de la campagne électorale de 2007 de Nicolas Sarkozy par la Libye » est en cours en France. Mais le pire concernant ces ingérences est que l'échec des interventions militaires des puissances occidentales a non seulement eu un impact négatif sur la sécurité intérieure des pays mais a également eu pour effet de générer de nouveaux flux migratoires en provenance cette fois des pays du Moyen-Orient.

Dans une interview avec le journaliste Jeffrey Goldberg[2], le président des Etats-Unis Barack Obama reconnaît que l'intervention militaire en Libye a été une erreur, il est à noter que cette intervention a davantage été le fait de Hillary Clinton, à l'époque secrétaire d'Etat. L'intéressée s'était déjà montrée favorable à l'intervention militaire en Irak tandis que Barak Obama, à l'époque sénateur, faisait campagne contre cette intervention. Depuis, Hillary Clinton a reconnu s'être trompée pour ce

---

[1] *"Former French Foreign Minister Alain Juppé, who introduced Resolution 1973, asserted in his speech to the Security Council that "the situation on the ground is more alarming than ever, marked by the violent re-conquest of cities". He stressed the urgency of the situation, arguing that "We have very little time left—perhaps only a matter of hours."*, Cf. Rapport parlementaire susmentionné paragraphe 18.
[2] Jeffrey Goldberg, « The Obama Doctrine », *The Atlantic*, April 2016.

qui concerne son soutien à l'intervention en Irak. Reconnaître ses erreurs après coup est une bonne chose mais mettre ses erreurs à profit pour ne pas en commettre de nouvelles est encore plus important. Les interventions militaires américaines en Afghanistan, en Irak et au Libye s'étant soldées par des échecs, le président Obama a eu la sagesse de tirer les leçons des défaillances américaines dans les guerres asymétriques. C'est ainsi que, mettant à profit l'opposition du parlement britannique concernant la proposition d'inspiration néoconservatrice d'intervention militaire en Syrie soutenue par François Hollande et David Cameron, Barak Obama a décidé de renoncer à une intervention militaire de plus au Moyen-Orient. Intervention militaire qui avait d'ailleurs été fortement dénoncée par le pape François comme il a été signalé plus haut. Si la position du président des Etats-Unis a pu être critiquée par François Hollande et certains néoconservateurs, c'est parce que les bienfaits de l'action intangible passent inaperçus. Comme il est vrai qu'on parle rarement des trains qui arrivent à l'heure, contrairement au récit qui en est donné par quelques-uns de ses détracteurs d'obédience néoconservatrice, il nous faut admettre que Barack Obama a raison de penser, sans toutefois l'exprimer en ces termes exactement, que *in fine* l'Histoire lui donnera raison.

Le journaliste américain Jeffrey Goldberg rapporte à ce sujet : « S'il n'y avait pas eu l'Irak, l'Afghanistan et la Libye, me dit Obama, il aurait peut-être été plus prompt à prendre des risques en Syrie. "Un président ne prend pas de décisions dans le vide. Il n'a pas une ardoise vierge. Tout président sensé, je pense, reconnaîtrait qu'après une décennie de guerre, avec des obligations qui requièrent encore aujourd'hui un important niveau de ressources et d'attention en Afghanistan, avec l'expérience de l'Irak, avec le stress que cela met sur notre armée – tout président sensé hésiterait à s'engager à nouveau dans la même région du globe avec les mêmes dynamiques à l'œuvre et probablement la même issue insatisfaisante". ». Le comble du cynisme des néoconservateurs européens tient au fait que ceux-ci ne semblent pas avoir pris conscience de la réalité qui consiste dans le fait que la situation chaotique de la Syrie est en grande partie une des inconséquences des interventions militaires passées, certes, mais pas exclusivement.

En effet, le soutien diplomatique et logistique, dont l'approvisionnement en armes des insurgés par des pays occidentaux et arabes, dont notamment la France et l'Arabie saoudite, a probablement contribué à aggraver sinon entretenir la guerre syrienne qui ne peut plus être décemment qualifiée de guerre civile. Il est à craindre que le soutien officiel au groupe djihadiste Jabhat al-Nosra de Laurent Fabius, alors ministre des Affaires étrangères, n'ôte toute crédibilité à la posture moralisatrice du pays de la déclaration des droits de l'homme. En effet, d'après le quotidien Le Monde[1], Laurent Fabius critiquant « la décision des Etats-Unis de placer Jabhat al-Nosra sur la liste des organisations terroristes », a déclaré « "tous les Arabes étaient vent debout" contre la position américaine "parce que, sur le terrain, ils font du bon boulot" ». Pour terminer glorieusement son parcours de ministre des affaires étrangères, l'intéressé a adopté une posture churchillienne en déclarant : « Il reste encore beaucoup de souffrance et beaucoup de travail pour que M. Bachar Al-Assad "dégage", comme on dit maintenant ».

---

[1] Cf. entre autres, l'article d'Isabelle Mandraud (avec Gilles Paris), « Pression militaire et succès diplomatique pour les rebelles syriens », le quotidien Le Monde, 13/12/2012.

Concrètement, la question importante qui se pose dans le cas de la Syrie consiste à savoir qui est l'ami du ministre des Affaires étrangères français Laurent Fabius : le groupe djihadiste Jabhat al-Nosra ? Qui est son ennemi : l'Etat syrien assurant la coexistence des communautés de confessions différentes ? Etat représenté en la personne du président Bachar El-Assad, représentant d'un régime brutal, certes, mais un président élu au suffrage universel qui est garant de la liberté confessionnelle et qui permet à la Syrie de disposer d'un d'Etat en bonne et due forme, sans quoi le pays devrait sombrer dans le chaos total et être livré pieds et mains liés à Daech, al-Nosra et autres groupes terroristes ? La France dispose d'une armée compétente, encore faut-il que celle-ci puisse être mise au service d'une stratégie cohérente à la hauteur de ses idéaux républicains.

Que dire alors du silence médiatique qui entoure le cas du Yémen, le seul pays du Golfe à ne pas disposer du pétrole, où les écoles, hôpitaux et marchés sont régulièrement bombardés par l'Arabie saoudite sans que l'opinion publique s'en émeuve pour autant, et pour cause ? Comment peut-on continuer à avoir foi dans une quelconque justice internationale lorsqu'on sait que ce sont avec des armes américaines, britanniques et françaises que les pays de la coalition menée par l'Arabie saoudite se battent contre les insurgés Houthis au Yémen, tandis que ces derniers subissent un embargo sur les ventes d'armes grâce à la résolution du Conseil de sécurité adoptée un mois après les bombardements saoudiens du mois de mars 2015 ? Un des effets de la posture moralisatrice à géométrie variable des pays occidentaux est le renforcement des groupes djihadistes en Afghanistan d'abord, puis en Irak, puis en Libye, puis en Syrie et au Yémen...

Ingérence d'intention machiavélique ou messianique ?

Ainsi que la *commun decency* nous y oblige, il nous faut admettre, qu'il soit le résultat d'une stratégie machiavélique délibérée ou le résultat de l'échec d'une stratégie d'ingérence messianique, le chaos actuel au Moyen-Orient est pour l'essentiel le résultat d'une ingérence externe et ne peut de ce fait même être assimilé à une catastrophe naturelle. Si, comme l'Histoire nous le révèle, le chaos actuel ne peut être considéré comme faisant partie du cours naturel des choses, le plus grave est qu'il nous faut admettre la triste réalité, à savoir que, sous couvert de droits de l'homme et de démocratie, ce sont en fait des considérations d'ordre géopolitique et des intérêts économiques et géostratégiques qui sont aux commandes.

Les principaux enjeux économiques et commerciaux pour les pays occidentaux sont l'accès sécurisé aux sources d'énergie et l'obtention de contrats de vente d'armements auprès des pétromonarchies, l'enjeu géostratégique essentiel est le contrôle des routes maritimes stratégiques, *last but not least* l'enjeu géopolitique majeur pour les Etats-Unis est de continuer à exercer le contrôle militaire et économique de la planète. Ces considérations d'ordre géopolitique et stratégique ont pour prix la défense des intérêts d'Israël tels qu'ils sont perçus par les dirigeants actuels de ce pays du Moyen-Orient. Le problème est que des intérêts mal compris n'augurent rien de bon à terme pour le peuple israélien. En effet, si l'exode du peuple juif a pu être qualifié de mythe par des historiens israéliens[1], le fait que le peuple juif a été victime d'holocauste est bel et bien une réalité historique. Reste à savoir maintenant si la guerre perpétuelle peut être une solution à ce drame.

---

[1] Cf. chapitre suivant.

CHAPITRE 4

Des entités en mal de reconnaissance internationale : "Etat palestinien", "Etat islamique", l'imposture de l'Histoire

Les Etats-Unis et les pays européens ne sont pas les seules entités occidentales à avoir instrumentalisé l'islam. Israël aujourd'hui n'a de cesse de se plaindre du Hamas mais on ne peut ignorer que, du vivant de Yasser Arafat, pour mettre en difficulté le leadeur de l'OLP, et affaiblir le mouvement nationaliste palestinien, les dirigeants israéliens ont apporté tout le soutien nécessaire aux islamistes, ce qui a contribué à l'essor de la nébuleuse islamiste en général et du Hamas en particulier. En effet, Fethi Benslama[1], à l'instar d'autres spécialistes du Moyen-Orient, rappelle que « Israël a soutenu les islamistes palestiniens au début pour affaiblir les laïques ». Les dirigeants israéliens n'étant pas nés de la dernière pluie, leur stratégie était évidemment plus subtile qu'elle n'en a l'air. En effet, l'OLP étant une organisation de nature aconfessionnelle, elle permettait de mobiliser sous une même bannière les nationalistes de toute confession. En renforçant les islamistes au détriment des nationalistes, les stratèges israéliens portaient un coup double : affaiblir le mouvement nationaliste et discréditer la cause palestinienne.

Autorité palestinienne ou Etat palestinien, le débat continue à faire rage mais, entre-temps, l'"Etat islamique"

---

[1] Fethi Benslama, « Les radicaux se sentent menacés de souillure », l'hebdomadaire *Le point* N° 2284, 16 juin 2016.

s'impose. Ironie ou imposture de l'Histoire ? De plus en plus complexe, le monde du XXIᵉ siècle connaît des changements majeurs, il n'en demeure pas moins que ce sont des concepts juridico-politiques des siècles passés qui continuent à définir et à conférer des droits et des devoirs aux entités collectives. C'est ainsi que, ironie du sort, après avoir pendant des décennies hésité à accorder le statut d'Etat à la Palestine[1], la communauté internationale s'est trouvée prise au piège d'un groupe terroriste qui s'est autoproclamé "Etat islamique", entité hybride (transnationale mais/et territorialisée) qui a mis au défi tout l'arsenal juridico intellectuel existant.

Le 29 novembre 1947, l'ONU vote le plan de partage de la Palestine[2]. En 1948, David Ben Gourion proclame la création de l'Etat d'Israël. Ce n'est que quarante ans plus tard, en 1988, que l'OLP proclame la création de l'Etat de Palestine selon le plan de partage de 1947. Mais il faudra attendre encore vingt-quatre ans pour que l'Assemblée générale des Nations Unies reconnaisse la Palestine comme Etat observateur non membre de l'ONU et assure ainsi la reconnaissance *de jure* du statut d'Etat à la Palestine en 2012[3]. La Palestine est légalement et légitimement reconnue comme une entité étatique empêchée dans l'exercice de sa souveraineté[4]. Il n'en demeure pas moins

---

[1] L'Etat palestinien n'a à ce jour été reconnu que par 136 des 193 Etats membres de l'ONU.
[2] Trente-trois pays votent pour, treize pays votent contre et dix pays s'abstiennent.
[3] Le 29 novembre 2012, jour du 65ᵉᵐᵉ anniversaire du plan de partage de la Palestine par l'ONU, l'Assemblée générale des Nations Unies reconnaît la Palestine comme Etat observateur non membre par 138 voix pour (dont la France), 9 contre (dont Israël et les Etats-Unis) et 41 abstentions (dont l'Allemagne et le Royaume-Uni).
[4] Nous sommes là en présence de la question qualifiée de « renforcement des capacités de l'État palestinien » dans la

qu'en l'absence d'une souveraineté effective nous sommes *de facto* en présence d'un proto-Etat et la complexité des questions en suspens, estimées inextricables par d'aucuns (statut de Jérusalem, retour des réfugiés, délimitation des frontières, occupation des terres, accès à l'eau), a généré une situation aux conséquences dramatiques pour tous.

Par souci de « neutralité », sinon de transparence, il a paru opportun de faire état de la position d'hostilité manifestée par certains juristes néoconservateurs américains[1] qui ont été jusqu'à dénier *de jure* et *de facto* toute réalité à l'entité "Etat palestinien". D'après ces juristes américains, « l'Assemblée générale ou le Conseil de sécurité de l'ONU n'ont pas le pouvoir de créer des Etats » ; de plus, pour qu'une entité puisse se proclamer « Etat », elle doit pouvoir remplir certaines conditions nécessaires telles que stipulées dans la convention de Montevideo de 1933 sur les droits et les devoirs des Etats. Il n'est évidemment pas question de savoir si l'Etat d'Israël a été astreint à la même exigence à sa création. Nous allons donc nous restreindre ici au cas de la Palestine.

En effet, même si devenir membre du système des Nations Unies équivaut *de facto* à une accession à la qualité étatique, la fonction de reconnaissance demeure une prérogative propre aux Etats, en raison du fait qu'en 1949, les Etats ont refusé de confier aux Nations Unies la fonction de reconnaissance. Il n'en demeure pas moins qu'il est classique de considérer que l'admission au sein du système des Nations Unies emporte la reconnaissance. Selon le professeur Joe Verhoeven, « On voit mal en effet comment

---

Déclaration conjointe de la Conférence pour la paix au Proche-Orient qui s'est tenue à Paris le 15 janvier 2017.
[1] David B. Rivkin et Lee A. Casey, tribune publiée le 20 novembre 2011 dans *The Wall Street Journal*, cf. Wikipedia.

contester une qualité étatique à celui auquel sont reconnus tous les droits – et toutes les obligations – d'un Etat membre »[1].

Venons-en maintenant au traité de Montevideo de 1933 sur les droits et les devoirs des Etats. A l'heure d'une mondialisation tous azimuts, aller déterrer un traité restreint de par sa portée et désuet à bien des égards laisse perplexe. Au fait du sort qui a été réservé à la Société des Nations par le sénat américain d'abord et par les grandes puissances d'avant la Seconde Guerre mondiale ensuite, le candide serait probablement étonné de constater qu'au XXIe siècle, alors même que l'Organisation des Nations Unies, qui a succédé à la défunte SDN, est riche d'une longue histoire et expérience en matière de gestion des conflits, des juristes aient eu l'idée apparemment saugrenue de se référer à un traité datant d'une époque a priori révolue ! En fait, il n'en est rien car l'article premier de cette convention est à la base de la conception classique de l'Etat. « A l'ordinaire, celui-ci naît ou meurt de la présence ou de l'absence d'éléments de pur fait : un territoire, une population, un gouvernement indépendant ». L'article premier du traité de Montevideo stipule : « L'Etat comme personne de Droit international doit réunir les conditions suivantes : I. Population permanente. II. Territoire déterminé. III. Gouvernement. IV. Capacité d'entrer en relations avec les autres Etats »[2].

---

[1] Joe Verhoeven, « La reconnaissance internationale, déclin ou renouveau ? », *Annuaire français de droit international*, vol.39, 1993, pp 7-43.
[2] Convention concernant les droits et les devoirs des Etats adoptée par la septième Conférence internationale américaine. Signée à Montevideo, le 26 décembre 1933.

La question se pose alors de savoir comment il se fait que plus de 89 Etats n'aient eu « aucune difficulté à reconnaître l'Etat palestinien au lendemain de sa proclamation, le 15 novembre 1988 », même s'il ne remplissait pas à l'époque les conditions requises (contrôle effectif du territoire, de la population, etc.). « Que peut bien signifier alors sa reconnaissance ? »[1] D'après les experts en droit international, « Sans doute rien de plus que l'affirmation du droit du peuple palestinien à créer aussitôt que possible cet Etat, dans les territoires - « illégalement » occupés par Israël - qui lui ont été promis »[2]. Comme on le constate du point de vue juridique, les néoconservateurs américains ont tort car empêcher une entité d'accéder à ses droits ne l'empêche pas d'y avoir droit. En effet, comme il ressort de la Déclaration conjointe de la Conférence pour la paix au Proche-Orient qui s'est tenue à Paris le 15 janvier 2017, il s'agit de respecter « le droit des Palestiniens à un État et à la souveraineté », et pour ce faire, d'œuvrer à créer les conditions permettant aux Palestiniens de bénéficier d'une souveraineté effective de façon à pouvoir « exercer leurs responsabilités étatiques par la consolidation de leurs institutions et de leurs capacités institutionnelles, y compris en termes de services ». Mais, en admettant que nous puissions donner raison à ces juristes d'obédience néoconservatrice, la question se pose alors de savoir si leurs conclusions invalident par voie de conséquence le vote de 1947 concernant le plan de partage de la Palestine.

Comme légitimité n'est pas effectivité et que, *in fine*, légalité non plus n'est pas effectivité, proto-Etat *de facto*, ni la proclamation, ni la reconnaissance *de jure*, de l'Etat de Palestine n'ont suffi pour qu'il en soit ainsi. Il n'en demeure pas moins que, nonobstant le cas apparemment

---

[1] Joe Verhoeven, op.cit.
[2] *Ibid.*

ardu sinon inextricable de la Palestine, depuis l'avènement de Daech sur la scène internationale, nous ne pouvons pas évacuer la question fondamentale de la reconnaissance, de droit et/ou de fait, du statut d'Etat à une entité politique. En effet, si la Palestine s'est trouvée confrontée à de multiples embûches l'empêchant d'accéder de fait au statut d'Etat, il n'en a pas été de même pour Daech. Non content de bénéficier du soutien financier des Etats arabes du Golfe lui permettant d'assurer son expansion internationale, Daech a été laissé libre de se doter de tous les attributs d'un Etat allant jusqu'à battre sa propre monnaie et a fini, au vu et au su de tout un chacun, par s'autoproclamer d'abord Etat islamique en Irak et au Levant, puis Califat, enfin "Etat islamique". Jusqu'au jour où, frappées en leur chair et leur cœur, les grandes puissances en viennent à lui déclarer la guerre, ce qui équivaut à lui conférer *de jure* le statut d'Etat, puisque, conformément au droit international, seules des entités étatiques se font la guerre.

C'est ainsi que, faisant fi du droit international, illégale du point de vue de la communauté politique et intellectuelle, illégitime au regard de la société civile mondiale, l'entité "Etat islamique" a réussi à s'imposer et à se faire nommer tel qu'elle en a décidé. Pris au piège d'une entité embarrassante mais bel et bien réelle, des intellectuels, des journalistes et des acteurs politiques se sont posé des questions quant à savoir s'il n'était pas nécessaire pour certains ou opportun pour d'autres de reconnaître le statut d'Etat à Daech alors que la question de la Palestine semble être mise de côté au regard des nouvelles priorités géopolitiques ! Nous allons tenter d'examiner ci-après un cas anecdotique, certes, mais ô combien parlant quant à l'état d'esprit ambiant.

D'après les experts en droit international[1], « l'Etat ne pourrait exister et, partant, être reconnu comme sujet de droit, lorsque sa naissance résulte d'actes, de conduites ou de situations contraires aux normes impératives du droit international général et notamment à celles qui entendent sauvegarder les droits de l'homme ou du peuple ». Dans le cas de Daech, ce ne sont pas seulement les modalités qui ont présidé à la naissance de cette entité politique, à savoir l'occupation illégale de territoires appartenant à des Etats membres de l'ONU qui font que Daech est illégal mais aussi et surtout sa manière et raison d'être, en l'occurrence le non-respect des droits de l'homme et de la démocratie, qui lui retire toute légitimité pour pouvoir perdurer.

Toute la panoplie de notre argumentaire n'y fait rien ou plutôt n'y peut rien. L'entité "Etat islamique" défie et la communauté internationale et la pensée géopolitique confrontée à une entité autoproclamée transnationale. D'apparence anachronique, si l'entité "Etat islamique" ne peut en aucune façon constituer la solution, elle n'en est pas moins tout à la fois le produit des dysfonctionnements des mécanismes de gouvernance locaux et internationaux et des mal-être de la communauté internationale des nantis et des exclus des bienfaits de la mondialisation. Après avoir tenté d'utiliser les acronymes de Daech pour les francophones et d'ISIL pour les anglophones, par conviction ou par résignation, certains ont fini par utiliser la désignation en vogue d'"État islamique" en oubliant, le plus souvent, d'y adjoindre le qualificatif d'« organisation ». Or, illégalement constituée du point de vue du droit international, l'entité autoproclamée « Etat islamique » ne peut donner lieu à une quelconque reconnaissance.

---

[1] *Ibid.*

En effet, pris au dépourvu par l'annonce de la création du califat, chacun a essayé au début d'utiliser la dénomination qui lui paraissait la plus idoine et aujourd'hui la communauté internationale est beaucoup plus préoccupée à détruire cette entité qu'à lui trouver une appellation de choix. Le présent essai n'ayant de cesse de rappeler l'importance qu'il y a à bien nommer les choses pour ne pas en ajouter au malheur des hommes et des femmes sur terre, il a paru utile de faire état ici de quelques exemples d'appellation dans le monde politique et d'essayer d'évaluer la portée et l'impact de ces dénominations. Car, comme nous allons le voir, le choix de la dénomination de ce groupe qualifié de terroriste n'est pas sans conséquence, ce qui explique pour partie les usages différenciés des médias, des autorités politiques et du monde de la recherche. Précisons d'emblée que les appellations en vogue dans les pays francophones sont Daech, Etat islamique en Irak et au Levant (EIIL) et "Etat islamique", tandis que, dans les pays anglophones, on a recours aux acronymes ISIS pour *Islamic State of Iraq and Syria*, ISIL pour *Islamic State of Iraq and the Levant* et IS pour *Islamic State*.

Nicolas Sarkozy, chantre de la guerre en Libye comme nous avons pu le constater précédemment, ne fait pas dans la dentelle. Indifférent aux mises en garde et remarques formulées par les chercheurs et hommes politiques, l'ancien président de la République fait usage de l'appellation « Etat islamique». Dans un entretien fait le jeudi 16 juin 2016, à six journaux européens dont le Figaro, il déclare que la France doit faire face à une guerre intérieure et extérieure : « une guerre extérieure, contre l'Etat islamique et al-Qaïda et une guerre intérieure contre ceux de nos compatriotes adeptes de l'islam radical». Le problème est que l'Organisation terroriste Daech en horreur avec l'acronyme Daech demande à être dénommée « Etat islamique». De plus, un des objectifs affichés de Daech consiste à générer une guerre civile au sein des pays occidentaux. Déclarer que le pays fait face à une guerre intérieure c'est reconnaître que Daech a gagné, qui plus est, en lui faisant l'honneur de le désigner comme il aime à être appelé !

Dans son livre *Vaincre le totalitarisme islamique* publé fin 2016[1], l'ancien Premier ministre, François Fillon, fait constamment usage de l'appellation « Etat islamique », pour désigner l'entité Daech. Dans un discours prononcé fin 2015[2], l'intéressé déclare : « Je refuse de l'appeler « Daech » qui n'est que l'acronyme arabe de l'Etat Islamique en Irak et au Levant. C'est encore une manière de ne pas nommer notre adversaire, de ne pas en prendre la dimension, de faire croire aux Français que c'est une organisation terroriste comme les autres, alors qu'il est aux portes de Damas et à portée de canons de Bagdad, prêt à réaliser son rêve de califat. » .. « Lorsqu'on fait la guerre, il faut se donner tous les moyens de la gagner ! Depuis la naissance du monstre qu'est l'Etat islamique en Irak et au Levant, je n'ai cessé de réclamer une coalition mondiale, intégrant la Russie, l'Iran et le régime syrien d'Assad. J'ai été longtemps seul, accusé de soutenir des dictateurs, jusque dans ma propre famille politique. Je n'ai aucune sympathie pour les dictateurs mais je vois monter une terrible menace pour le monde : une nouvelle folie totalitaire. Après le nazisme qui voulait dominer le monde et imposer la supériorité d'une race sur toutes les autres, après le communisme qui voulait façonner par la force l'homme à l'image d'une créature idéale, voilà le

---

[1] François Fillon, *Vaincre le totalitarisme islamique*, Paris, Albin Michel, 2016.
[2] François Fillon, « Discours prononcé à Montluçon dans le cadre du meeting de soutien à Laurent Wauquiez en Auvergne Rhône-Alpes », le 20 novembre 2015.

totalitarisme islamique qui veut asservir le monde au nom d'un islam dévoyé. »

Faisant preuve d'un réalisme assumé, la position de François Fillon paraît bien plus censée du point de vue de la politique de lutte contre le terrorisme que celle de Laurent Fabius[1], mais pour ce qui est de la dénomination de Daech, elle a l'inconvénient de satisfaire cette entité terroriste qui, non content d'avoir agrandi le territoire sous son contrôle, ne souhaite plus être dénommé « Etat Islamique en Irak et au Levant » (ISIL, en anglais) mais « Etat islamique ». Plus restrictive, l'appellation « Etat Islamique en Irak et au Levant » n'est pas des plus satisfaisantes pour autant. Cette appellation pose un double problème. Le premier est de conférer à cette entité le statut d'Etat et le second inconvénient, et non des moindres, est de lui reconnaître un territoire beaucoup plus vaste que celui qu'elle a occupé de fait, lorsqu'elle était au sommet de sa gloire. En effet, géographiquement, le Levant couvre le Liban, la Syrie, la Palestine, Israël, la Jordanie et le Sinaï égyptien.

Il est à noter ici que le recours à l'acronyme ISIS dans le monde anglo-saxon a l'avantage de mieux circonscrire le champ d'action de cette organisation terroriste, avant d'en avoir été "chassée" fin 2017[2] ; « champ d'action » compris au sens conventionnel du terme, puisque cette entité terroriste a réussi, grâce à ses adeptes, qui ne sont pas exclusivement des locaux, à semer la terreur un peu partout dans le monde[3]. Il est à noter que, dans son interview

---

[1] Cf. supra, chapitre 3.
[2] Nous disons bien chassée car Daech a été chassé des territoires illégalement occupés ce qui ne veut pas dire qu'il a été définitivement vaincu, cf. chapitres 3, 7 et 9 du présent essai.
[3] Il s'agit là d'une des raisons pour lesquelles Deach souhaite être dénommé « Etat islamque ».

d'avril 2016 avec le journaliste Jeffrey Goldberg, le président des Etats-Unis, Barak Obama, fait davantage usage de l'acronyme ISIS que d'ISIL. Fin politique, l'intéressé n'a pas attendu deux ans pour prendre conscience du danger qu'il y a d'utiliser l'appellation « Etat islamique ». Dans sa prise de parole du mercredi 10 septembre 2014, Obama[1] déclare « Ce groupe se fait appeler "Etat islamique" mais il faut que deux choses soient claires : ISIL n'est pas islamique. Aucune religion ne cautionne le meurtre d'innocents et la majorité des victimes de l'ISIL sont des musulmans. ISIL n'est certainement pas un Etat. Il était auparavant la branche d'al-Qaïda en Irak ».

Comme on le constate, Barak Obama balaye d'un revers de main l'appellation d'« Etat islamique ». « Proactif », il clarifie la position de la première puissance mondiale : refus de la diabolisation de l'islam, d'une part et refus de la culpabilisation des musulmans qui sont les premières victimes de Daech, d'autre part. Brève, précise et percutante, cette citation, qui, de plus, a l'avantage de mettre les pendules à l'heure, fournit un rappel historique quant à l'origine de ce groupe qui n'est rien d'autre qu'une émanation d'al-Qaïda. En effet, c'est en 2006 que Zarkaoui, alors chef d'al-Qaïda en Mésopotamie, décide de se débarrasser de la tutelle d'al-Qaïda qui, suite à une bévue de taille de Zarkaoui, n'a aucun scrupule à le traiter pour ce qu'il est à ses yeux, à savoir : un ancien petit caïd délinquant bardé de tatouages. Pour ce faire, il s'associe avec des tribus et des groupes djihadistes irakiens et ils fondent ensemble un groupe qu'il dénomme « Etat islamique en Irak ». C'est la fusion en 2013 de ce groupe avec l'entité syrienne Front al-Nosra, créée en 2010, qui donnera naissance au groupe « Etat islamique en Irak et au Levant ». ISIS ou ISIL ? L'acronyme ISIS est sans conteste

---

[1] Le quotidien *Le Figaro* du 12 septembre 2014.

plus exact. Il convient toutefois de préciser que, ainsi qu'il a été signalé plus haut dans le chapitre 3, l'acronyme ISIS (*Islamic State of Iraq and Syria*) ne tient malgré tout pas compte de l'incursion de cette organisation en Libye.

Revenons au cas de la France. Conscient des implications politiques des appellations utilisées par François Fillon et Nicolas Sarkozy, treize jours après les horribles attentats du 13 novembre 2015, le quotidien *Le Figaro*[1] fait étalage d'un vocabulaire apparemment consensuel : « Même revirement sémantique à gauche : "ces deux appellations (Daech et Etat islamique), je les assume, parce que c'est une guerre, et qu'il faut bien comprendre qui nous combattons" explique le Premier ministre Manuel Valls sur le plateau du *Petit Journal*. "Nous sommes en guerre contre un Etat", abonde Patrick Kanner, ministre de la Ville, de la Jeunesse et des Sports. ». Si le recours à l'appellation "Etat islamique" relève de l'erreur stratégique, la déclaration gouvernementale « nous sommes en guerre contre un Etat » peut être assimilée à une erreur tactique.

Créé illégalement et illégitime dans ses pratiques et finalités, il n'est pas possible de considérer Daech comme un Etat et encore moins de lui déclarer la guerre. En effet, qui dit guerre, pense « traité de paix ». Or, il est difficile d'imaginer un traité de paix avec une organisation terroriste qui cherche à instaurer une société mondiale wahhabite par des moyens criminels. Contrairement à ce qui a été traduit sur le net, le président des EU Barak Obama déclare « nous allons détruire ISIL » et non pas comme ceci a été traduit en français par les journalistes « nous allons détruire l'Etat islamique ». En effet, après avoir annoncé « la création d'une coalition internationale », Barak Obama déclare très exactement « notre objectif est clair : nous allons détruire

---

[1] Le quotidien *Le Figaro* du 26 novembre 2015.

ISIS. Conformément à notre stratégie anti-terroriste. Nous allons traquer ISIS où qu'il soit en Irak, en Syrie. Si vous nous menacez, vous n'aurez pas de répit où que vous soyez ». Il était important de faire état de l'essentiel de son communiqué de presse pour lever tout risque d'ambiguïté quant aux raisons qui nous amènent à disqualifier l'usage des concepts d'« Etat » et de « guerre » dans le cas de Daech. Comme on le constate, le président des Etats-Unis n'utilise aucun de ces termes. Ce qui ne l'empêche pas d'être percutant, tout au contraire.

Compte tenu du fait que, concernant l'organisation Daech, nombre de chercheurs, d'experts et d'acteurs politiques continuent et risquent même après sa disparition de continuer à raisonner en termes d'« Etat » et de « guerre », il nous paraît important d'expliciter encore une fois pourquoi, tant d'un point de vue théorique que pratique, il y a lieu d'éviter ce mode de raisonnement. En réalité, avec la coalition internationale contre Daech, nous avons affaire à une intervention militaire conjointe de plusieurs Etats mais cette intervention ne peut être qualifiée de guerre puisque l'objectif n'est pas d'obtenir un traité de paix avec l'adversaire mais de l'anéantir en tant qu'entité indésirable. En effet, étant donné que nous sommes en présence d'une intervention militaire conjointe de plusieurs Etats contre une entité illégale au regard du droit international et considérée comme moralement illégitime au regard des principes des droits de l'homme, on peut, dans une perspective post-westphalienne, raisonner en termes d'intervention policière destinée à démanteler une bande d'activistes pour certains, de terroristes pour d'autres, étant entendu que les forces de l'ordre luttent toujours contre le banditisme, la criminalité organisée ou le terrorisme, mais qu'elles ne peuvent lui faire la guerre.

Reconnaître le statut d'Etat à une organisation terroriste n'est pas un fait nouveau, mais, consciemment ou par maladresse, faire des déclarations qui pourraient laisser entendre ne serait-ce qu'un soupçon d'une telle reconnaissance à une organisation terroriste qui promeut des institutions et des pratiques estimées archaïques car discriminatoires et barbares en est une autre. Anachronique ? Anodine ? Il n'en est rien, bien au contraire, en phase avec la réalité des transformations en cours, de sorte que la banalisation du concept d'"Etat islamique" peut s'avérer bien plus dangereuse que celle du prétendu califat. Il y a en effet bien longtemps que, vidée de tout pouvoir réel, la fonction de califat avait perdu de son lustre pour n'être plus et ce, dès l'an mille, qu'un simple titre honorifique[1]. Il en va tout autrement du concept d'« Etat » qui, en principe, est rejeté comme « héritage colonial »[2] par le courant de pensée panislamiste favorable à la restauration du *khalifat*.

Il n'en va pas de même avec le concept « Etat islamique ». Concept de nature hybride, contrairement aux apparences, avec l'appellation « Etat islamique » nous n'avons pas affaire à une entité étatique au sens classique et westphalien du terme. Ce qu'il importe de souligner et qui n'a semble-t-il pas été relevé à ce jour, est que la dénomination « Etat islamique » confère à cette réalité politique le statut d'entité post-westphalienne. Avec ses visées expansionnistes, Daech raisonne en termes d'empire mondial et a pour volonté d'instaurer un califat "wahhabite sunnite" couvrant l'ensemble du monde musulman d'abord et la planète entière ensuite[3].

---

[1] Voir, entre autres, Bernard Lewis, *Le langage politique de l'Islam*, Paris, Gallimard, 1988.
[2] Cf. Bruno Étienne, « L'islamisme comme idéologie et comme force politique », *Cités* 2/ 2003 (n° 14), p. 51.
[3] Cf. chapitre 7.

Le terrorisme n'est pas un fait nouveau. Israël, la Palestine et bien d'autres entités politiques aujourd'hui respectables et respectées, s'y étant adonnées dans le passé, disposent d'un passif en la matière. Optant pour le spectaculaire dans l'acte terroriste, Daech, entité politique illégale au regard du droit international et illégitime au regard des principes des droits de l'homme et de la démocratie, a sciemment fait sombrer la communauté internationale dans l'horreur absolue. Il est vrai que les séries télévisées américaines ont banalisé les scènes de violence extrême au point où l'on se demande si Daech n'a pas voulu leur faire concurrence. De toute façon, réelle ou virtuelle, la violence récréative est une réalité et, si la fascination des hommes pour la violence est bien réelle, elle ne date pas d'aujourd'hui, pour preuve le pouvoir d'attraction qu'ont eu dans le passé les exécutions en place publique, exutoire pour certains, distraction pour d'autres.

Mais, de nos jours, habitués que nous sommes via les différents médias à la vue des scènes d'extrême violence fictives ou réelles (le mime réel des atrocités des films d'horreur pour la plupart inspirés de faits divers, les scènes de carnages occasionnés par l'usage du LSD et autres drogues synthétiques, etc.), la question se pose de savoir pourquoi nous sommes tant affectés par les images de mise à mort de ce pseudo "Etat islamique" qu'est Daech.

N'est-ce pas, en partie, parce qu'on croit avoir affaire à l'entité « Etat » qui continue de nos jours à disposer *de jure* du monopole de la violence légitime ?

Comme l'usage du terme « islamisme » pour désigner l'islamisme radical n'a pas été sans conséquence, la dénomination "Etat islamique" n'en est pas moins dangereuse. Mais plus grave de conséquence encore est la reconnaissance du statut d'Etat à un mouvement terroriste à des fins insidieuses telles que la diabolisation d'Etats *de jure* que l'on considère comme des « ennemis », bien plus pour des raisons d'alliance géostratégique que pour des raisons d'antinomie religieuse. Parmi les messages de propagande capturés aux fins de la présente étude, prenons le cas du psychanalyste Daniel Sibony[1], invité du Journal de 12h30 de France Culture le 12 mars 2015 pour la promotion de son livre intitulé « le Grand Malentendu Islam, Israël, Occident ».

Avant de lui donner la parole, le présentateur du Journal de midi, qui clôt sa présentation du journal d'information proprement dit en fournissant des nouvelles du Front en Irak, annonce que la ville de Tikrit en Irak, qui était occupée par Daech depuis juin 2014, vient tout juste d'être libérée par une milice chiite. L'organisation "Etat islamique" avait dû battre en retraite grâce à l'appui apporté par l'Iran. Se tournant alors vers son invité, le journaliste introduit son livre en l'interrogeant sur le malentendu qui existerait selon l'auteur entre l'Occident et l'islam. Et demande à savoir en quoi « l'Occident a du mal à réagir au surgissement de l'Etat islamique au Moyen-Orient. »

Au psychanalyste Daniel Sibony de répondre : « Parce qu'il y a comme "un secret" qui est bien maintenu en Occident. Car il y a dans l'islam deux parties : une partie pacifique et une partie agressive »… « La partie pacifique construite à partir du message judéo-chrétien (ce sont les valeurs de charité, de justice, de pèlerinage, de prière) et la partie agressive qui dénonce on peut dire même qui maudit, les gens du Livre : les juifs et les chrétiens. Il y a un vrai problème mais dont on ne parle pas, c'est dommage parce que c'est un mépris pour les différents acteurs, c'est un

---

[1] On est évidemment tout à fait en droit d'être étonné de la réaction de cet invité de nationalité française sur les antennes publiques et de vouloir comprendre, sinon trouver une explication plausible à ses propos qui pourraient porter à faire croire qu'il fait allégeance à un mouvement terroriste.

problème que vivent les musulmans partout dans le monde ».

Il est nécessaire d'ouvrir ici une parenthèse pour signaler que la philosophe Marie-José Mondzain, invitée sur les ondes de France Culture[1] pour débattre de son livre qui traite de la question des erreurs et dérives sémantiques graves entretenues par ceux qui ont fini par s'y habituer, déclare : « je m'insurge contre des syntagmes totalement banalisés du type "judéo-christianisme" alors que le monde chrétien a quand même maltraité les juifs et tué infiniment plus de juifs pour l'instant que n'en ont tué les gens de l'islamisme ».

Pour surmonter la posture belliciste, il y aurait peut-être lieu de raisonner à l'instar de Henry Corbin[2] en termes de judéo-chrétien-islamique puisque le judaïsme, le christianisme et l'islamisme[3] sont les trois religions révélées et que, comme l'islamologue Christian Jambet[4] le rappelle, « l'islam se présente comme la religion du Livre qui parachève le judaïsme et le christianisme ».

Pour revenir sur le cas de Daniel Sibony, comme on le constate à partir de ses dires du mois de mars 2015, le psychanalyste compatit avec les musulmans qui se trouvent être victimes de l'islam ! A lui de préciser : « On a peur d'en parler car ça serait stigmatisant »... « C'est se voiler la face et c'est passer à côté de problèmes réels, de problèmes concrets »... « Avec la ville de Tikrit libérée avec l'aide de l'Iran, on assiste à une substitution d'un Etat islamique par un autre. On s'obnubile sur l'opposition des chiites et des

---

[1] Emission du 6 mars 2017 de la grande table ayant pour titre « Comment réhabiliter la radicalité ? ».
[2] Henry Corbin, *Le paradoxe du monothéisme*, Paris, L'Herne, 1981.
[3] Islamisme compris en son sens originel (voir chapitre 1), c'est-à-dire avant que ce terme ne soit malencontreusement dévoyé.
[4] Jean Birnbaum, *Un silence religieux*, Paris, Seuil, 2016, p.44.

sunnites. On ne voit pas qu'à terme c'est un Etat islamique chiite qui remplace un Etat islamique sunnite. On ne voit pas que, du point de vue du rapport à l'Europe, à l'Occident, à Israël, ça sera pareil : un Etat intégriste très fort qui affiche des intentions exterminatrices ». Alors que le journaliste, donnant des nouvelles du front en Irak, fait état de la contribution de l'Iran dans la libération des villes conquises par Daech, Daniel Sibony rappelle à qui veut bien l'entendre que les Iraniens sont chiites. Et comme le journaliste fait état du fait qu'avec l'Iran on a affaire au moins à un Etat (Etat théocratique certes mais Etat *de jure*) et que le journaliste préoccupé par la réalité des faits ne semble pas aller dans le sens de Sibony, à lui d'insister... Iran ceci cela... toute une panoplie d'arguments destinés à diaboliser les chiites qui, comme les sunnites, sont victimes d'une guerre confessionnelle sciemment et habilement fomentée et entretenue à des fins géopolitiques. Car, *in fine*, si agressivité il y a, celle-ci relève davantage du wahhabisme (secte sunnite apparue au XVIIe siècle) qui se caractérise par un rigorisme moral, une exaltation du djihad et une agressivité contre les autres branches du sunnisme en général et les chiites en particulier en raison de la pratique de l'« *ijtihâd* » par ces derniers[1].

En résumé, comme on le constate avec le cas Sibony cité à titre d'exemple, l'exercice de désinformation destinée à générer une islamophobie a pour procédé de dévoiler une prétendue vérité historique gardée « secrète » à ce jour en Occident qui consiste dans le fait que l'islam est fait de deux parties : une partie pacifique qui est d'origine judéo-chrétienne et une partie agressive avec des intentions

---

[1] Interprétation et établissement de normes dans la perspective d'une conception dynamique de la charia, cf. chapitres suivants et plus spécifiquement chapitre 9.

exterminatrices qui a de tout temps constitué l'essence de l'islam. L'opération de sape auprès des Occidentaux ne suffisant pas, elle sera complétée par une opération de manipulation des musulmans en diabolisant les chiites. Alors qu'en fait, avec le chiisme, nous avons, à l'instar du christianisme, affaire à la doctrine de l'intercession des imams comme des saints chez les chrétiens. Il y a de plus une affinité spécifique entre le chiisme et le christianisme tenant à la croyance dans un messianisme : l'Imam caché et Jésus doivent venir sur terre avant la fin des temps.

Comme nous avons pu le constater, avec le psychanalyste Daniel Sibony, nous avons affaire à une version brouillon de la thèse de la « menace islamiste » développée par l'orientaliste Bernard Lewis qui devrait à terme conduire à une « guerre de civilisations » selon une interprétation radicale de la thèse « du choc des civilisations » de Samuel Huntington et que Shlomo Sand[1], historien israélien réputé pour avoir dénié au mythe de l'exode du peuple juif le statut de réalité historique, qualifie plus simplement de « guerre de religions ». Compte tenu du fait que la guerre entre deux religions, toutes deux à vocation universelle, n'a pas lieu de prendre fin sauf extermination des tenants de l'une des deux religions, nous serions en présence d'une « guerre perpétuelle » entre chrétiens et musulmans. Ce qui a l'avantage de permettre aux juifs de jouir d'une paix bien méritée après avoir subi, tout au long de leur histoire, des persécutions.

---

[1] Travail de sape ou lui-même victime de manipulations, selon Shlomo Sand « Nous ne savons pas si l'Europe va s'unifier contre l'Islam, à la fin dans cent ans, on va peut-être retrouver aussi le terme « peuple chrétien ». Pour le moment on ne l'autorise pas ». Cf. Shlomo Sand, « L'exode du peuple juif : mythe ou réalité ? » in Michel Collon (dir.), *Israël, parlons-en !*, Bruxelles, écitions Investig'Action - Couleur Livres, 2011, pp. 29-43.

Le problème est que, théologiquement, l'islam n'a pas pour vocation d'entrer en conflit avec les autres religions monothéistes. En effet, ainsi que Georges Corm[1] le souligne à très juste titre, « L'islam, dernier né des monothéismes, aura une attitude moins exclusive que les deux premiers. En effet, il reconnaît dans les prophètes, d'Abraham au Christ, ses propres ancêtres et garantit aux « gens du Livre » le libre exercice de leur culte s'ils ne montrent pas d'hostilité à la nouvelle religion. Celle-ci est censée venir compléter et achever définitivement l'aventure monothéiste débutée par Abraham. Le pouvoir chrétien à Byzance ou à Rome n'aura pas une telle vision lui permettant d'accepter l'existence de juifs et de musulmans au sein de son territoire sans les pousser à embrasser la « vraie foi ». »

Reste à savoir maintenant quels sont les soubassements de ce courant de pensée « pan-occidentaliste » fondé sur l'alliance judéo-chrétienne. Selon Noam Chomsky[2], « Les Etats-Unis possèdent un mouvement chrétien évangéliste d'une force inhabituelle, avec lequel sympathisent – pour ne pas dire plus – des figures haut placées comme George W. Bush, et aussi Reagan. Soutenir Israël est un élément clé de leur théologie (laquelle est en fait extrêmement antisémite, mais ça c'est une autre histoire) ». Se basant sur les thèses des stratèges néoconservateurs américains, Sibony veut créer une affinité entre juifs et chrétiens français qui devraient tout naturellement s'allier pour se défendre contre l'ennemi commun constitué par les musulmans, en faisant croire que les juifs et les chrétiens partagent une même idéologie, la « partie pacifiste », tandis que les musulmans ont une « partie » en plus qui, elle, est agressive.

---

[1] Georges Corm, « Religion et géopolitique : une relation perverse », *Revue internationale et stratégique* 2009/4 (n° 76), p. 23-34.
[2] Noam Chomsky, « Pourquoi les Etats-Unis protègent-ils Israël ? » in Michel Collon (dir.), *Israël, parlons-en !*, op.cit., pp. 167-171.

Si on fait appel à la théorie des ensembles comme Daniel Sibony semble le faire implicitement, avec les trois religions monothéistes (islam, christianisme et judaïsme), nous n'avons pas deux mais trois parties qui s'emboîtent. Si nous devions, à l'instar du psychanalyste Sibony, nous placer dans une posture conflictuelle (affinité / discorde), on constate que, d'un point de vue mémoriel, il y a en réalité une affinité au sens d'acceptation et/ou de reconnaissance entre islam et christianisme, entre islam et judaïsme, puis entre christianisme et judaïsme. Mais il ne peut y avoir d'affinité originelle entre le judaïsme et les deux autres religions : affinité au regard du christianisme et/ou de l'islam par le judaïsme puisqu'il s'agit là de la première des trois religions monothéistes. Ceci explique d'ailleurs pourquoi il est si difficile, pour ne pas dire quasi impossible, pour un chrétien ou un musulman, de se convertir au judaïsme. Mais, lorsqu'on se libère de la posture conflictuelle et que l'on se réfère à ce qu'il y a d'essentiel dans ces religions, on constate que, sur certains aspects --qu'il n'y a pas lieu de développer ici-- il y a plus d'affinités entre juifs et musulmans qu'entre juifs et chrétiens et que, d'une façon générale, ce qui unit juifs, chrétiens et musulmans est bien plus important que ce qui les différencie. Et vouloir tranformer un conflit territorial en conflit de religions n'augure rien de bon pour personne.

Ce psychanalyste, qui semble souffrir quelque peu du mal de l'islamophobie, n'est évidemment pas seul à vouloir discréditer l'islam ou à surfer sur l'opposition chiites-sunnites. Dans l'émission *Mots croisés* du 9 mars 2015 consacrée à un débat intitulé « La guerre des religions aura-t-elle lieu ? » diffusée sur France 2, nous avons eu droit à deux apartés du même calibre.
En plein milieu de l'émission, alors que le débat porte sur la question de savoir si la priorité serait d'éradiquer

Daech ou se « débarrasser de Bachar El Assad », Pascal Brucker déclare : « Dans une guerre il faut diviser ses ennemis et pour nous la division chiites-sunnites aussi est une bénédiction. La guerre entre les deux factions (chiites-sunnites) –qui provoque des centaines de milliers de morts, c'est malheureux– est salutaire, salvatrice pour nous ». Un peu plus tard, lorsqu'Odon Vallet déclare : « Il faut respecter à la fois la liberté d'expression et les religions », l'ancien ministre des Affaires étrangères français Bernard Kouchner réplique : « Il faut respecter les religions lorsqu'elles sont respectables. Les religions ne sont pas toutes respectables ». Tâche désespérée ou pas, pour départager les religions qui méritent notre respect, il va falloir nous mettre à la recherche d'un « Juge suprême ». Le problème est que, si chacun doit être libre dans la sphère du religieux de décider de « sa » vérité, nul hormis Dieu n'est habilité en la matière à se poser en Juge suprême pour autrui.

Il y a maintenant bien plus de trente ans que certains acteurs et pays occidentaux s'emploient, au gré du temps, à attiser des conflits au sein et entre pays musulmans sans tenir dûment compte de l'impact négatif de ces conflits dans les pays occidentaux en général, et sur le devenir d'Israël en particulier. Pour ce qui concerne Israël, comme Dominique Moïsi[1] le souligne, « depuis que le monde arabo-musulman est entré dans une période de guerre civile, Israël n'apparaît plus comme un foyer de tension. Mais, tant qu'un Etat palestinien ne sera pas créé, ce calme ne restera que provisoire : si la guerre civile actuelle au sein de l'islam évoque la guerre de Trente Ans, la référence à la guerre de Cent Ans s'imposera bientôt dans le cas du conflit Israël-Palestine ».

---

[1] Dominique Moïsi, « Israël : la possibilité d'un îlot de stabilité », le quotidien *Les Echos* du 5 décembre 2016.

CHAPITRE 5

Des orientalistes aux islamologues : pour en finir avec le concept d'« islamisme », trou noir de la géopolitique mondiale

Aucune puissance ni gouvernement ne semble aujourd'hui apprécier Daech, mais ce dernier est pourtant « un sous-produit du choix américain de jouer l'islamisme contre l'Union soviétique » comme tout un chacun le sait et qu'un professeur de relations internationales à Sciences Po[1] se plaît à le rappeler.

La question de l'instrumentalisation de l'islam à des fins géopolitiques a été longuement évoquée dans les chapitres précédents mais le problème est que, comme il a déjà été précisé, cette instrumentalisation n'est pas le seul fait des Occidentaux. Après l'échec de la tentative panislamiste du sultan Abdül Hamid II et l'effondrement de l'empire ottoman, les pays du Moyen-Orient étaient majoritairement engagés dans des projets de libération politique et économique areligieux. Mais, après la crise du panarabisme consécutive à la défaite des pays arabes « laïcisants » dans la guerre des Six jours de 1967 d'abord et la crise du tiers-mondisme consécutive à la crise des dettes souveraines ensuite, vint l'ère du « consensus de Washington » qui a eu pour corollaire l'abandon de l'idéologie développementaliste. L'univers géopolitique ayant horreur du vide idéologique, c'est ainsi que le vide, créé par la crise de l'idéologie tiers-mondiste d'inspiration marxiste, l'abandon

---

[1] Pierre Grosser (prof.), « Après les réactions épidermiques, penser Trump et le monde », le quotidien *Libération* du 21 novembre 2016.

de l'idéologie développementaliste d'inspiration rostowienne et l'échec du panarabisme dans les pays arabes, sera pour partie comblé par la montée de l'islam comme idéologie politique. Ce courant de pensée politique sera soutenu et/ou sublimé par une partie de l'élite intellectuelle et médiatique occidentale et la mise en pratique de l'islam politique fortement encouragée par les puissances occidentales, les activistes fondamentalistes étant conséquemment soutenus et/ou financés par des pétromonarchies d'abord et ensuite par des régimes islamiques ou en voie d'islamisation.

Si, au XIXe siècle, des intellectuels islamologues moyen-orientaux à l'image de Djamâl ed-Dîn al-Afghani[1] admettent la dimension idéologique des religions en général dont celle de l'islam qu'ils cherchent à réformer[2], au XXe, siècle ce sont des activistes de différents bords, intellectuels et religieux, qui vont œuvrer à l'idéologisation et à la politisation de l'islam. Le fait que des idéologues tels que Ali Shariati aient considéré l'islam comme une idéologie et que des religieux tels que l'ayatollah Khomeiny aient

---

[1] Nous avons cité Afghani (1838-1898) car il est de loin le précurseur des réformistes. Olivier Roy, dans son ouvrage *Généalogie de l'islamisme*, prend soin de faire l'inventaire de l'ensemble des intellectuels qualifiés d'islamistes. Pour ce qui est des premiers idéologues de tendance réformiste, contemporains d'Afghani, il y a lieu de citer : l'Egyptien Mohammad Abduh (1849-1905) et le Libanais Rashid Ridâ (1865-1935). Plus contemporains, les seconds de tendance franchement radicale sont : Hassan al Banna (1906-1949) fondateur de l'association des Frères musulmans en Egypte et Abu Ala Maududi (1903-1979) fondateur de Jama'at-i Islami.

[2] D'après Afghani, « la science, si belle qu'elle soit, ne satisfait pas complètement l'humanité qui a soif d'idéal et qui aime à planer dans des régions obscures et lointaines que les philosophes et les savants ne peuvent ni apercevoir ni explorer ». Comme on le constate, Afghani croit à la pérennité de la religion pour la très bonne raison qu'elle répond à un besoin éternel inhérent à la psychologie humaine.

considéré l'islam comme une réalité politique sont des faits avérés que des islamologues français trouveront commode de dénommer « islamisme ». Si, d'un point de vue pragmatique, la question consiste à savoir comment il se fait que ceux-ci, les dénommés « islamistes », ont pu, plus que d'autres, avoir droit au chapitre au point d'accéder au pouvoir, d'un point de vue théorique, la question se pose de savoir si l'« islamisme » des islamologues français ne fait pas pendant à l'Orient des orientalistes anglo-saxons.

En effet, si les islamologues français, dont Bruno Etienne en tête, prennent soin de faire un distinguo entre islam et « islamisme », ceci n'a pas été le cas des orientalistes anglo-saxons comme Bernard Lewis, qui n'a eu de cesse d'invoquer pour ce faire l'unicité de l'islam basée sur la tautologie islam égale culture fondée sur l'hypothèse selon laquelle l'islam annihile toute idée de culture d'abord. Ensuite, le présupposé caractère entier pour ne pas dire totalitaire de l'islam démontré à partir de la situation en vigueur dans le passé lointain, nous sommes là en présence d'un « holisme décontextualisé »[1] malencontreusement dénommé *holisme islamique*. Et, pour finir, le caractère définitif et inamovible du Coran[2] qu'il invoque, la fidélité à la lettre et non à l'esprit du Coran servant de soubassement à ses diverses assertions.

Dans son ouvrage « Le langage politique de l'islam »,

---

[1] Il ne s'agit pas, par souci de réalisme, d'opter pour une approche holistique dans le présent mais d'opérer un holisme par anachronisme, c'est-à-dire par un retour à des temps passés qui ne peuvent être puisqu'il s'agit d'une époque révolue. Le point de référence pour les sunnites étant la situation en vigueur du temps du prophète et des quatre califes orthodoxes.
[2] Selon Bernard Lewis, « l'origine divine et l'intangibilité du Coran sont des axiomes pour les musulmans », cf. Bernard Lewis, *Islam et démocratie* in Bernard Lewis, *Islam*, Gallimard (Quarto), 2005, p.824.

Bernard Lewis[1] se réfère à une période aux contours flous qu'est le passé en général, et l'avant occidentalisation en particulier, pour définir ce qu'est le bon c'est-à-dire le « vrai » islam. Les situations et expériences récentes sont analysées à l'aune de cette norme du véritable islam. C'est ainsi que, d'après Bernard Lewis, en 1979, la révolution a permis à l'Iran de mettre fin à une « anomalie... qui pourrait bien aussi approcher de sa fin dans d'autres pays islamiques »[2]. Les déviations au regard de ce qui est vrai en islam sont tolérées par l'intéressé pour autant qu'elles demeurent dans la sphère ou plus exactement sous la coupe d'un régime islamique. Il en va ainsi par exemple de la doctrine de « l'Autorité du faqih ». Selon Bernard Lewis, « L'aspect le plus novateur du système politique conçu par l'ayatollah Khomeiny est sa doctrine de l'« Autorité du *Faqih* » qui prévoit l'établissement d'un seul *faqih* comme autorité légale suprême de l'Etat – sorte de cour suprême constitutionnelle, ayant pouvoir d'invalider toute action ou tout décret du gouvernement, jugé contraire à l'islam. Il n'y a –précise-t-il– aucun précédent d'une pareille fonction dans la doctrine et la pratique islamiques passées »[3].

Il n'est évidemment pas question de frayer avec des concepts et idées supposés selon lui inconciliables avec l'islam vrai tels que la laïcité, la modernité, la démocratie, et autres « drogues offertes par divers colporteurs étrangers »[4]. Prenons le cas de la démocratie. Dans son étude *Islam et démocratie*[5], Bernard Lewis qui est, encore faut-il le préciser, viscéralement contre toute idée ou pensée

---

[1] Bernard Lewis, *Le langage politique de l'Islam*, in Bernard Lewis, *Islam, op.cit.*, pp. 671-811.
[2] *Ibid.*, p. 676.
[3] *Ibid.*, p. 708.
[4] *Ibid.*, pp 810-811.
[5] *Ibid.*, pp 813-838.

marxisante, après avoir fait état des vicissitudes de la démocratie libérale au Moyen-Orient où elle n'a, selon lui, «jusqu'à aujourd'hui jamais produit d'effet durable», en arrive à se poser la question de fond qui consiste à savoir si islam et démocratie sont compatibles. La réponse est évidemment négative pour ce qui est du fondamentalisme islamique et circonspecte pour ce qui est de l'islam : «sous l'angle politique, l'islam paraît le moins adaptable à une démocratie libérale»[1]. Selon l'intéressé, «Pour les fondamentalistes, la démocratie n'a évidemment aucun sens... Leur attitude à l'égard des élections démocratiques a été résumée par le slogan "un homme, une voix, une fois". Cela n'est pas tout à fait exact, au moins pour les Iraniens. La République islamique d'Iran organise des élections disputées et offre plus de liberté de débat et de critique dans la presse et au Parlement qu'il n'en existe habituellement dans la plupart des pays musulmans. Mais la candidature, la formation d'un groupe, l'expression des idées font l'objet de réglementations strictes et rigoureusement appliquées. Il va sans dire qu'on ne saurait mettre en doute les principes de base de la révolution ou de la République islamique»[2]. Comme on le constate, pour Bernard Lewis, la démocratie et l'islam sont incompatibles, puisqu'en terre d'islam, au pire la démocratie est un moyen d'accès au pouvoir et au mieux, elle est restreinte par les limites naturelles d'un régime théocratique.

Afin d'éviter de se perdre dans les méandres de la pensée de Bernard Lewis, il convient de préciser ici que son analyse est la plupart du temps fondée sur des conclusions douteuses qui font figure de prémisses, à savoir : «les talismans du mystérieux Occident («gouvernement constitutionnel», «indépendance» et autres « remèdes

---

[1] *Ibid.*, p.829.
[2] *Ibid.*, p. 825.

importés ») n'opèrent aucun miracle »[1] en terre d'islam. Ce qui, pour faire bref, équivaut à la tautologie : la modernité et la démocratie sont incompatibles avec l'islam. Etant donné qu'il fait preuve d'une grande érudition, son analyse philologique qui est savante à bien des égards donne une illusion de scientificité à sa démarche et de vérité à ses assertions. En bon philologue, il accomplit l'exploit de faire ressortir, dans les interstices d'un examen minutieux des racines historiques d'une longue liste de mots, des affirmations qui font figure de vérités scientifiques concernant ses interprétations partisanes des événements politiques actuels. Ses prises de position en faveur ou en défaveur des pratiques et/ou acteurs politiques se font à l'aune de ses préjugés, orientations et desseins politiques.

Bernard Lewis rejette le terme « fondamentalisme »[2], car l'islam par essence ne peut être que fondamentaliste. A croire que, lorsqu'il défend quelqu'un ou quelque chose en « terre d'islam », c'est pour mieux l'enfoncer aussitôt après. Pour lui, le terme de « culture » est inopérant en terre d'islam. Valable pour la culture avec un grand « C », puisque la musique, la peinture et la sculpture sont, d'après lui, bannies sinon interdites en terre d'islam[3]. Ceci est tout aussi valable pour la culture comprise au sens anthroposociologique du terme[4], c'est-à-dire la culture avec un petit c, puisque d'après lui en terre d'islam « la culture se définit par l'islam »[5], simple tautologie. Lorsque B. Lewis s'adonne à un discours apologétique à l'égard d'un acteur

---

[1] Le langage politique de l'islam, op.cit., p.810.
[2] D'après Bernard Lewis, « tous les musulmans, dans leur attitude à l'égard du texte du Coran, sont, en principe, pour le moins fondamentalistes », Ibid., p.678, note 3.
[3] Ibid., pp 686-687.
[4] Pour les définitions de la culture cf. Ninou Garabaghi, Les espaces de la diversité culturelle, Paris, Karthala, 2010, p.54.
[5] Bernard Lewis, op.cit., p. 676.

politique ou un pays musulman ce n'est pas parce qu'il est d'accord avec la politique prônée par cet acteur politique ou la voie suivie par le pays concerné, mais bien souvent parce que ceux-ci fournissent une assise à ses stratagèmes et/ou dénigrements. Il n'est pas question de dresser ici la liste des remarques désobligeantes pour ne pas dire péjoratives et méprisantes. Un simple exemple suffit à résumer la haine et l'esprit d'animosité de ce stratège du président Bush junior. Selon Bernard Lewis, « L'islam est perçu, dès ses origines, comme une religion militante, à vrai dire comme une religion militaire, et ses adeptes comme des guerriers fanatiques, enrôlés pour propager leur foi et leur loi par la force armée »[1].

La construction de la nouvelle figure de l'ennemi de l'Occident est une tâche ardue et semée d'embûches. Selon Bernard Lewis[2], « l'alliance de l'adjectif « saint » et du substantif « guerre » ne se trouve pas dans les textes islamiques classiques. Son usage, en arabe moderne, est d'origine récente et étrangère ». D'après l'intéressé, le mot arabe *djihad*, aujourd'hui compris comme signifiant « faire la guerre », en fait, signifie, au sens littéral, « tentative », « effort », ou « lutte ». Dans le Coran, il est invariablement suivi des mots « dans la voie de Dieu ». Ce qui ne l'empêche pas de conclure : « L'obligation du djihad se fonde sur l'universalité de la révélation musulmane. La parole de Dieu et le message de Dieu s'adressent à l'humanité ; c'est le devoir de ceux qui les ont acceptés de peiner (*djahada*) sans relâche pour convertir ou, à tout le moins, soumettre ceux qui ne l'ont fait. Cette obligation n'a pas de limites ni dans le temps ni dans l'espace. Elle doit durer jusqu'à ce que le monde entier ait rallié la foi

---

[1] *Ibid.*, p.760.
[2] *Ibid.*, p. 761.

musulmane ou se soit soumis à l'autorité de l'Etat islamique. »[1].

Pour en revenir à la question du développement et de la modernisation des pays du Moyen-Orient, selon B. Lewis, « La deuxième moitié du XXe siècle suscita une grande déception et beaucoup d'interrogations. Les talismans du mystérieux Occident n'opérèrent aucun miracle, les drogues offertes par divers colporteurs étrangers ne guérirent pas les maux des pays et des peuples islamiques. Le gouvernement constitutionnel, contrairement à toute attente, ne leur donna ni la santé, ni la richesse, ni la force. L'indépendance résolut peu de problèmes et en engendra plus encore... La révolution en Iran a montré une voie, et il y a aujourd'hui, dans tous les pays musulmans, des hommes et des femmes qui essaient de suivre la voie iranienne, ou d'en trouver une autre meilleure, afin de revenir à l'islam vrai, original et authentique du Prophète et de ses compagnons ». Ce qu'il importe de bien comprendre ici est que, pour Bernard Lewis, la voie suivie par l'Iran n'est pas bonne parce qu'elle apporte « la santé, la richesse et la force » ou l'« indépendance » au peuple iranien ; cette voie n'est pas bonne en soi par rapport à la morale, à l'éthique ou au salut. Cette voie est bonne par rapport à ce que cet orientaliste définit comme étant le « bon » islam : « l'islam vrai, original, authentique du Prophète et de ses compagnons »[2].

Comme on le constate à la lecture de l'ouvrage de Bernard Lewis, nous sommes, au mieux, en présence d'un des multiples visages de ce qui est aujourd'hui qualifié d'islamisme radical : un islam essentialisé, figé dans des pratiques et institutions révolues qui, aujourd'hui, s'incarne dans la figure du wahhabisme. En effet, comme nous serons amenés à le voir plus loin, il s'agit là d'un « islam original

---

[1] *Ibid.*, p. 763.
[2] *Ibid.*, p.811.

dans sa forme », alors que l'islam vrai et authentique du Prophète est d'esprit progressiste. Et, au pire, nous sommes en présence d'un islam dévoyé sous la forme qualifiée aujourd'hui de djihadisme qui, après les heures de gloire d'al-Qaïda, s'est incarné en la figure de Daech et qui risque, si rien n'est fait pour assainir la situation géopolitique, de se métastaser demain en des configurations beaucoup plus dangereuses et quasi ingérables.

Si, pour les islamologues occidentaux il existe deux conceptions de l'islam : l'islam et l'islamisme, pour les orientalistes de la veine de Bernard Lewis, il n'y a pas deux perceptions possibles de l'islam. Il n'y a en fait qu'une seule conception possible de l'islam : « l'islam vrai, original, authentique du Prophète et de ses compagnons » qui n'est autre que l'« islamisme radical » des islamologues. C'est ainsi que la mauvaise perception de l'islam, celle de la guerre, a fini par chasser la bonne, celle de la paix. On peut évidemment être surpris de constater que des lois économiques puissent opérer dans la sphère du religieux. Il n'en demeure pas moins que l'adage économique selon lequel « la mauvaise monnaie chasse la bonne »[1] s'avère des plus pertinents au regard des concepts d'islam et d'islamisme. Lorsqu'on examine la situation dans toute sa complexité, on constate que c'est via un double mécanisme de dévalorisation, que la nouvelle perception en mal du concept d'islamisme a rendu inopérante la "neutralité" du terme islam, "neutralité" si l'on considère que l'islam a autant droit que les autres religions au respect ni plus ni moins.

---

[1] Selon la loi de Gresham, lorsque deux monnaies, avec un taux de change légal fixe, se trouvent simultanément en circulation, les agents économiques thésaurisent la « bonne » monnaie et utilisent la « mauvaise » monnaie pour s'en débarrasser au plus vite.

En effet, avec l'extension du terrorisme islamique à l'échelle mondiale et la montée de la terreur au sein des populations des différentes sociétés, le concept d'islamisme a scotomisé celui de l'islam, en ce sens que, dans l'esprit des gens, les termes « islam » et « islamisme » sont devenus quasi identiques pour ne pas dire synonymes. C'est ainsi que, de plus en plus mal perçu, le concept d'islamisme, assimilé ici à la mauvaise perception de l'islam en tant que religion belliciste, a chassé la conception pacifique et pacifiante de l'islam dans l'esprit d'une grande partie de la population occidentale. Le problème est que le concept d'« islamisme » a non seulement fini au fil du temps par banaliser l'instrumentalisation de l'islam par les acteurs politiques mais a conféré à ce phénomène une certaine légitimité, ce qui est pire encore.

La question se pose dès lors de savoir pourquoi et comment la nouvelle définition du terme « islamisme » a pu faire l'unanimité. Il faut évidemment se garder, à la faveur de la ténébreuse stratégie du chaos, de sombrer dans le conspirationnisme. La science étant censée être neutre, la question se pose d'abord de savoir pourquoi les chercheurs "non partisans", c'est-à-dire ceux qui cherchent à être aussi indépendants et objectifs que possible, ainsi que la « *commun decency* » l'exige[1], ne se sont pas élevés contre ce concept. Ceci semble tenir au fait que, à l'origine, avant de se référer à une « lecture politique et radicale du fondamentalisme »[2], ce concept se limitait à désigner une idéologie, celle de l'islam politique. Etant donné que le monde musulman, loin d'être homogène, est de surcroît particulièrement complexe, solution de facilité peut-être, ce concept semblait en tout

---

[1] Posture de l'observateur « neutre ».
[2] Olivier Roy, *Généalogie de l'islamisme*, Paris, Editions Pluriel, 2011, p.30.

cas faciliter la compréhension de la situation, alors qu'il n'en a rien été, bien au contraire.

Le plus important maintenant consiste à comprendre pourquoi ce nouveau concept défini en Occident n'a pas été rejeté, ni remis en cause dans le monde musulman : qu'il s'agisse des pays musulmans et/ou des ressortissants de pays occidentaux de confession musulmane quel que fût leur engagement religieux. Il est utile pour ce faire de rappeler au préalable la définition originelle du concept d'islamisme telle qu'elle a été établie par des chercheurs occidentaux dans les années 80. Il est communément admis que le concept a été consacré par l'ouvrage « L'islamisme radical » de Bruno Etienne paru en 1988. Dans un article dédié en 2003 au concept d'islamisme, Bruno Etienne fait preuve de modestie en partageant la paternité de ce concept avec des collègues, nommément : « Kepel, Tosy et Burgat ». Consacré, ce terme mal défini et flou à maints égards est depuis devenu un fourre-tout utile que chacun, du chercheur au politicien, en passant par le journaliste, se permet de définir de façon circonstancielle selon ses besoins du moment. Souci de pertinence ou quête de rigueur, il n'est pas surprenant de voir un même chercheur dans un même ouvrage recourir à différentes définitions toutes de son propre cru.

*

En cherchant à sélectionner les définitions historiquement les plus emblématiques, nous avons été amenés à remonter dans le temps et nous avons, semble-t-il, réussi à circonscrire les origines et la paternité exacte du concept d'« islamisme ». Décédé depuis, son concepteur a, semble-t-il, lui-même oublié comment ce concept a pu voir le jour et s'imposer sur la scène géopolitique mondiale. Nous allons pour commencer fournir ci-après plusieurs définitions consacrées du terme « islamisme », puis nous ferons état de la polémique entre les chercheurs concernant la vie et la mort de ce concept. Après quoi nous ferons état de l'origine du concept et chercherons à voir pourquoi personne n'a cherché à remettre en cause le processus de dévoiement de ce terme à l'origine inoffensif, devenu depuis anxiogène.

Commençons par faire état de quelques définitions, étant entendu qu'aucune de ces définitions ne nous paraît entièrement satisfaisante au regard du phénomène qu'elles cherchent à circonscrire et à rendre intelligible, à croire que plus les chercheurs s'approchent du phénomène plus il leur échappe ! Et ce pour la bonne et simple raison que nous pensons que ce concept n'a pas lieu d'être. Faisant office de trou noir, ce concept décrit un ensemble de phénomènes divers, voire disparates, sans réussir à en faire une synthèse capable de résister à l'usure du temps et à la réalité des situations qu'il cherche à dévoiler. Avatar de la recherche en sciences politiques des islamologues, à l'instar de l'« Orient » des orientalistes des premières heures, ce concept, qui n'avait pas lieu d'être, disparaîtra. Mais, entre-temps, combien de malentendus et de dégâts occasionnés ! On ne peut en effet ignorer que ce concept a, entre autres, fortement contribué à légitimer l'instrumentalisation de l'islam par des acteurs internes et externes qui ont œuvré à l'instauration de régimes islamiques dont une théocratie cléricale, mode de gouvernement inédit dans l'histoire du monde musulman[1].

---

[1] La République islamique d'Iran, moins fondamentaliste que bien d'autres Etats islamiques, n'en demeure pas moins le seul Etat islamique où la séparation entre le pouvoir temporel et le pouvoir spirituel est quasi inexistante. Ce qui ne présage rien de bon quant à l'avenir de la religion d'Etat dans ce pays.

En 2003, Bruno Etienne[1] nous rappelle « que le terme « islamisme » a changé de sens deux fois en deux siècles : avant la période coloniale il signifie tout simplement l'islam comme « mahométisme ». Il déclare que « Ce n'est que plusieurs décennies après les indépendances nationales qu'il va prendre le sens actuel » sans spécifier pour autant qui lui a donné son sens actuel et pour quelle raison on a choisi ce terme. Bruno Etienne précise : « Nous avions tous (Kepel, Tosy, Burgat, quelques autres et moi-même) décrit l'« islamisme radical » comme étant l'utilisation politique de thèmes musulmans mobilisés en réaction à la "westernization" considérée comme agressive à l'égard de l'identité arabo-musulmane, réaction perçue comme une protestation antimoderne. » Comme on le constate, nous avons droit là à une ébauche de définition sans pouvoir affirmer pour autant qui en est l'auteur.

Bruno Etienne précise s'en tenir à sa terminologie de 1987 qui a consacré l'expression d'« islamisme radical » pour la très bonne raison qu'il exprime selon lui le fait que « la doctrine orthodoxe est prise au sérieux sur tous les plans y compris dans le passage à la violence ». D'après Bruno Etienne, « François Burgat a donné la meilleure définition de cet « islamisme radical » : « Le recours au vocabulaire de l'islam opéré (initialement mais non exclusivement) au surlendemain des indépendances, par les couches sociales freinées dans leur accès aux bénéfices de la modernisation pour exprimer (contre ou, le cas échéant, depuis l'Etat) un projet politique se servant de l'héritage occidental comme d'un repoussoir mais autorisant, ce faisant, sa réappropriation ». Bruno Etienne conclut : « L'islamisme est l'utilisation politique de l'islam par les acteurs d'une contestation, antimoderne perçue comme portant atteinte à leur identité à la fois nationale et

---

[1] Bruno Étienne, « L'islamisme comme idéologie et comme force politique », *Cités* 2003/2 (n° 14), pp. 45-55.

religieuse » et prend soin de préciser qu'il est d'accord avec G. Kepel et O. Roy quant à l'échec du « projet politique » des contestataires. Etant donné que ces deux islamologues ont chacun écrit un livre faisant le constat de cet échec (O. Roy : « *Echec de l'islam politique* » et G. Kepel : « *Expansion et déclin de l'islamisme* »), on en vient à se poser la question suivante : avait-on le droit de dévoyer un terme destiné à identifier une religion pour définir un mirage ? A la lecture de ces ouvrages on peut arriver à la conclusion que le terme islamisme comme religion a été vainement sacrifié sur l'autel d'une dystopie.

Polysémique, le concept d'islamisme, comme la réalité géopolitique nous le rappelle à longueur de journée, a la vie dure. Ceci nous amène à passer en revue quelques-unes des différentes définitions qu'Olivier Roy en donne. Nous avons opté pour cet islamologue pour la très bonne raison que, selon nous, il est le chercheur qui, à la suite de Bruno Etienne, a le plus travaillé à définir ce concept.

Dans son ouvrage *Généalogie de l'islamisme*[1], l'islamisme est défini comme « une idéologie qui veut faire de l'islam et du respect intégral de la charia un modèle politique alternatif à la démocratie ». Dans *Echec de l'islam politique*, Olivier Roy[2] déclare : « j'appelle comme d'autres, « islamisme » ce mouvement contemporain qui pense l'islam comme une idéologie politique ». Ailleurs, il précise que l'islamisme est « la lecture politique et radicale du fondamentalisme »[3]. Confronté au « mythe de l'islamisme », il pose la question : « Quel est le contenu de l'islamisme ? » et répond : « c'est un moment, l'expression idéologique de mouvements sociaux et d'une crise

---

[1] *op.cit.*
[2] *Ibid.*, p.9.
[3] *Généalogie de l'islamisme,* op. cit. p.30.

identitaire profonde et violente »[1] et précise : « nous n'en avons pas fini avec l'islamisme comme contestation sociale, recherche d'identité fondée sur un code normatif, faute de véritables racines, mais aussi comme stratégie de pouvoir pour une *intelligentsia* exclue des avenues du pouvoir »[2]. Au sens propre ou figuré, le terme « moment » est utilisé à bon escient. Il s'agit dans l'esprit d'Olivier Roy d'un moment spécifique : le moment révolutionnaire. C'est cette insurrection contre l'assignation de sens qu'a été la révolution iranienne qui sert de point de référence. Il y a le fond de ce moment historique : l'expression du politique et la forme qu'elle va prendre : le fondamentalisme. « Le fondamentalisme n'est pas en soi politiquement radical, ou révolutionnaire, il le devient quand il exprime en termes politiques la volonté de réforme de la société » écrit Olivier Roy[3]. « Or, cette politisation, ou plus exactement idéologisation du fondamentalisme, est récente » précise-t-il. A la lecture de ces définitions on peut déduire qu'il y a lieu de faire une distinction entre le fondamentalisme quiétiste et contestataire et le fondamentalisme radical et révolutionnaire. Le premier a été dénommé « islamisme » et le second « islamisme radical ». D'après Gilles Kepel[4] qui se vante d'avoir contribué à la mise en vogue du concept dans les années 80, l'islamisme est un mouvement qui lutte pour le changement du pouvoir, c'est « un mouvement qui lutte pour l'instauration d'un Etat qui soit basé sur les injonctions contenues dans les textes sacrés ». Comme on le constate avec la définition de Gilles Kepel nous avons

---

[1] *Ibid.*, p.117.
[2] *Ibid.*, p. 118.
[3] *Ibid.*, p.29.
[4] France Culture, *émission Répliques d'Alain Finkielkraut*, « Le terrorisme en face » avec François Burgat et Gilles Kepel, le 7 janvier 2017.

affaire à un mouvement qui ne dure qu'un moment : l'« islamisme » s'arrête à la porte du pouvoir. Une fois au pouvoir les islamistes deviennent de simples musulmans !

S'il est vrai que l'islamisme, en tant qu'instrument de délégitimation du pouvoir[1], est par essence la négation de la politique, la question qui se pose alors est de savoir pourquoi les islamologues français ont ressenti le besoin de dévoyer le terme islamisme pour exprimer cette réalité brûlante qu'est l'aporie de la révolution permanente sousjacente à l'idéologie islamiste. Utiliser un même terme pour définir une idée théologico-politique et ce qui s'est passé en Afghanistan, en Iran, en Algérie et ailleurs tout autant que ce qui se passe aujourd'hui en Irak, au Mali et dans les banlieues parisienne et bruxelloise n'est pas chose aisée. Olivier Roy n'est pas responsable du dévoiement de l'acception première du sens du mot « islamisme », fidèle à la doxa, il a fait sien le nouveau concept d'islamisme et a assidument travaillé à le définir. Le problème est que les phénomènes, qu'il cherche à circonscrire et à définir à partir du concept d'*islamisme* sont multidimensionnels, complexes et dynamiques. Olivier Roy nous paraît optimiste car, contrairement à ce qu'il pense, avec « islam et islamisme » nous n'avons pas affaire à un « grand malentendu »[2] mais à une multitude de malentendus qui font que le monde va de mal en pis.

*

Après les terribles attentats qui ont frappé la France en 2015, nous avons eu droit à la polémique « islamisation de

---

[1] Islam signifie « soumission à Dieu » et à Dieu seul. Et, d'après le Coran, chaque musulman est en droit de se lever et de demander des comptes à celui qui le gouverne.

[2] *Ibid.*, « *Islam et islamisme : le grand malentendu* ».

la radicalité » ou « radicalité de l'islam » : Olivier Roy défendant la thèse de l'islamisation de la radicalité et Gilles Kepel s'obstinant dans l'idée essentialiste de la radicalité de l'islam. Mais, bien avant cette polémique, la revue Esprit avait en 2001 fait étalage d'une polémique autrement célèbre concernant le concept « islamisme » qui n'a malencontreusement pas pu aboutir pour des raisons que, faute de mieux, nous qualifierons ici de *paradoxe épistémique*. En effet, il est communément admis aujourd'hui que l'islamisme se réfère à l'islam politique (titre de l'ouvrage d'Olivier Roy *Echec de l'islam politique*). Or, si l'islam des « islamistes » est atemporel, il ne peut être politique, c'est-à-dire temporel.

Il n'est évidemment pas question d'entrer dans les détails de la polémique entre François Burgat, Alain Roussillion et Olivier Roy concernant l'islam politique[1].

Dans son article intitulé « De l'islamisme au post-islamisme : vie et mort d'un concept », François Burgat soulève une question intelligente : « Qui a réellement évolué ? Les islamistes ou le discours de ceux qui ont fait profession d'en parler ? ». En vérité, le problème est que François Burgat qui a pris la cause des « islamistes » (acteurs locaux, militants et activistes) n'apprécie guère les annonces de « déclin de l'islamisme » (thèse de Kepel) ou de son dépassement par le « post-islamisme » (thèse d'O. Roy). Optant pour la posture tiers-mondiste, Burgat feint de défendre la cause des contestataires (les exclus de la mondialisation, les dominés, les populations ayant subi le joug du colonialisme, etc.) mais le problème est qu'en pensant défendre la juste cause des contestataires, il prend fait et cause pour le fondamentalisme comme l'« unique » solution aux problèmes du mal-développement du monde musulman. Avec François Burgat, nous sommes dans le

---

[1] « Polémique entre chercheurs à propos de l'islam politique », in revue *Esprit*, août-septembre 2001.

registre du « ressentiment » qui pousse les ex-colonisés embourbés dans le ressentiment à faire de l'islam une « identité » à défendre.

D'un autre bord, Alain Roussillon, dans son article intitulé *Les islamologues dans l'impasse*, met à juste titre en garde contre le risque que comporte la tentation de *totaliser*, à l'enseigne de l'islamisme, le déchiffrement des processus en cours dans les sociétés musulmanes (cette critique concerne tout autant G. Kepel qu'O. Roy) et le risque qu'il y a « de réinterpréter « l'islam » à la lumière de ses avatars contemporains les plus « extrémistes » » (critique limitée à Kepel). Mais le problème de fond demeure intact car, *in fine*, Roussillon scotomise la question des risques que comporte l'utilisation même de la catégorie « islamisme ». Car, au fond, la question ne consiste pas à savoir si l'islamisme doit, à l'instar de l'islam, être considéré comme un phénomène pluriel pour que le concept d'« islamisme » retrouve toute sa pertinence mais de savoir si ce concept a lieu d'être. C'est pourquoi, lorsque O. Roy soulève la question : « *Les islamologues ont-ils inventé l'islamisme ?* » (titre de son article, déjà évoqué en début du présent essai), la réponse est non pour le(s) phénomène(s) en tant que réalité factuelle mais oui en tant que réalité conceptuelle. Ce qui a eu pour conséquence, entre autres, d'essentialiser cette réalité.

Si, pour Gilles Kepel, l'islamisme est un phénomène inhérent à la réalité de l'islam, il n'en va pas de même pour Olivier Roy et Alain Roussillon qui tous deux font une distinction entre islam et islamisme. Mais A. Roussillon refuse l'idée d'une « matrice conceptuelle » commune de l'islamisme attribuée à O. Roy. Pour Roussillon, si islamisme il y a, comme pour l'islam il se conjugue au pluriel. Contrairement à Kepel pour qui l'islam est par essence une religion figée, Roussillon estime que l'islam, à l'instar des autres religions, est une religion apte à se réformer. De sorte que lorsqu'on examine les perspectives de changement explorées, on constate que Roussillon et Roy sont plus complémentaires qu'opposés. Pour enrayer la montée de l'intégrisme et du fondamentalisme islamiques, A. Roussillon mise sur les intellectuels réformistes musulmans : les penseurs locaux ou expatriés car persécutés chez eux. O. Roy, pour sa part, croit à l'autorégulation de l'islamisme par les acteurs sociaux (les acteurs islamistes au pouvoir ou en quête du pouvoir, c'est-à-dire les activistes islamistes) sous la contrainte du réel. Roussillon semble plus « interventionniste » en ce sens qu'il estime qu'il faut soutenir les intellectuels musulmans réformistes. Tandis que Roy a une posture d'« observateur » en ce sens plus attentiste pour ne pas faire usage du terme de « neutralité » qui pose problème. L'accent mis sur la nécessité de la réforme de l'islam par des acteurs « internes » de la part d'O. Roy[1] peut prêter à confusion. Ceci pourrait laisser croire que, sous couvert de pragmatisme, O. Roy ne ferait en fait pas suffisamment confiance en la capacité des musulmans européens ou américains à réformer l'islam. Nous sommes là en présence de la question des fondements de la légitimité de la pensée religieuse musulmane à l'ère de la mondialisation et du paradoxe de la dichotomie « *Dar al-islam – Dar al-harb* », qui sert de fondement au stéréotype d'un islam djihadiste par essence, stéréotype défendu et propagé par Bernard Lewis et ses disciples.

Comme on le constate, si ces islamologues ne s'entendent pas toujours, aucun ne remet en cause le nouveau paradigme constitué. De sorte que, au pire de sa virulence, la polémique entre islamologues demeure circonscrite dans le cadre du paradigme de l'islamisme. O. Roy et F. Burgat, en l'an 2000 comme aujourd'hui, demeurent dans le

---

[1] *Ibid.*, p.112.

registre politique ; dans cette perspective, l'islam cesse d'être une véritable religion et devient une idéologie politique. Le problème est que Burgat, à l'instar de Bruno Etienne et de Kepel, fait le choix d'une « altérité islamiste » qui s'avère *in fine* être irréductible. Si O. Roy et G. Kepel ont une position radicalement opposée, ceci tient pour partie au fait que Gilles Kepel est plus proche de la réalité wahhabite tandis que Olivier Roy, qui a davantage frayé avec le monde chiite, semble moins pessimiste pour ne pas dire hostile à l'égard de cette réalité qui n'est finalement pas aussi étrangère au monde occidental que d'aucuns le croient et/ou aiment à le faire croire. En effet, que l'on se situe au niveau du monde des idées ou de celui de l'action, l'islam participe de l'Occident. Eh oui, et n'en déplaise aux mouvements populistes européens, l'islam a « quelque chose à voir » avec l'Occident.

Ainsi qu'il a été signalé, l'islam n'est pas une réalité étrangère à la pensée occidentale puisqu'il reconnaît faire suite aux deux religions monothéistes qui le précèdent et qu'il parachève. De même, les pays musulmans et les pays occidentaux ne forment pas deux mondes parfaitement distincts et indépendants l'un de l'autre. Si les soi-disant « islamistes » ne sont pas tous des Occidentaux, loin s'en faut, ils sont en revanche presque tous formés à la doctrine wahhabo-salafiste, grâce essentiellement aux pétromonarchies soutenues par des pays occidentaux et leur accès au pouvoir est souvent facilité par le soutien de la « main invisible » de l'Occident et l'apport financier des pétromonarchies. Concrètement, les talibans, al-Qaïda, Daech sont des réalités propres au monde islamique mais, comme nous l'avons vu précédemment, l'accès au pouvoir des talibans en Afghanistan ou la formation de l'entité Daech sont en grande partie le produit de l'ingérence occidentale. Et, si l'on réfléchit bien, on constate que le pétrole, dont l'Occident a tant besoin, a été une véritable malédiction pour le monde musulman[1].

*

La remise en cause du paradigme de l'islamisme nous amène à chercher à mieux circonscrire l'origine exacte du concept d'« islamisme ». Lorsqu'on remonte dans le temps et que l'on cherche à identifier la source de ce nouveau concept, on constate que c'est via la création du nouveau concept d'« islamiste » que le processus de dévoiement du terme « islamisme » a commencé. En effet, c'est dans un article paru dans la revue du Tiers Monde en 1982 que Bruno Etienne[2] déclare que « Tosy et moi-même avons proposé d'appeler « islamistes » cet ensemble de militants de l'islam ; puisque c'est ainsi qu'eux-mêmes se nomment : al-islamiyin ». C'est dans ce même article qu'il fait usage du terme « islamisme » compris dans sa nouvelle acception. Il convient d'ouvrir ici une parenthèse pour souligner à cet égard que, dès le départ, Maxime Rodinson préfère utiliser le terme intégrisme[3], ce qui a le double avantage d'éviter le dévoiement du terme islamisme et le risque de fournir un outil conceptuel aux fins de la légitimation théorique de l'instrumentalisation de l'islam ; ce qui n'a pas manqué d'être le cas comme nous serons amenés à le constater.

Revenons pour l'heure à notre enquête sur la genèse du concept d'« islamisme ». Dans l'article susmentionné qui fait figure d'hymne à la faveur des « islamistes » à

---

[1] Cf. supra, chapitres 3 et 7.
[2] Bruno Etienne, « La vague « islamiste » face aux nations arabes », in revue *Tiers Monde*, tome 23, n° 92, 1982, p.915.
[3] Le terme intégrisme se justifie par la fidélité à des pratiques en vigueur à l'époque des Rashidun (les califes bien guidés) dans l'islam sunnite. Le terme fondamentalisme porte sur la fidélité à une conception littérale du Coran, ce terme peut être utilisé indistinctement pour des sunnites ou des chiites fondamentalistes.

l'encontre des « nationalistes » arabes qui, d'après l'auteur de l'article, « ne produisent plus assez de légitimité après vingt-cinq ou trente ans d'expérience », Bruno Etienne aborde la question de l'islamisme en ces termes : « Les groupes humains ont besoin de produire de la *Weltanschauung* pour persister dans leur être. Toute idéologie organique donne un sens au monde, c'est le cas de l'*islamisme* »[1]. Cette définition fait penser à la définition que Henry Corbin entend donner au terme « iranisme ». Le problème est que, si l'iranisme de H. Corbin est circonscrit dans le champ du spirituel comme nous allons le voir plus loin, l'islamisme de B. Etienne opère dans le champ du politique. Il importe de préciser à cet égard que Bruno Etienne[2] déclare soutenir avec d'autres que « l'essence du politique se trouve, ou tout au moins peut être recherchée, dans le champ religieux ».

Le sort en est jeté, c'est ainsi que le terme « islamisme » va progressivement changer de sens pour finir par signifier « islam politique ». Si, d'après la définition implicite dans la citation susmentionnée, le terme islamisme ne définit plus une religion mais une idéologie, la question se pose alors de savoir si toutes les religions ne doivent pas être considérées comme des idéologies. Ce qui conforterait la thèse d'Elie Barnavi[3] selon qui « toute religion est politique ». La phrase susmentionnée de Bruno Etienne pourrait dès lors être reformulée comme suit : « Les groupes humains ont besoin de produire de la *Weltanschauung* pour persister dans leur être. Toute idéologie organique donne un sens au monde, c'est le cas du *judaïsme*, du *christianisme*, de l'*islamisme, etc.* ». Pourquoi devrait-on réserver cette spécificité à l'islam ?

---

[1] *Ibid.*, p.923.
[2] Bruno Etienne, *L'islamisme radical*, Paris, Hachette, 1987, p.41.
[3] Elie Barnavi, *Les religions meurtrières*, *op. cit.*, pp 27-37.

Le fait que l'islam soit traité comme une « idéologie politique » n'est pas chose nouvelle. Le problème est que le concept « islamisme radical » a progressivement changé de sens pour finir par signifier « islam politique », le terme politique étant désormais compris au sens de « la politique » et non plus « du politique ». C'est ainsi que, à trop vouloir frayer avec le monde politique, instrumentalisée, la religion a perdu de sa sacralité. Alors, la porte a été ouverte à toutes les dérives... Si nous faisons abstraction pour un moment du monde politique, pour nous atteler à ce qui nous intéresse plus spécifiquement ici, à savoir le monde de la recherche, on constate alors que le problème réside dans la capacité de distanciation du chercheur vis-à-vis des idéologies politiques. Maxime Rodinson[1], qui a longuement travaillé sur l'islam et le capitalisme, met en garde contre la mise en tutelle de l'activité scientifique par une idéologie politique quelle qu'elle soit.

Comme nous avons pu l'illustrer, source à l'appui, c'est à Bruno Etienne qu'incombe la paternité du nouveau concept d'*islamisme*. La question qui se pose maintenant consiste à savoir pourquoi Bruno Etienne a choisi de dévoyer ce terme plutôt que de recourir à la panoplie des concepts mis à sa disposition pour définir l'objet de sa recherche. C'est dans son ouvrage *Islamisme radical* que nous pensons avoir trouvé la réponse à notre question. Pour bien comprendre les motivations de Bruno Etienne, il faut savoir qu'avec ce chercheur nous avons affaire à un intellectuel marxisant déstabilisé par l'échec de l'idéologie

---

[1] Selon ce sociologue et orientaliste, « Ce fut un grand malheur pour l'activité scientifique (on l'appelait alors philosophie) d'avoir été longtemps la servante de la théologie. Le malheur ne serait pas moindre d'en faire maintenant la servante de l'idéologie politique qui a succédé à la théologie ». Maxime Rodinson, *Islam et capitalisme*, Paris, Seuil, 1966, p.9.

communiste, pris de court et fasciné par la dimension religieuse de la révolution iranienne. A l'affût d'alternatives, il n'a d'autre choix que de se laisser emporter par la vague d'enthousiasme, de toute façon il déclare écrire pour ses « concitoyens, et non les Arabes, ni pour les musulmans, qui n'ont que faire de (ses) analyses »[1], ses dires ne peuvent donc porter à conséquence.

« La lutte des classes, comme l'avait pressenti Engels, écrit Bruno Etienne[2], ne débouche sur la révolution que lorsqu'elle peut se présenter en termes religieux : les victimes de la « colère de Dieu » élucident le sens de l'Histoire. Le but de l'islamisme radical est bien terrestre : créer un royaume égalitaire qui mette à bas la morgue des possédants ». Le communisme est en agonie, vive l'islamisme radical ! Il est à noter à cet égard qu'une fatwa, en date du 28 mars 1948, de l'université d'Al Azhar proclame avec énergie : « pas de communisme dans l'Islam »[3]. On ne peut en effet ignorer que « le Coran n'a rien contre la propriété privée », qu'« il recommande de ne pas mettre en cause les inégalités », que « le salariat est une institution naturelle contre laquelle rien n'est à objecter » et que, *last but not least*, « le Coran considère avec faveur l'activité commerciale »[4].

Comme on le constate, B. Etienne n'arrive pas à s'extraire de son contexte pour aborder avec un esprit indépendant et critique la réalité du fait islamique tel qu'il se présente à ses yeux. De sorte qu'il pense que, s'il n'arrive pas à percevoir la réalité telle qu'elle est, c'est

---

[1] « J'écris pour mes concitoyens, et non les Arabes, ni pour les musulmans, qui n'ont que faire de mes analyses ». Bruno Etienne, *Islamisme radical*, op.cit., p.331.
[2] *Ibid.*, p.327.
[3] Maxime Rodinson, *Islam et capitalisme*, Paris, Seuil, 1966, p.42.
[4] *Ibid.*, p.31.

parce qu'il ne dispose pas de concept adapté à cette réalité qui lui échappe. Comme il ne peut y avoir d'accord de la pensée au réel sans sujet, à trop vouloir être objectif, B. Etienne finit par perdre toute subjectivité "naturelle". Or, c'est via l'usage du langage et de l'outillage intellectuel commun à sa culture qu'un chercheur fait preuve d'une subjectivité naturelle ou commune. Lorsque le chercheur introduit un nouveau concept ou un objet étranger à sa culture il fait preuve de créativité, nous avons alors affaire à une subjectivité personnelle ou intuitive. Ce nouveau concept doit alors faire l'objet d'une validation et non pas être accepté d'emblée par la communauté intellectuelle pour finir par s'imposer tel quel au commun des mortels. Que s'est-il passé ? Examinons comment et pourquoi le terme « islamisme » a pu être dévoyé, sans que l'acte de dévoiement de ce terme ne fasse l'objet d'une analyse critique conséquente.

*

Les causes et mobiles véritables qui sont à la base de la création du nouveau concept d'*islamisme* se trouvent explicités dans les « quelques remarques méthodologiques » que B. Etienne formule en début de son ouvrage *Islamisme radical* et qui se résument en fait dans la question centrale qui le taraude et qu'il met en exergue en italique : « *comment parler de la religion de l'Autre, comment analyser l'Islam avec des concepts européens ?* ». Que faire ? La solution est toute trouvée, il en vient à forger un concept bâtard né du mariage de deux termes : « islam » et « politique ». On ne peut en effet ignorer que Bruno Etienne est un « enseignant de sciences politiques », ainsi qu'il tient à le souligner lui-même. Cinq ans après, il revient à son idée phare qui faisait figure de conclusion dans son article publié dans la revue Tiers Monde en 1982 : l'islamisme est une idéologie organique qui donne sens au monde. Comme toutes les religions peuvent être considérées comme des idéologies organiques qui donnent sens au monde, Bruno Etienne va plus loin, selon lui, l'islam dès le départ est une religion politique. Mais, a contrario, il prend soin de préciser : « je n'utilise pas l'ensemble « islam politique » parce que le terme *politique*, tel que nous le concevons en Occident depuis les Grecs, n'existe pas dans le Coran »[1]. On croirait entendre Bernard Lewis. A ceci près que Bruno Etienne a quelques sympathies pour le marxisme : « Le but de l'islamisme radical est bien terrestre : créer un royaume égalitaire qui mette à bas la morgue des possédants »[2]. Comme on le constate, si « royaume » il y a, ce royaume ressemble fort à celui du monde communiste.

Le premier problème avec la remarque d'ordre méthodologique de Bruno Etienne est que ce n'est pas parce qu'on prend un terme conçu à l'origine pour définir l'islam, en l'occurrence l'« islamisme », qu'on a affaire à une catégorie islamique. Ce n'est pas parce qu'on ne se réfère pas nommément à la politique en ayant recours à la terminologie « islamisme radical » qu'on évacue la dimension politique du concept. Pourquoi ne pas nommer un chat, un chat ? Nous sommes en présence du phénomène communément connu sous le nom d'instrumentalisation de la religion par des activistes fondamentalistes ou par des religieux. Ce que d'aucuns dénomment politisation de l'islam. Pourquoi les musulmans n'auraient-ils pas le droit d'être fondamentalistes et activistes à la fois et d'être nommés en tant que tels ? Bruno Etienne est un Français qui, d'après ce qu'il dit, écrit pour « ses concitoyens », pourquoi ne pas utiliser les concepts propres au paradigme ambiant ? Pourquoi créer un concept bâtard qui n'appartient

---
[1] *Ibid.*, p. 21.
[2] *Ibid.*, p. 327.

ni au monde islamique, ni au monde occidental ? Pourquoi ne pas faire appel à des concepts existant dans le système d'intelligence en usage sans nier pour autant la « spécificité » de l'objet ? Au lieu de dévoyer le terme islamisme, pourquoi ne pas faire usage du terme fondamentalisme en y adjoignant le qualificatif islamique ou wahhabite ? Pourquoi ne pas faire usage du terme fondamentaliste ou insurgé ou activiste musulman ou, mieux encore, wahhabite ou salafiste, au lieu de faire usage du terme « islamiste » ?

Il est à noter que, plutôt que de dévoyer le terme « islamisme », l'islamologue Maxime Rodinson[1] opte pour l'appellation « intégrisme islamique ». En effet, abordant les questions de terminologie dans un entretien inédit réalisé par le politologue Gilbert Achcar, Maxime Rodinson précise que, si l'appellation « intégrisme islamique » n'est pas très bonne, celle de « fondamentalisme » l'est encore moins ; quant au terme « islamisme », il entraîne une confusion avec l'Islam ; « Islam radical » n'est pas si mal, ajoute-t-il, mais aucune appellation ne correspond tout à fait à l'objet ». Il convient de préciser que, sous la rubrique « intégrisme islamique », M. Rodinson regroupe « tous les mouvements qui pensent que l'application intégrale des dogmes et pratiques de l'Islam, y compris dans les domaines politique et social, mènerait la communauté musulmane, voire le monde entier, vers un État harmonieux, idéal, reflet de la première communauté musulmane idéalisée, celle de Médine entre 622 et 632 de l'ère chrétienne ».

Le second problème avec la remarque d'ordre méthodologique de Bruno Etienne est que son concept d'« islamisme » est fondé sur l'hypothèse de

---

[1] Gilbert Achcar, « Maxime Rodinson : sur l'intégrisme islamique », *Mouvements* 2004/6 (no 36), pp. 72-76.

l'incommunicabilité entre l'Islam et l'Occident : « les catégories occidentales sont incompatibles avec la réalité islamique ». Nous sommes en présence d'une aporie. Posée sous forme de tautologie, cette hypothèse se présente d'emblée comme une assertion donc en principe inaccessible à la réfutation. Il suffirait de rejeter cette hypothèse pour que le château s'effondre. Or, ainsi qu'il a déjà été signalé, l'islam comme religion est une réalité distincte mais non étrangère à la pensée occidentale puisqu'il se fonde sur les deux religions monothéistes qu'il fait siennes.

En effet, comme Eric Geoffroy[1] le souligne : « Être musulman implique de reconnaître l'authenticité de toutes les religions révélées avant l'islam. Le Coran est explicite sur cet héritage. Dites : "Nous croyons en Dieu, à ce qui a été révélé à Abraham, à Ismaël, à Isaac, à Jacob et aux tribus ; à ce qui a été donné à Moïse et à Jésus ; à ce qui a été donné aux prophètes, de la part de leur Seigneur. Nous n'avons de préférence pour aucun d'entre eux ; nous sommes soumis à Dieu" ». En tout état de cause, il n'y a pas de « culture » pure : culture occidentale d'une part et moyen-orientale d'autre part. La culture occidentale a été forgée par la culture moyen-orientale et continue à s'en nourrir, comme la culture moyen-orientale a été et continue à être façonnée par la culture occidentale. Ces deux cultures sont constitutives l'une de l'autre. A l'ère de la mondialisation, d'internet et des réseaux sociaux, plus que jamais, toute culture est nécessairement hybride.

Même en admettant que les hypothèses qui sont à la base de la création du concept d'*islamisme radical* soient valides et que ce concept puisse être considéré comme pertinent à tous points de vue, un autre problème d'ordre épistémologique se pose à nous, qui consiste à savoir si

---

[1] Éric Geoffroy, *Le pluralisme religieux en islam, ou la conscience de l'altérité*, Fondation pour l'innovation politique, janvier 2015.

nous avons affaire à un concept rigoureux d'un point de vue scientifique. Honnêtement parlant, il faut reconnaître qu'avec ce concept nous sommes dans un flou artistique complet. Il suffit de se référer au *signifiant* tel que défini par son concepteur pour s'en convaincre. L'*islamisme radical*, écrit Bruno Etienne, « Je le prends au sens premier du terme, la doctrine de l'Islam *à la racine*, et au sens américain du terme, l'Islam politiquement radical, presque révolutionnaire »[1]. Comme on le constate, on retrouve les problèmes évoqués dans le cadre de la polémique entre chercheurs exposée précédemment. Mais, qui plus est, cette définition permet de saisir le problème à sa racine. Le problème est que ce concept crée un monde imaginaire idéalisé à outrance où l'insurgé fondamentaliste est hissé au rang de héros libérateur. Tous deux fardés d'une certaine arrogance, à la figure dégradée du musulman de l'orientaliste anglo-saxon, s'oppose ici l'image glorifiée de l'islamiste de l'islamologue français. C'est tout juste si, avec l'activiste fondamentaliste dénommé « islamiste », nous n'avons pas affaire à un surhomme.

*

Un des deux propos primordiaux du présent chapitre de cet essai se situe au plan de la connaissance et consiste dans le rejet de l'hypothèse de l'incompatibilité des systèmes de pensée que nous avons dénommée *incompatibilité épistémique* qui est à la base de la conception du concept d'islamisme. Ce qui échappe à B. Etienne en la matière est que si la réalité (« objet » selon ses propres termes) lui paraît « insaisissable », c'est parce qu'il a affaire à un phénomène nouveau en gestation qui est une réponse

---

[1] Maxime Rodinson, *op.cit.*, p. 21.

d'acteurs fondamentalistes à la crise de la modernité[1]. En d'autres termes, l'« objet est insaisissable » parce qu'il s'agit d'une nouvelle réalité en gestation qui échappe à l'entendement de B. Etienne et non parce qu'il a affaire à une altérité irréductible et que les concepts mis à sa disposition ne lui permettent pas de saisir cette autre culture dans son essence. Mais admettons que Bruno Etienne ait raison, lorsqu'on se situe dans le monde des idées, la question qui se pose alors consiste à savoir, si incompatibilité épistémique il y a, si nous avons affaire à deux mondes ou espaces culturels ontologiquement distincts, en quoi ce concept a pu résoudre cette aporie. En rien, qui plus est, au-delà des multiples problèmes inhérents à l'approche culturaliste, non moins grave est que, lorsqu'on se situe dans le monde de l'action politique, le concept d'islamisme a légitimé le phénomène qu'il était censé décrire et a, entre autres, contribué à la consolidation des régimes fondamentalistes.

Avant d'aborder le second propos essentiel, il nous faut ici revenir sur le problème que nous avons évoqué ci-dessus qui a trait au fait que si l'« objet est insaisissable » pour Bruno Etienne c'est parce qu'il s'agit en vérité d'une nouvelle réalité en gestation. En effet, ainsi qu'il a été signalé précédemment à la fin des années 70, les religions séculières, dont le marxisme, sont en crise : « certains disent que les grandes idéologies sont en train de mourir, d'autres qu'elles nous submergent par leur monotonie ». Fortement inspiré par la révolution iranienne, le philosophe Michel Foucault ne désespère pas pour autant car, si les grandes idéologies meurent, « Le monde contemporain, à l'inverse, fourmille d'idées ».

---

[1] Attention, nous disons bien réponse à la crise de la modernité et non pas réponse à la crise de la modernisation du pays qui a été évoquée précédemment.

En vérité, « Il y a plus d'idées sur la terre que les intellectuels souvent ne l'imaginent », écrit M. Foucault[1]. « Il faut assister à la naissance des idées et à l'explosion de leur force : et cela non pas dans les livres qui les énoncent, mais dans les événements dans lesquels elles manifestent leur force », précise-t-il. Et pour cause, lui, pris d'enthousiasme pour la révolution iranienne, il est allé chercher ses idées sur le terrain en Iran en s'improvisant comme « reporter » d'idées ou plus exactement reporter des « points de croisement des idées et des événements ». C'est lui qui, en 1978, avant tout autre, utilise le terme « mouvement » pour qualifier l'événement auquel il participe sur place en qualité, non pas de simple journaliste, mais de véritable « reporter » opérant pour le compte du quotidien italien *Corriere della Sera*.

Il est important de noter que Michel Foucault[2] brandit le « mouvement iranien » comme la plus grande insurrection qui va soulever le poids formidable de l'ordre du monde qui pèse sur chacun de nous : « C'est l'insurrection d'hommes aux mains nues qui veulent soulever le poids formidable qui pèse sur chacun de nous, mais particulièrement sur eux, ces laboureurs du pétrole, ces paysans aux frontières des empires : le poids de l'ordre du monde entier. C'est peut-être la première grande insurrection contre les systèmes planétaires, la forme la plus moderne de révolte..». Inversion des ordres !? Comme on le constate, les fondements de l'« islamisme » comme idéologie alternative

---

[1] Du 16 septembre 1978 au 11 février 1979, Michel Foucault va écrire une série d'articles qu'il dénomme « reportages » d'idées pour le compte du quotidien italien *Corriere della Sera*. Cf. Michel Foucault, « Les « reportages » d'idées », in Michel Foucault, *Dits et Écrits II*, 1976-1988, Gallimard, 2001, texte n° 250, pp. 706-707.
[2] Michel Foucault, « Le chef mythique de la révolte de l'Iran », in *Dits et écrits*, op. cit., texte n° 253, pp. 713-716

sont posés et les musulmans ne sont pas les seuls à y avoir cru.

Pour qui en douterait, il est à noter à cet égard que, dans sa « lettre ouverte à Mehdi Bazargan », Michel Foucault reproche au Premier ministre du gouvernement de l'ayatollah Khomeiny de lui avoir fait croire que, sur le point spécifique des droits de l'homme « l'islam, dans son épaisseur historique, dans son dynamisme d'aujourd'hui, était capable d'affronter... le redoutable pari que le socialisme n'avait pas mieux tenu – c'est le moins qu'on puisse dire – que le capitalisme »[1]. Posté à l'avant-garde, c'est lui, Michel Foucault, qui, fasciné par le soulèvement populaire, fera, avant tout autre, référence au « gouvernement islamique » en termes d'« utopie » et d'« idéal »[2] que les islamologues auront vite fait d'ériger en « Cité idéale ».

Il serait utile de rappeler que l'idée que Michel Foucault se fait de la spiritualité politique est bel et bien ancrée dans l'ordre de l'immanence : « un mouvement traversé par le souffle d'une religion qui parle moins de l'au-delà que de la transfiguration de ce monde-ci »[3]. Or, comme Olivier Roy[4] le souligne à très juste titre, il ne suffit pas que les musulmans, à commencer par leurs dirigeants, soient vertueux pour que la société soit juste et islamique. Mais, au-delà des contraintes d'ordre systémique que nous serons amenés à examiner plus loin, la véritable aporie de la Cité idéale est ailleurs. Elle est dans le fait que la Cité idéale est un mythe. La Cité idéale au sens d'une société parfaite exempte de tous maux est une utopie et en aucune façon une réalité concrétisable. C'est une « réalité impossible », le versant négatif de l'utopie c'est-à-dire une réalité qui ne

---

[1] *Ibid.*, texte n° 265, p. 781.
[2] Michel Foucault, « A quoi rêvent les Iraniens ? », *op.cit.*, texte n° 245, p.691.
[3] Michel Foucault, *op.cit.*, texte n° 253 susmentionné, p.716.
[4] Olivier Roy, *L'échec de l'islam politique*, Paris, Seuil, 2015, p.11.

peut être sur terre. Ce n'est pas pour rien qu'utopie signifie « en aucun lieu ». Car si, comme Paul Ricœur l'a parfaitement théorisé, le versant positif de l'utopie est de permettre de défier l'ordre présent avec l'idée de le transformer à la lumière du champ des possibles, son versant négatif est la chimère et/ou la nostalgie, la fuite dans un futur imaginaire ou, pire encore, dans un passé idéalisé qui n'a jamais existé et qui ne peut en conséquence être reproduit[1].

Mais, tout aussi paradoxal que cela puisse paraître, c'est Michel Foucault, et encore lui mais lui seul, qui parlera de « spiritualité politique »[2], expression que, plus tard, les islamologues, Bruno Etienne en premier, au fait et au feu de la réalité, qualifieront d'« islam politique »[3]. Ainsi qu'il a été souligné précédemment, Michel Foucault, examinant le soulèvement iranien à un moment précis de l'histoire, ne peut pas ne pas avoir été conditionné par le contexte idéologico-politique ambiant et le climat intellectuel et spirituel qui le caractérise (crise des idéologies, annonce du retour du religieux comme repère identitaire). *Last but not least*, sa vision ne peut pas ne pas avoir été biaisée par son genre et sa nationalité : c'est un homme, qui plus est Occidental, qui n'a donc pas à vivre les conséquences de la révolution islamique iranienne. Au grand dam du peuple iranien en général et des femmes en particulier, Foucault évacue le risque d'un régime théocratique : « il n'y aura pas de parti Khomeyni, il n'y aura pas de gouvernement

---

[1] Pour les différentes fonctions de l'utopie, voir, entre autres, Paul Ricœur, *L'idéologie et l'utopie*, Paris, Seuil, 1997.
[2] Michel Foucault, « A quoi rêvent les Iraniens ? » in *Dits et écrits*, *op.cit.*, texte n° 245.
[3] Il est vrai, comme ceci a été signalé plus haut, que Bruno Etienne déclare ne pas vouloir faire usage de l'ensemble « islam politique », il n'en demeure pas moins qu'il est le premier à s'y référer.

Khomeyni »[1]. Fasciné, il ne peut s'empêcher d'être enthousiasmé par « la révolution contre la politique » car, comme Olivier Roy[2] le souligne, Michel Foucault est parfaitement conscient qu'il n'assiste pas à une révolution politique mais à une révolution contre la politique. Foucault, en tant qu'intellectuel progressiste ayant très longuement travaillé sur la question du « pouvoir », est fasciné par le soulèvement de toute une population qui se présente à ses yeux comme une négation du pouvoir politique. Mais, comme la négation du pouvoir n'est qu'un moment, se pose alors la question de la responsabilité de l'intellectuel quant à la perspective des lendemains qui déchantent et qui aurait dû l'amener à tempérer son enthousiasme. Mais ceci est difficile car l'élan spirituel est contagieux, de sorte qu'il est possible d'affirmer que l'élan spirituel est tout aussi dangereux sinon plus que l'élan révolutionnaire qui est au clair, pour ne pas dire au fait, de ses objectifs qui relèvent de l'ordre du temporel. En tout état de cause, l'idée que se fait Foucault de la fonction de l'intellectuel est radicalement autre puisque ce dernier a eu le mérite de fustiger, deux ans avant la révolution iranienne, l'idée de l'intellectuel "universel" « maître de vérité et de justice... un peu conscience de tous »[3]. Mais, si l'on s'en tient à ses propres catégories, il n'en demeure pas moins un intellectuel "spécifique", le « savant-expert » selon la définition qu'il en donne.

Fausse conscience de l'intellectuel « spécifique » peut-être, mais, comme il s'en est expliqué lui-même après

---

[1] Michel Foucault, « Le chef mythique de la révolte de l'Iran », in *Dits et écrits*, *op.cit.*, texte n° 253, pp 713-716.
[2] Olivier Roy « L'énigme du soulèvement » - *Foucault et l'Iran*, Vacarme 29, automne 2004, pp 34-35.
[3] Michel Foucault, « La fonction politique de l'intellectuel », in *Dits et écrits*, *op.cit.*, texte n° 184 (pp. 109-114).

l'instauration d'un régime théocratique en Iran et la pratique d'exécutions sommaires qui s'en est suivie, il pensait être resté fidèle à sa « propre morale théorique » qu'il qualifie d'« antistratégique » : « Me demanderait-on comment je conçois ce que je fais, je répondrais –écrit Foucault[1]–, si le stratège est l'homme qui dit « Qu'importe telle mort, tel cri, tel soulèvement par rapport à la nécessité de l'ensemble et que m'importe en revanche tel principe général dans la situation particulière où nous sommes », eh bien, il m'est indifférent que le stratège soit un politique, un historien, un révolutionnaire, un partisan du chah ou de l'ayatollah ; ma morale théorique est « antistratégique » : être respectueux quand une singularité se soulève, intransigeant dès que le pouvoir enfreint l'universel ».

Comme on le constate, Foucault ne parle pas d'éthique mais de morale, ce qui est tout autre[2]. Lourde de la charge du passé, la morale peut être d'ordre théorique mais, au feu de l'action, l'éthique est nécessairement d'ordre pratique. De sorte que, si Foucault fait une nette distinction entre le moment du soulèvement qui est celui de la négation du pouvoir politique et l'après-soulèvement qui est celui de l'institution du nouveau pouvoir, en tant que philosophe antisystème, il n'a aucune affinité pour le second moment. Or, comme l'histoire nous le rappelle constamment, c'est ce second moment qui est décisif, de sorte que, si l'on n'y prend pas garde dès le premier moment, après c'est souvent déjà beaucoup trop tard.

---

[1] Michel Foucault, « Inutile de se soulever ? », in *Dits et écrits, op.cit.*, texte n° 269, p.794.
[2] David Macey utilise le terme éthique alors que Foucault parle de morale. Cf. David Macey, *Foucault et l'Iran*, in la revue *La Rose de Personne* de 2005, passage extrait de l'ouvrage : David Macey, *Michel Foucault*, Gallimard (biographie), 1994 (traducteur Pierre-Emmanuel Dauzat) pp 415-421.

Mais, pour ce qui est de la dimension spirituelle à proprement parler de l'expression de « spiritualité politique », il importe de noter que Foucault parle de spiritualité politique et non d'islam politique. Reste à savoir dans quelle mesure il n'a pas été influencé par Malraux[1]. Ainsi que Marc-Alain Ouaknin, à l'instar de beaucoup d'autres, l'a rappelé sur France Culture[2], « On répète à satiété la formule si souvent citée et attribuée à Malraux : le XXIe siècle sera religieux ou ne sera pas ou encore le XXIe siècle sera métaphysique ou ne sera pas ». En effet, « Qu'a dit exactement André Malraux sur le retour du religieux au XXIe siècle ? » interroge Henri Tincq[3] parmi tant d'autres. « Dès l'année 1955, il avait déclaré, dans une interview restée célèbre : *« Le problème capital de la fin de siècle sera le problème religieux. »* Plus tard, invité à préciser son propos, il a ajouté : *« On m'a fait dire que le XXIe siècle sera religieux. Je n'ai jamais dit cela bien entendu, car je n'en sais rien. Ce que je dis plus incertain, mais je n'exclus pas la possibilité d'un événement spirituel à l'échelle planétaire. »* Effectivement, le doute n'est pas permis, si le changement de régime parait inéluctable, c'est le changement spirituel qui intéresse au premier chef Michel Foucault. Sa grille de lecture de ce que pensent les Iraniens est influencée par les « dires » d'André Malraux certes mais aussi et surtout conditionnée par les « écrits » d'Henry Corbin sur l'« iranisme ».

En effet, dans un de ses articles phares, « Esprit d'un monde sans esprit »[4], Michel Foucault décrit ainsi « l'âme du soulèvement » : « il nous faut changer, bien sûr, de

---

[1] Michel Foucault, « Ils ont dit de Malraux », *Dits et écrits*, op.cit., texte n° 183, p.108.
[2] Emission Talmudique du 29 janvier 2017.
[3] Henri Tincq, « La grande crispation religieuse », *Slate*, 16.02.2015.
[4] Michel Foucault, « Esprit d'un monde sans esprit », in *Dits et écrits*, *op.cit.*, texte n° 259 (pp. 743-775).

régime et nous débarrasser de cet homme, il nous faut changer ce personnel corrompu, il nous faut changer tout le pays, l'organisation politique, le système économique, la politique étrangère. Mais surtout, il nous faut changer nous-mêmes. Il faut que notre manière d'être, notre rapport aux autres, aux choses, à l'éternité, à Dieu, etc., soient complètement changés, et il n'y aura de révolution réelle qu'à la condition de ce changement radical dans notre expérience. Je crois –ajoute Foucault– que c'est là où l'islam a joué son rôle ». C'est là où l'expérience religieuse iranienne semble avoir agi en profondeur sur Foucault pour finir par devenir, osons-le, une expérience religieuse foucaldienne.

Nous sommes là pour reprendre l'idée de Cavagnis[1] au « point de fusion entre religion et politique » qui devient alors « le lieu de la conduite subjective de l'éthique ». Mais selon nous, le recours à l'expression de « spiritualité politique » qu'il faut mettre au crédit de Michel Foucault, a été fortement conditionné par sa lecture des écrits d'Henry Corbin sur les aspects spirituels et philosophiques de l'islam iranien. En effet, la lecture foucaldienne du soulèvement iranien a été idéalisée par l'influence qu'a eue sur lui son immersion dans la pensée d'Henry Corbin de sorte que, submergé dans l'« iranisme » sublimé, il cherche à annihiler « la politique », le versant négatif pour d'aucuns et réaliste pour d'autres de la situation que les islamologues à la suite de Bruno Etienne finiront par dénommer « islamisme ». Dans sa lettre ouverte à Mehdi Bazargan, Premier ministre du gouvernement de Khomeyni, Foucault écrit : « Dans l'expression « gouvernement islamique »,

---

[1] Julien Cavagnis, « Michel Foucault et le soulèvement iranien de 1978 - Retour sur la notion de « spiritualité politique » in Foucault, une politique de la vérité, *Cahier Philosophique*, n° 130, 3e trimestre 2012, pp 51-71.

pourquoi jeter d'emblée la suspicion sur l'adjectif « islamique » ? Le mot « gouvernement » suffit, à lui seul, à éveiller la vigilance » mais là où il faut faire pour partie crédit à Foucault c'est que, avant la mise en place d'un gouvernement islamique, il ne peut s'empêcher de mettre en exergue sa « méfiance à l'égard du légalisme » tout en avouant sa « foi dans la créativité de l'islam »[1].

Fortement influencé par la pensée d'Henry Corbin, Foucault pense en termes de « liberté créatrice », concept clef de la pensée iranienne selon Corbin. Il est nécessaire de faire état ici de la définition qu'Henry Corbin[2] donne du terme iranisme pour mieux saisir l'état d'esprit de Foucault. Selon ce grand érudit, « l'*iranisme* est le principe spirituel créateur, religion de la liberté morale ». Il a pour pendant « le *kouschisme* (Koukh est le nom biblique de l'Ethiopie) » qui est « la religion de la nécessité physique aussi bien que de la nécessité logique. Le *kouschisme* ne connaît que l'enchaînement nécessaire du raisonnement logique ; en ce sens, il trouve son expression la plus achevée dans le système de Hegel. L'*iranisme,* la pensée iranienne, « se fonde sur la tradition et ne se restaure point par une action purement logique, car le concept de liberté créatrice ne s'enferme pas dans les formules et ne s'en déduit pas. Elle peut être discernée seulement par une intuition supérieure, dépassant les étroites limites du raisonnement, ou par le travail des siècles, ayant parcouru tous les degrés de la négation ».

Pour résumer la parenthèse foucaldienne de la fin des années 70, disons que le concept d'islamisme, cet « objet

---

[1] Michel Foucault, « A quoi rêvent les Iraniens ? », in *Dits et écrits*, op.cit., texte n° 245, p. 691.
[2] Henry Corbin, *En Islam iranien – Aspects spirituels et philosophiques*, Paris, Gallimard, 1971, tome II : « Sohrawardî et les platoniciens de Perse », pp 335-346.

insaisissable » de Bruno Etienne, devait répondre à deux des fonctions essentielles qui incombent à la religion. Il y a, d'une part, le « prophétisme politico-religieux » qui permet de canaliser les espoirs en une société idéale et, d'autre part, l'iranisme en tant que religion de la liberté morale et éthique, tremplin à la quête du changement de soi. Les islamologues ne retiendront que la première mais Foucault, judicieusement inspiré par Henry Corbin, misera sur les deux. C'est pourquoi sa pensée, à l'instar de celle du visionnaire et éminent érudit qu'a été Henry Corbin, a pu et devrait perdurer à sa mort. Pour clore la parenthèse, il convient de rappeler qu'avec l'idée du « changement de soi », nous sommes en présence de l'idée du grand djihad qui est le djihad sur soi. Le grand djihad, c'est cet effort qui permet d'acquérir un pouvoir sur soi-même. Une personne qui a acquis un tel pouvoir est invincible.

*

Le second propos essentiel du présent chapitre se situe au plan du langage et consiste dans la remise en cause d'une réalité factuelle qui est l'acte de dévoiement du terme islamisme. Si le dévoiement de l'islam est le fait de fondamentalistes musulmans (intellectuels, activistes, religieux, etc.), le dévoiement du terme islamisme est le fait des islamologues français. Pour ce qui est du dévoiement de l'islam, entendons-nous bien : si le dévoiement de l'islam est le fait de fondamentalistes musulmans (intellectuels, activistes, religieux, etc.), les causes du dévoiement ne sont pas toutes endogènes au monde musulman. L'ingérence occidentale y est pour quelque chose (voir chapitre précédent et suivants). Mais pour ce qui est du dévoiement du terme islamisme (objet du présent chapitre), les acteurs locaux et/ou internes au monde musulman n'y sont pour rien. Nous parlons de dévoiement parce que nous sommes en présence d'un changement de sens du mot. Contrairement à la coutume, ce mot a cessé de se référer à l'islam en tant que religion pour vouloir désigner un « objet insaisissable ».

Avec le terme « islamisme », nous avons, d'ores et déjà, affaire à un mot qui appartient au langage commun mais il s'agit en réalité d'un mot intelligible pour personne, pas même pour Bruno Etienne, puisqu'il sert à désigner un « objet insaisissable ». Insaisissable depuis son dévoiement, ce terme demeure insaisissable aujourd'hui ; à croire que le mot « islamisme » a communément acquis le statut de mot à « usage privé » selon les catégories de Wittgenstein[1], chacun l'utilisant à sa guise dans le langage commun. Il n'est pas exagéré de dire que le mot change de sens au gré du locuteur, du contexte et des circonstances. Nous avons fourni ci-dessus quelques définitions émanant d'islamologues mais ils ne sont pas seuls à faire un usage « moderne » du terme. Dans un article paru dans le quotidien Le Monde, l'écrivain Kamel Daoud[2] écrit : « L'islamisme est un attentat contre le désir ». Pourquoi pas après tout, au point où nous en sommes ! La question consiste maintenant à savoir pourquoi Bruno Etienne a décidé de dévoyer le terme « islamisme ». Pourquoi a-t-il choisi ce terme plutôt qu'un autre ? Il s'agit ensuite de savoir pourquoi il a été suivi.

Commençons par la première question, pourquoi dévoyer un terme au lieu de créer un nouveau mot ? Le problème de Bruno Etienne consiste à « intégrer un objet nouveau au système d'intelligence en usage », encore faut-

---

[1] Linsky Leonard, « Wittgenstein, le langage et quelques problèmes de philosophie », in *Langages*, 1e année, n°2, 1966. Logique et linguistique. pp. 85-95.
[2] Kamel Daoud : « Cologne, lieu de fantasmes », le quotidien *Le Monde* du 31 janvier 2016.

il que le corps étranger greffé soit compatible pour qu'il n'y ait pas de rejet. Si l'on se sert d'un terme d'usage courant en en changeant le sens, on ne diminue pas le risque de rejet à proprement parler, mais on facilite l'assimilation de l'objet nouveau par le système existant en écartant le risque d'un rejet immédiat du fait de l'intrusion d'un corps étranger. L'on comprend mieux ainsi l'avantage qu'il avait à dévoyer un terme d'usage courant pour exprimer une réalité autre.

Le problème avec le nouveau concept, qui procède par dévoiement du terme islamisme, réside dans la nature de la réponse que B. Etienne fournit au problème du rapport du sujet à une réalité estimée inaccessible avec les concepts appartenant au monde de la pensée du sujet. Nous sommes supposés être en présence d'une réalité supposée insaisissable du fait d'une altérité absolue, irréductible, ce qui n'est pas le cas puisque les trois religions monothéistes ont des fondements communs. Le problème est que, avec l'« objet insaisissable », nous n'avons plus affaire à une « religion », en l'occurrence l'islam, mais à un « événement » : une réalité nouvelle qui est l'instrumentalisation de l'islam par des musulmans (activistes, religieux, etc.). De sorte que ce nouveau concept a créé des confusions dans le champ sémantique : si, avant le dévoiement du terme islamisme, les termes islam et islamisme signifiaient la même chose, maintenant, ces deux termes sont considérés comme synonymes, à ceci près que c'est la nouvelle définition d'islamisme qui s'impose.

Nous en arrivons maintenant à la seconde question qui consiste à savoir pourquoi le dévoiement du terme islamisme n'a suscité aucun émoi dans les pays musulmans. Cette définition de l'islamisme, qui permet de consacrer l'« islam politique » en lui dédiant une des deux dénominations réservées à la religion musulmane, n'est pas sans déplaire à certains pays musulmans et plus spécifiquement aux régimes islamiques. Pour les dirigeants de ces pays, les termes « islam » et « islamisme » demeurent à l'identique mais cette définition de l'islamisme permet de légitimer aux yeux des Occidentaux leur lecture et conception de l'islam qui ne peut être qu'un islam politique dans sa version totalisante pour ne pas dire totalitaire. Pour les pays musulmans où les mouvements islamiques commencent à cette époque à poser des problèmes, comme l'Egypte et d'autres pays du Moyen-Orient, lorsqu'il n'est pas possible de composer avec les militants et/ou d'instrumentaliser le phénomène, il vaut mieux ne pas faire trop de bruit pour ne pas envenimer la situation.

Mais qu'en est-il alors des pays musulmans, alignés ou non alignés, mais en tout état de cause plus ou moins bien engagés dans la voie de la sécularisation, sinon de la laïcisation, de la société à la fin des années 70 ? Qu'en est-il de la population occidentale de confession musulmane, pourquoi cette frange de la population occidentale n'a-t-elle pas émis d'objection au dévoiement du terme « islamisme » ? Qu'advient-il des réfugiés politiques musulmans hostiles aux régimes théocratiques ? Pourquoi ces différents groupes, qui sont presque unanimement en faveur de la séparation des pouvoirs religieux et politique, n'ont-ils pas réagi contre cette définition de l'islamisme qui équivaut à la consécration de l'islam politique ? Il est vrai que les réfugiés politiques sont astreints à l'obligation de réserve et occupés à trouver les moyens de gagner leur vie. Mais est-ce là la véritable raison d'absence de réaction de la part des intellectuels, exilés inclus ? Admettons que oui, mais, dans ce cas, qu'advient-il alors de la classe des intellectuels de confession musulmane de nationalité européenne ou américaine ? Pourquoi ceux-ci ne se sont-ils pas opposés au dévoiement du terme islamisme et continuent-ils à garder le silence ?

Entendons-nous bien, les exilés iraniens de confession musulmane n'ont pas tous été frappés de mutisme mais la bataille s'est déroulée à un niveau, qui à première vue, paraissait refléter l'essentiel, à savoir le champ de la politique. Des voix se sont élevées pour se faire entendre mais lorsque la contestation n'a pas été étouffée, elle a été circonscrite dans le champ du politique, les questions de sémantique étant le cadet des soucis des protagonistes. Et, dans les rudes combats souvent perdus d'avance, les dissidents (l'élite ayant gardé un brin de lucidité) étaient le plus souvent accusés de s'être occidentalisés : certains islamologues vont jusqu'à utiliser le qualificatif de « renégat ». Nous sommes en présence d'une actualisation à la lumière du retour du fait religieux, de la rhétorique marxiste de « la bourgeoisie compradore »[1].

Ceci dit, pour bien comprendre ce qui s'est passé réellement, il faut se mettre dans le contexte de l'époque et examiner la définition susmentionnée de l'islamisme à l'aune du paradigme en vogue à ce moment. Selon cette définition, l'islam politique est assimilé à une idéologie. Or, à l'époque, le terme idéologie n'a pas bonne presse. En effet, comme Maxime Rodinson[2] le souligne dans son livre *Islam et capitalisme* : « Ce fut un grand malheur pour l'activité scientifique (on l'appelait alors philosophie) d'avoir été longtemps la servante de la théologie. Le malheur ne serait pas moindre d'en faire maintenant la servante de l'idéologie politique qui a succédé à la

---

[1] Voir, entre autres, les répliques de Michel Foucault, de Bruno Etienne, de François Burgat, etc. Pour le qualificatif de « renégat », cf. Bruno Etienne, « Les vagues islamistes ». op.cit., p.915.
[2] *Ibid.*, p.9.

théologie ». Les essais dans ce sens (auxquels j'ai participé) ont mal tourné pour la science et même pour la politique. » Sans entrer dans le détail des débats, il est utile ici de rappeler que, pour Ludwig Feuerbach, auteur de « *L'essence du christianisme* », « l'idéologie est l'ensemble plus ou moins systématique, plus ou moins cohérent des représentations, des « valeurs » et des principes que sécrète une société globale pour puiser dans cet imaginaire l'apaisement et le réconfort rendus nécessaires par les contradictions, les obscurités et les déchirements de sa réalité »[1].

C'est en ce sens que Marx déclarera que la religion est « l'opium du peuple » : « expression de la vraie détresse, protection contre cette vraie détresse ». Mais, après avoir mis à profit la critique de l'idéalisme de Hegel au profit du matérialisme de Feuerbach qu'il considère comme « le seul à avoir constitué un progrès »[2], Marx critique à son tour le pseudo-matérialisme de Feuerbach[3] qui, selon lui, a le grand défaut de continuer à « croire, dans le monde existant, au règne de la religion, des concepts et de l'Universel »[4] pour bâtir un matérialisme « historique » de son propre cru. C'est ainsi que Marx sera amené à mettre en exergue l'opposition « idéologie et matérialisme

---

[1] François Châtelet, *Idéologie* in André Akoun (dir.), « La philosophie », La bibliothèque du CEPL, 1977.
[2] Karl Marx et Friedrich Engels, *L'idéologie allemande*, Paris, Editions sociales, 1968, p. 20.
[3] « Dans la mesure où il est matérialiste, Feuerbach ne fait jamais intervenir l'histoire, et dans la mesure où il fait entrer l'histoire en ligne de compte, il n'est pas matérialiste. Chez lui, histoire et matérialisme sont complètement séparés », Marx, *Idéologie allemande*, *op. cit.*, p.72.
[4] Selon Marx, « La seule différence est que les uns combattent comme une usurpation cette domination que les autres célèbrent comme légitime », *Idéologie allemande*, *op. cit.*, p.22. Il convient de préciser que les uns sont ceux que Marx qualifie de jeunes-hégéliens et les autres sont les vieux-hégéliens.

historique », laquelle opposition sera perçue et interprétée comme l'opposition de l'idéologie et de la science.

Si, dans les années 70, le terme idéologie est « surchargé de significations », comme le précise à l'époque François Châtelet[1], en fait, c'est la connotation péjorative héritée de Marx qui semble l'emporter, de sorte que ce philosophe qui a pignon sur rue, après avoir tenté de définir le concept, déclare : « reste qu'actuellement, le mot *idéologie* est devenu une sorte de « réceptacle » où l'on entasse pêle-mêle les idées confuses de l'adversaire, étant entendu que l'on possède soi-même l'idéologie guidée par la science. Il y a de ce fait –précise-t-il– à se demander s'il ne serait pas raisonnable d'en bannir l'usage, et de se demander, chaque fois que ce terme survient, de quelle réalité on veut effectivement parler »[2].

En effet, depuis l'avènement de l'impitoyable idéologie marxiste, les sociétés occidentales vivent une crise des idéologies. C'est contre l'idéologie capitaliste que les idéologies communiste d'abord et maoïste ensuite ont vu le jour. Le concept d'idéologie ayant une connotation marxiste trop marquée, dans le titre de l'ouvrage de Max Weber « *Esprit protestant et Esprit du capitalisme,* le terme « idéologie » sera remplacé par « Esprit »[3]. Si les deux idéologies dominantes du XXe siècle, à savoir le communisme et le maoïsme ont fait leur temps, aujourd'hui on commence à dénigrer les droits de l'homme en les taxant d'idéologie des « droits-de-l'hommisme ». Mais ironie du sort, le capitalisme continue à faire ses ravages envers et contre tout et ce n'en déplaise à l'anthropocène. En effet,

---

[1] François Châtelet (dir.), *Histoire des idéologies,* (3 tomes), Hachette, Paris, 1978, tome 1, p.10.
[2] François Châtelet, *Idéologie,* op.cit.
[3] Christian Godin, *La totalité,* Paris, Ed. Champ Vallon, 1998, tome 2, p.448.

comme Fernand Braudel l'a si bien illustré : le capitalisme, on le chasse par la porte, il rentre par la fenêtre. Il en va de même pour la démocratie. Maintenant, s'il est difficile de savoir par quel miracle des intellectuels du calibre de Michel Foucault ou de Bruno Etienne ont pu croire que l'« islamisme » comme nouvelle idéologie aurait pu servir d'alternative au capitalisme, il est en revanche plus aisé de comprendre comment l'« islamisme » peut continuer à servir de tremplin au vide idéologique actuel.

Ceci dit, on comprend mieux ainsi pourquoi, en qualifiant l'islam politique d'idéologie, on ne le gratifiait pas, mais, qui plus est, on abondait dans le sens d'une critique des régimes politiques islamiques où la religion fait office d'idéologie lorsqu'elle n'est pas instrumentalisée à des fins politiques. Si les musulmans non-pratiquants et les agnostiques pour ne pas dire athées[1] étaient et demeurent par définition hostiles à tout régime islamique, qu'il soit monarchique ou républicain, pour les pratiquants vivant dans un environnement sécularisé, la religion doit au mieux se cantonner à la vie privée et au pire guider l'action sociale des individus. Les premiers, partisans de l'islam spirituel (religion de l'éthique du salut), se situent plus dans le registre de la « foi », tandis que les seconds, partisans du salafisme dit quiétiste, se situent davantage dans le registre de la « loi »[2]. Mais, dans tous les cas, l'islam ne doit et/ou ne peut pas être instrumentalisé à des fins de prise du pouvoir temporel.

---

[1] L'apostasie peut entraîner des conséquences assez graves pour la personne.
[2] Aujourd'hui, trois courants se réclament du salafisme : quiétiste, politique et djihadiste. Cf. Samir Amghar, « Le salafisme en Europe. La mouvance polymorphe d'une radicalisation », *Politique étrangère* 2006/1 (Printemps), p. 65-78.

En effet, dans un article intitulé « L'idéologie de l'Islam » publié dans les années 70, Mohammed-Allal Sinaceur précise que : « Islam signifie « soumission totale à Dieu » ; nul pouvoir ne peut donc se légitimer que par la foi en lui. Cette foi, dont les docteurs en la loi aiment parer leur science, explique le *primat de l'éthique sur la politique*, la priorité du dessein communautaire et de l'intention qui l'anime sur les formes de son incarnation et sur son organisation, et justifie l'absence de théorie du pouvoir en tant que tel, puisque le vrai pouvoir est à Dieu »[1].

Comme on le constate, nous sommes en présence de cette fameuse soumission totale mais volontaire à Dieu, qui a tout aussi brillamment/bruyamment été fustigée par Houellebecq, qu'elle a fasciné Foucault de par son pouvoir de négation du pouvoir politique ainsi que nous avons pu le constater plus haut.

Pour ceux qui sont au fait de l'islam spirituel, il y a lieu d'ouvrir une parenthèse pour signaler que la « soumission volontaire et totale à Dieu » consacre en fait le « croire libre », c'est-à-dire la liberté d'interprétation : le travail sur les textes comme support amène à une transformation intérieure. En effet, si, théoriquement, du fait du principe d'absence d'intermédiation, nul n'est habilité à imposer sa vision de la vérité, en réalité, c'est-à-dire pratiquement, nul n'est en mesure d'imposer « la » vérité à qui pense l'incarner pour la très bonne raison que celui-ci ne pense pas, il « est » la vérité. Ceci est magistralement illustré dans le « ana al hagh » de Hallaj : « je suis la Vérité (Dieu) ». Ce n'est pas la pensée de Hallaj qui a été transformée par sa quête spirituelle mais son être même. Phénomène à appréhender en termes de « naissance à soi-même ».

---

[1] Mohammed-Allal Sinaceur, *L'idéologie de l'Islam* in François Châtelet (dir.), *Histoire des idéologies*, op.cit., tome 1, p.263.

Ceci dit, pour rester fidèle à l'esprit et à la lettre de cette religion, il convient toutefois de souligner que le mot islam est polysémique. Selon l'islamologue Tayeb Chouiref, invité de l'émission « Questions d'islam » de Ghaleb Bencheik sur France Culture[1], le terme islam dans son sens premier signifie « entrer dans la paix, faire la paix, s'en remettre à une puissance supérieure ». Comme il a déjà été signalé, c'est via le grand djihad qui permet la maîtrise de son ego qu'une pacification intérieure se produit.

Le problème maintenant est que, assimilé au salafisme au moment de son dévoiement, l'islamisme a été défini comme « une idéologie qui veut faire de l'islam et du respect intégral de la charia un modèle politique alternatif à la démocratie ». Aujourd'hui, le terme islamisme est utilisé pour définir la radicalité de mouvements ou d'organisations tels que Daech, de sorte que, *in fine* le terme islamisme sert maintenant à définir les actes de violence et de terrorisme perpétrés au nom de l'islam et que d'aucuns qualifient plus simplement de terrorisme djihadiste[2]. C'est ainsi que, dans le glossaire figurant à la fin de l'ouvrage intitulé « *Qui est Daech ?* » publié au lendemain de la tragédie du 13 novembre 2015 qui a endeuillé la France, il est spécifié que le terme « islamisme »... « a aujourd'hui tendance à qualifier l'islam fondamentaliste, traditionaliste et prosélyte, adepte de la violence »[3]. Aujourd'hui, si l'on veut éviter

---

[1] Ghaleb Bencheikh, « Considérations sémantiques à propos du vocable islam », émission *Cultures d'Islam*, France Culture, 08.05.2016.
[2] Dans l'émission Répliques du 11 février 2017 intitulée « Les nouveaux défis de la France », Alain Finkielkraut fait usage du terme « islamisme » pour se référer aux actes de violence terroristes tandis que Jean-Pierre Chevènement essaie en vain de rectifier le tir en invitant à bien nommer les choses.
[3] Eric Fottorino (dir.), *Qui est Daech ? Comprendre le nouveau terrorisme*, Paris, Ed. Le1, 2015, cf. Glossaire, pp 92-93.

d'envenimer la situation qui pourrait à terme conduire à un état de terrorisme permanent, il y aurait lieu de faire appel à l'éthique de responsabilité afin que, dans un souci de pacification des esprits, tout un chacun prenne conscience au quotidien du pouvoir des mots.

Les concepts sont des outils nécessaires pour autant qu'ils participent à une meilleure perception et/ou plus grande intelligibilité de la réalité. Comme Olivier Roy le souligne à juste titre, le concept de post-islamisme est incompréhensible sans le concept d'islamisme. Mais comme le post-islamisme est fondé sur un constat d'échec et non de dépassement de l'islamisme, l'islamisme étant compris comme un état idéal à atteindre et non comme un moment révolutionnaire, on est en droit de s'interroger sur la pertinence même du concept du fait même que l'islamisme, en tant qu'un état idéal[1], se présente comme une aporie.

En tout état de cause, ce concept n'est d'aucune utilité pour les chercheurs qui, par conviction ou pour des raisons plus ou moins obscures, que par souci de civilité nous qualifierons de simple « parti pris », défendent les thèses du courant de pensée essentialiste. Pour ces derniers, islam et islamisme sont une seule et même chose. Pour les personnes qui adhèrent à ce courant de pensée, sans toujours l'avouer publiquement, le wahhabisme et/ou le salafisme sont la réalité de l'islam. Ce sont en général ces mêmes personnes, qu'elles appartiennent au monde intellectuel, politique ou médiatique, qui adhèrent aux thèses du « clash » des civilisations de Samuel Huntington. Ces personnes déclarent qu'il faut regarder la réalité en face et ils ont raison. Le problème est qu'il faut s'assurer qu'en

---

[1] Y compris l'idée de révolution permanente qui a tant fasciné Foucault au point de l'amener à faire corps avec la révolution iranienne.

regardant la réalité : l'on perçoit juste et l'on nomme juste. Il s'agit, en d'autres termes, de bien nommer les choses pour ne pas ajouter au malheur des hommes, comme Camus le préconisait.

Trou noir de la géopolitique mondiale, l'islamisme a cessé d'être un concept opératoire pour autant qu'il l'ait jamais été. Mais, au-delà des questions d'ordre géopolitique, le problème majeur avec le concept d'islamisme est qu'il produit à longueur de journée des « clashs » de « signifiés » qui deviennent des chocs d'identités entre musulmans et non-musulmans. Lorsque l'émetteur d'information – journaliste, intellectuel, politique, ou artiste – critique des comportements, actes ou événements éminemment critiquables, voire même blâmables, en faisant usage du terme « islamisme », le musulman a la désagréable impression d'être visé et fait ainsi une bonne provision de ressentiments. Le non-musulman pour sa part croit entendre le son d'une sirène d'alarme qui suscite en lui l'inquiétante impression de vivre à proximité de dangereux individus qui n'ont rien à faire "chez lui".

De sorte qu'il y a lieu, en paraphrasant François Châtelet, de se demander s'il ne serait pas raisonnable de bannir l'usage du terme « islamisme » et de se demander, chaque fois qu'on veut en faire usage, de quelle réalité on veut parler pour faire usage du terme idoine ou lorsqu'on entend ce terme, de quelle réalité son émetteur veut effectivement parler pour percevoir la réalité à laquelle elle est censée se référer et non point celle à laquelle on a instinctivement pensé.

CHAPITRE 6

De la peur du djihadisme à la haine de l'islam : construction de la figure du nouvel ennemi

Le mythe de la menace islamique fait place aujourd'hui à celui de la menace extrême du terrorisme islamiste. C'est ainsi que, non contents d'avoir trop longtemps posé la question de savoir si l'islam et la démocratie sont compatibles, d'aucuns en arrivent aujourd'hui à se poser la question de savoir si la démocratie et le terrorisme sont compatibles. Comme poser la bonne question, c'est bien souvent contribuer pour partie tout au moins à la résolution du problème, au vu des apories, à bien y réfléchir, on est en droit de se demander si l'on n'est pas là en présence de problèmes inhérents à la nature des questions posées.

Pour les islamologues concepteurs du nouveau concept d'«islamisme», les termes «islam» et «islamisme» désignent deux réalités distinctes : l'islam est une religion et l'islamisme se réfère au mieux à l'islam politique que d'aucuns qualifient d'islam radical. Comme on ne se débarrasse pas si facilement des automatismes mentaux générés par une définition parce qu'on a décrété sa désuétude, la vieille définition de l'islamisme reste présente dans les esprits. De sorte que, les tragiques évènements aidant, à la longue, les deux termes finissent par se confondre pour désigner au mieux deux aspects d'une même réalité : le « bon islam » et le « mauvais islam ». Et, comme Nietzsche l'a si bien dit, « ceux qui se nomment les bons et les justes, il ne leur manquait que le pouvoir pour devenir des pharisiens », on sait ainsi à quoi s'en tenir.

Le nouveau concept d'islamisme a été prolifique. Une simple enquête généalogique révèle comment il a donné naissance au terme islamiste, banalisé son usage et a contribué au fil du temps à réactualiser le terme d'islamophobie qui était tombé en désuétude. Ironie du sort, ce sont ceux qui tiennent le plus aux termes islamisme et islamiste qui sont synonymes pour eux de terrorisme et terroriste, qui s'insurgent contre le terme islamophobie. Conséquence logique du cercle vicieux de la peur, de la haine et du mépris de l'autre, le terme islamophobie ne serait qu'une pure illusion ? Apparemment pas puisque, d'après Jacques Julliard[1], « le peuple et les élites n'ont plus les mêmes valeurs ni les mêmes priorités. Pour le peuple, le danger principal est le terrorisme islamiste. Pour les élites, c'est le fascisme d'extrême droite ». Est-ce à croire que l'islamophobie serait limitée au peuple ? Hormis l'extrême droite, l'élite, tout d'un bloc guidée par les valeurs républicaines issues des Lumières, serait objective, neutre, maître de ses affects, exempte de toute haine à l'égard des musulmans, exempte de tout préjugé à l'égard de l'islam, libre de toute mauvaise foi, et elle n'aurait en aucune façon contribué à la montée de l'islamophobie.

Fidèle à sa propre devise « qui ne prend pas de risque est mis à l'écart »[2], Pascal Bruckner persiste et signe : « Il est des mots qui contribuent à infecter la langue, à en obscurcir le sens, «islamophobie» fait partie de ces termes à bannir

---

[1] Jacques Julliard, « Le peuple et les élites : pourquoi ce divorce ? », le journal *Marianne*, samedi 26 décembre 2015.
[2] Interviewé par Philippe Plassart, Pascal Bruckner déclare : « A un moment, il faut savoir taper. Je donne des coups et, en retour, il m'arrive de me faire insulter. Celui qui ne prend pas de risques est vite laissé à l'écart » voir « Pascal Bruckner : L'impératif de bien nommer les choses », *Le nouvel économiste*, le 5 février 2015.

d'urgence du vocabulaire »[1]. Bruckner, comme tout un groupe identifiable d'intellectuels, d'écrivains, de journalistes et de leaders politiques de pensée néo-conservatrice, n'a pas de problème avec le dévoiement du terme « islamisme », ce qu'il récuse, en revanche, c'est le terme islamophobie sans se rendre compte du lien de causalité existant entre les deux phénomènes que définissent ces termes. Le binôme Pascal Bruckner et Caroline Fourest manifeste un même mépris de l'islam, affiché plus par intérêt que par conviction. Si le premier a un passif à se faire pardonner, il tient aussi et surtout à demeurer sur le devant de la scène. Et, comme le mépris n'empêche pas le déni, en déclarant « l'islamophobie n'existe pas »[2], il dénie la réalité non par compassion vis-à-vis des victimes mais par mépris vis-à-vis de l'islam.

Reprenant les dires de Caroline Fourest dont la réputation en matière de contrevérités et de manipulation de l'information n'est plus à faire, selon Pascal Bruckner[3], « Forgé par les intégristes iraniens à la fin des années 70 pour contrer les féministes américaines, le terme «islamophobie», calqué sur celui de xénophobie, a pour but de faire de l'islam un objet intouchable sous peine d'être accusé de racisme ». Si les intégristes iraniens ont contribué à la montée de l'islamophobie, ils n'ont en revanche pas le privilège d'avoir inventé le terme islamophobie. Contrairement aux dires de Caroline Fourest repris par Pascal Bruckner et d'autres membres de l'élite, dirigeants politiques inclus, ce ne sont pas des mollahs iraniens mais

---

[1] Pascal Bruckner, « L'invention de l'« islamophobie » », Tribune de *Libération* du 23 novembre 2010, déclaration reprise dans son interview de 2015 susmentionnée.
[2] Pascal Bruckner, « Islamophobie ça n'existe pas », *Le Causeur*, 29 octobre 2012.
[3] Pascal Bruckner, « L'invention de l'« islamophobie » », *op.cit.*

un groupe d'«administrateurs-ethnologues»[1] spécialisés dans les études de l'islam ouest-africain qui ont en 1910 utilisé pour la première fois le terme islamophobie.

A la question « Avez-vous peur de l'Islam ? », Pascal Bruckner répond : « Absolument pas »[2]. S'il est vrai qu'à l'origine l'islamophobie savante était davantage motivée par la haine et/ou le mépris que par la peur, depuis les atrocités commises par les terroristes affiliés à Daech, la donne a changé. Les élites ne sont plus à l'abri de la peur, de sorte que, plus par souci de sécurité que par égard vis-à-vis des fondamentalistes, ils doivent, eux aussi, comme tout un chacun, « se surveiller pour ne pas offenser les islamistes » pour reprendre les termes utilisés par l'écrivain Boualem Sansal[3]. Si, à quelques nuances près, Boualem Sansal a raison lorsqu'il déclare que « Les musulmans ont tout perdu, ils ont perdu leurs pays "colonisés" par les dictateurs et/ou par les islamistes, et ils ont perdu leur religion, que l'islamisme a phagocytée et dont les dictateurs ont fait la religion d'État, autrement dit leur religion puisque l'État c'est eux. Sans pays et sans leur religion, il se pose à eux un sérieux problème d'identité et de dignité ». Il oublie, au passage, de faire état de la part de responsabilité des grandes puissances dans l'état actuel du monde.

Comme il ressort de la lecture de l'histoire, aucune ingérence n'est parfaitement innocente. Il faut effectivement, en limitant autant que faire se peut

---

[1] Maurice Delafosse (1870-1926), Paul Marty (1882-1938) et Alain Quellien. Voir citations et références in Marwan Mohammed et Abdellali Hajjat, *Islamophobie : comment les élites françaises fabriquent le problème musulman*, Paris, La Découverte, 2013.
[2] Pascal Bruckner, « Islamophobie ça n'existe pas », op.cit.
[3] Alexandre Sulzer, « Boualem Sansal : "A ce point, la passivité des musulmans est mortelle" », *L'express.fr*, 13 janvier 2015.

l'impératif "missionnaire", permettre aux populations non occidentales de confession musulmane d'exercer leur droit à l'autodétermination, et aux Occidentaux de confession musulmane de se sentir chez eux là où ils sont et tels qu'ils sont, c'est-à-dire des citoyens pluriculturels et libres de l'être. En effet, comme Amartya Sen[1] le souligne à très juste titre, aujourd'hui plus que jamais, « notre identité étant nécessairement plurielle, il nous appartient de décider de l'importance relative de ces différentes associations et affiliations dans un contexte donné ».

Pour ce qui est de la composante religieuse des identités qui nous intéresse plus spécifiquement ici, il s'agit de refuser que celle-ci soit phagocytée par les « fanatiques intégristes musulmans » et par les islamophobes qui font le jeu de ces derniers en les qualifiant d'« islamistes ». Il faut oser nommer mais « nommer juste » pour voir juste, penser juste et agir juste. Il est difficile, sinon quasi impossible, de pouvoir en toute circonstance faire preuve d'une pensée, d'une parole et d'une action justes, mais chacun doit pouvoir se sentir « libre » de « vouloir » pouvoir et/ou faire l'effort et aussi et surtout, il est à la portée de tout un chacun de mettre de côté la mauvaise foi qui tue aux sens propre et figuré du terme.

« La critique de l'islam n'est pas une agression contre lui ou contre les musulmans, l'islam ne peut pas, lui seul, être hors du champ de la critique », comme le souligne à juste titre Boualem Sansal à l'instar de bien d'autres musulmans. Mais déclarer, comme Bernard Kouchner le fait au sujet de l'islam, qu'« il faut respecter les religions lorsqu'elles sont respectables. Les religions ne sont pas toutes respectables », ce n'est pas critiquer l'islam mais s'ériger en « Juge suprême ». Adopter une posture d'affrontement, c'est

---

[1] Amartya Sen, *Identité et violence – l'illusion du destin*, Paris, Odile Jacob, 2007, pp. 11-12.

refuser de débattre. Si tout un chacun est en droit de débattre sur les systèmes de croyance et de les critiquer, nul n'est en droit de s'arroger le rôle de « juge suprême » en la matière, car aucun individu ne peut prétendre à la fois connaître et savoir tout et être en mesure d'être parfaitement objectif et neutre à tous points de vue. D'après le sociologue Raymond Boudon qui a longuement travaillé sur la question de l'objectivité des valeurs et de la connaissance (le juste et le vrai) et a forgé une « Théorie générale de la rationalité », « le respect à l'égard de tous les systèmes de croyances est le seul principe compatible avec la valeur fondamentale du respect de l'autre ».

Pascal Bruckner s'afflige en déclarant solennellement : « L'islamisme nous a fait oublier la grandeur de l'islam d'autrefois ». Il ne semble pas être conscient que critiquer ce n'est pas mépriser. Pascal Bruckner, comme bien d'autres, est en droit de critiquer comme il est en droit de mépriser, mais il est, en tant qu'« intellectuel médiatique », tenu à distinguer entre mépris et critique. Si, d'après la définition en vogue, les termes islam et islamisme désignent deux réalités distinctes, on est en droit de s'interroger pour savoir en quoi l'islamisme peut retirer quoi que ce soit à l'islam. Qui plus est, comme il a été signalé dans le chapitre précédent, l'islamisme est pluriel, tout comme le mode de gouvernement des pays qualifiés de « musulmans » est divers. Les républiques islamiques sont de différentes formes, certaines sont de nature théocratique, d'autres pas. Il en va de même des monarchies : l'Arabie saoudite, la Jordanie, le Maroc sont tous trois des pays musulmans à régime monarchique mais la composante religieuse de l'identité est différemment vécue dans ces pays.

Dans son étude sur la conquête et l'exercice du pouvoir en islam suite à la place vide laissée par le Prophète,

Ahmad Hasnâvi[1] déclare : « il y a quelque chose d'irritant dans le problème de la détermination du pouvoir en islam. Depuis l'image à la fois théorique et littéraire du « despote oriental » jusqu'aux thèmes, plus élaborés, de la confusion du spirituel et du temporel, le terrain est encombré de clichés, de stéréotypes et de prénotions dont il est difficile de se défaire »... « les constructions idéologiques des musulmans eux-mêmes, sous forme d'interprétations du passé qui « louchent » vers le présent, sont encore plus difficiles à défaire... Ici comme ailleurs, précise-t-il, l'histoire est un enjeu politique, un certain type d'arme dans un certain type de stratégie. » A lui de conclure que « l'Islam, si l'on peut parler d'*un* Islam, n'apporte pas de restrictions positives sur la nature du pouvoir. Tout ce qui est déterminé... c'est une place vide (la succession du prophète)... Ensuite, il faut se demander si nous ne sommes pas atteints d'une certaine forme de myopie quand nous nous fixons sur les formes les plus évidentes du pouvoir. »

Il est de bon sens de lutter contre tout prosélytisme et toute dictature, quels qu'ils soient, mais ce qui importe avant tout est de savoir dans quelle sphère on se situe. Pour ce faire, il convient d'éviter de confondre réductionnisme et simplification. Le rejet du terme « islamisme » permet de clarifier la situation, de saisir la réalité dans sa complexité pour pouvoir agir en conséquence. Il faut libérer la parole mais parler libre a ses exigences. Il faut savoir parler vrai, c'est-à-dire parler « juste ». Parler vrai permettrait de démasquer tous ceux qui, sous couvert de liberté de la parole, stigmatisent plus d'un milliard et demi d'individus. Pour parler juste, il est bon d'éviter la confusion des genres. Par exemple, lorsqu'on veut dresser la liste des religions

---

[1] Ahmad Hasnâwi, *Islam : la conquête, le pouvoir* in François Châtelet (dir.), *Histoire des idéologies*, Paris, Hachette,, 1978, tome 1, pp. 310-349.

pratiquées dans le monde, on peut tenter de les nommer : le judaïsme, le christianisme, l'islam, le mazdéisme (ou zoroastrisme), le bouddhisme et l'hindouisme, le sikhisme, le shintoïsme, le taoïsme, etc.

Aujourd'hui il ne viendrait à l'idée de personne, fût-il le pire des islamophobes, de mentionner l'islamisme comme religion. Ce qui signifie qu'avec le terme « islamisme », on n'est plus dans la sphère du religieux mais au mieux de la politique. Comme on le constate, Pascal Bruckner, Brice Couturier et d'autres membres de l'élite néo-conservatrice ont raison de vouloir nommer librement les choses pour autant qu'il soient fidèles à la devise d'Albert Camus : « mal nommer les choses, c'est ajouter au malheur du monde ». Par exemple, réuni à la matinale de France Culture le 25 mai 2015, le trio : Pascal Bruckner, Brice Couturier et Caroline Fourest, s'insurge contre le terme islamophobie. Caroline Fourest aborde la question en termes de désinformation. Pascal Bruckner propose de bannir l'utilisation du mot. Et Brice couturier avoue ouvertement être islamophobe mais rectifie le tir en précisant que sa haine est essentiellement tournée vers les « islamistes ». Il propose donc de remplacer l'expression « islamophobie » par « islamistophobie » ; "manque de chance", le terme est estimé trop long par Bruckner.

Au-delà de la question de terminologie, la question de fond, qui se pose ici, consiste à savoir à quoi l'on se réfère exactement lorsqu'on déclare être contre l'islamisme. Lorsqu'on déclare haïr l'islamiste, qui hait-on exactement ? Non moins importante, se pose alors la question de savoir qui va répondre à ces questions ? Qui a voix au chapitre, comme dirait Noam Chomsky ? On constate que, lorsqu'on ose se poser les questions de fond, lorsqu'on a le courage de chercher à éviter l'amalgame, tout devient plus limpide. La réalité demeure complexe mais on est mieux à même de l'examiner et de la critiquer en évitant la confusion des genres.

Est-ce à dire que la compréhension des actes de violence perpétrés avec un label islamique doit se limiter à l'examen critique des causes socio-économiques et/ou des motivations d'ordre géopolitique ? Évidemment que non, mais il faut commencer à rendre à César ce qui est à César et à Dieu ce qui est à Dieu. L'islam doit pouvoir être critiqué, la critique interne de l'islam ne date pas d'aujourd'hui. A l'instar des autres religions, elle s'est faite parfois aux dépens de la vie de ceux qui s'y sont livrés. Les personnes de confession musulmane, qu'elles soient d'Orient ou d'Occident, n'ont pas attendu les orientalistes[1] pour s'adonner à la tâche de l'exégèse. L'islam ne doit pas seulement être critiqué, il doit être réformé dans sa pratique. La pratique de l'islam doit pouvoir évoluer en fonction des exigences de chaque époque et des besoins propres à chaque société. Plus d'un milliard et demi d'individus sur terre sont musulmans.

Pour être réaliste, stratégiquement parlant, une critique constructive de l'islam ne peut être une critique essentialiste ayant pour objet de stigmatiser l'islam comme religion violente et intolérante. La critique doit pouvoir ouvrir la porte à la réforme des pratiques. Mais, même dans une perspective belliciste, la critique, y compris celle effectuée dans le registre d'une approche essentialiste, devrait être faite selon les règles de l'art. Comme nous serons amenés à le voir, c'est parce que la liberté n'est pas la licence, que la "parole libérée" ne peut s'exempter de toute éthique de responsabilité.

Quelques exemples tirés de la littérature scientifique et/ou médiatique permettent d'étayer notre propos. Mars

---

[1] Islamologues de confession chrétienne, juive ou autres.

2015, *Philosophie magazine*, qui consacre un dossier à l'islam, ouvre le débat par la question suivante : « Nous abordons un point sensible : quelle est la nature des liens entre islam et islamisme ? Y a-t-il, entre eux, une différence de nature ou simplement de degré ? »[1]. Le langage manipulateur ne décrit pas la réalité du monde existant, il la crée. Le 17 janvier 2015, invité de l'émission « On n'est pas couché » de Laurent Ruquier[2], le philosophe Michel Onfray déclare : « La question que l'on devrait pouvoir poser, sans être assimilé à Marine Le Pen, c'est : est-ce qu'il y a une différence de nature entre un musulman pacifique et un terroriste, ou une différence de degré ? »[3]. Dans son livre « Une France antijuive ? Regards sur la nouvelle configuration judéophobe », Pierre-André Taguieff, qui se présente comme un intellectuel engagé, déclare : « Il est légitime de poser la question ainsi formulée par Michel Onfray : "est-ce qu'il y a une différence de nature entre un musulman pacifique et un terroriste, ou une différence de degré ?" il faut en débattre ».

Telle que formulée, cette question étant lourde de conséquences, nous allons nous soumettre à l'injonction de Taguieff et tenter d'en débattre. Comme on le constate, contrairement à *Philosophie magazine*, il n'est plus question, dans cette phrase interrogative, de religion (« islam » et « islamisme »), mais de personnes : « un musulman pacifiste et un terroriste ». Mettant à profit le contexte de la « parole libérée » après les tragiques attentats terroristes, Michel Onfray, riche de sa notoriété, fait usage dans cette phrase de sa maîtrise des règles et usages du code discursif : le récepteur de l'information (auditeur/lecteur) est invité à percevoir le réel non tel qu'il est mais tel que l'émetteur d'information veut lui faire croire qu'il est, tout en créant

---

[1] Philosophie Magazine, N° 87, mars 2015, p.40.
[2] Emission de la chaîne publique : France 2.
[3] Citation reprise le lundi 19 janvier 2015 dans *Marianne.fr*.

chez lui une illusion de liberté. L'auditeur / le lecteur est d'emblée mis devant un choix : différence de nature ou de degré. Il ne peut évidemment pas opter pour une différence de nature ; il s'agit donc d'une différence de degré.

Le plus intéressant dans cette phrase est que la première partie est posée comme « non problématique », de sorte que l'auditeur/lecteur est acculé à un seul choix : il doit choisir entre différence de « nature » ou de « degré ». En d'autres termes, comme on le constate, l'auditeur / le lecteur n'est pas habilité à se poser des questions quant au choix du binôme : « un musulman pacifique et un terroriste ». Nous sommes là au cœur de la stratégie discursive du langage manipulateur.

La première remarque, concernant ce binôme explosif, est relative au terme de « terroriste ». Celui-ci est délesté du qualificatif islamique, à croire que tous les terroristes sont musulmans ou vice versa. Cette remarque n'est pas aussi anodine qu'il n'y paraît et pour preuve, d'après un article publié le 24 février 2015 par l'hebdomadaire "*L'Obs*"[1], « les attentats meurtriers contre "Charlie-Hebdo" et l'Hyper Casher ont marqué les esprits. Ils ont aussi laissé place aux amalgames : tous les terroristes sont des musulmans radicaux, voire l'inverse ». Pour démêler le vrai du faux et relativiser les déclarations, l'hebdomadaire procède à une analyse statistique rigoureuse du terrorisme en France et en Europe qui montre, entre autres choses, que « seulement 1% des terroristes en France sont des islamistes ». Peine perdue, la rhétorique anti-islam a de beaux jours devant elle, et ce n'est pas l'arrivée au pouvoir de Donald Trump qui va assainir le climat.

---

[1] Article du journaliste Boris Manenti intitulé « Les terroristes sont tous islamistes ? Des chiffres pour relativiser ». Comme on le remarque dans ce titre, on ne parle pas de musulmans mais d'islamistes.

145

Ceci dit, les discours de haine sont plus ou moins réussis : moins un propos est attaquable en justice, plus il est réussi. A titre d'exemple, nous avons « les propos du président du Crif, Roger Cukierman, qui --d'après l'hebdomadaire *L'Obs*-- a estimé que "tous les terroristes qui ont commis des meurtres dans la période récente se réclamaient de l'islam" et ceux du patron d'un restaurant en Seine-Saint-Denis qui, refusant de servir deux femmes voilées, leur a déclaré : « tous les terroristes sont musulmans et tous les musulmans sont terroristes. Voilà, cette phrase veut tout dire. Analysez-la »[1]. Il est intéressant de noter que l'intéressé a pris soin de se présenter comme un « laïc » et comme un « raciste qui ne tue pas les gens », ce qui selon les catégories de Michel Onfray, équivaut à un raciste pacifique. Plus maladroit que bien d'autres, le restaurateur a été poursuivi et contraint à présenter ses excuses : « J'ai pété un plomb, et je m'en excuse. J'ai un ami qui est mort au Bataclan, j'ai tout mélangé », a-t-il précisé. Il n'en va pas de même pour tous : Onfray persiste et signe sur Marianne.net, lui aussi a perdu des amis et il n'est d'ailleurs pas le seul à avoir été atteint dans sa chair et son âme, mais beaucoup pour ne pas dire la majorité, refusent les amalgames.

La seconde remarque, concernant le binôme « un musulman pacifique et un terroriste », se rapporte au choix de la personne du « musulman pacifique ». Le choix du qualificatif « pacifique » est moins anodin qu'il n'y paraît, il est censé faire pendant au qualificatif de « terroriste ». Nous sommes en présence de deux termes antinomiques. On n'a pas affaire à un musulman lambda mais à un pacifique, donc en principe exempt de tout danger. Mais, même dans ce cas, le musulman pacifique a quelque chose à voir avec le terroriste. En effet, dans sa réponse à ses détracteurs,

---

[1] Propos relayés par le quotidien *Le Parisien* du 28 août 2016.

Onfray écrit : « Le soleil n'a rien à voir avec le jour, ni la lune avec la nuit ; la casserole n'a rien à voir avec la cuisine ; la salle de bain n'a rien à voir avec la propreté ; ni la bibliothèque avec l'intelligence ; le Christ n'a rien à voir avec le christianisme ; ... ni *Le Meilleur des mondes* avec Huxley ; le réel n'a jamais rien à voir avec ce qui a lieu – *pas d'amalgames*, vous risqueriez de faire le jeu du réel ». Comme on le constate, le tour de passe-passe de Michel Onfray a pour objet de nous faire admettre qu'il n'y a pas d'amalgame, nous sommes donc acculés à devoir reconnaître, comme l'intéressé le déclare, que, même pacifique, le musulman a quelque chose à voir avec le terroriste ! Reste à savoir ce que Michel Onfray et Pierre-André Taguieff pensent au sujet de la question suivante : « Est-ce qu'il y a une différence de nature entre un musulman pacifique et un juif pacifique, ou une différence de degré ? ». En attendant la découverte par un savant farfelu d'un "gène de l'islam" expliquant ce qui différencie les musulmans du reste de l'humanité civilisée[1], il convient de noter que la première femme médaillée *fields* est une Iranienne[2].

Comme il ressort de l'exemple susmentionné, le problème consiste à avoir la décence de poser de bonnes questions car, avec une fausse question, on ne peut espérer avoir une bonne réponse. Venons-en maintenant à la question qualifiée de sensible de *Philosophie magazine* que nous avons citée plus haut et qui a été formulée comme suit : « Quelle est la nature des liens entre islam et islamisme ? Y a-t-il, entre eux, une différence de nature ou simplement de degré ? ». Si, avec une bonne question, on peut espérer trouver une réponse, avec une mauvaise, on ne

---

[1] Alain Gresh, « Bernard Lewis et le gène de l'islam », *Le Monde diplomatique*, août 2005.
[2] Equivalent Nobel de la discipline mathématique.

peut au mieux que s'égarer dans des débats futiles, à moins que, jouant le rôle d'un bon kôan, ladite question n'assure le décentrement nécessaire à l'éveil. Qu'en est-il de la question « Islam et islamisme : différence de nature ou de degré ? ». Sommes-nous en présence d'une bonne ou d'une mauvaise question ? En invitant à opter pour une solution parmi deux options alternatives dégagées, cette question peut conduire à des errements dans des débats interminables qui peuvent être d'une certaine utilité mais qui n'apportent pas de solution au problème de fond. Comme Michel Onfray aime à le souligner, le journaliste vit d'émotion, tandis que « le philosophe vit de réflexion » et le dossier sur l'islam de *Philosophie magazine* a d'ailleurs été motivé par le « besoin de prendre le temps de la réflexion »[1].

Nous allons donc tenter de réfléchir à la question « Islam et islamisme : différence de nature ou de degré ? ». Pour ce faire, nous allons tenter d'évaluer les deux options alternatives qui s'offrent à nous. On constate alors que l'on ne peut opter pour la différence de nature car ceci reviendrait à dire ou à croire qu'il existe deux vérités en islam, ce qui est impossible. Opter pour une différence de degré, c'est passer du registre de la discontinuité au registre de la continuité en reconnaissant une pluralité d'interprétations de la vérité. Dans ce cas, nous ne sommes plus dans le monde de la dualité mais de la diversité. Comme on le constate, nous sommes confrontés à l'inévitable tension entre vérité et réalité.

Qui dit religion monothéiste et plus spécifiquement « islam », dit principe d'Unicité (Tawhîd). Eric Geoffroy[2] relève à cet égard « les affinités du paradigme holistique

---

[1] *Ibid.*
[2] Eric Geoffroy, *L'islam sera spirituel ou ne sera plus*, Paris, Seuil, 2009, p.92.

avec le principe de l'Unicité : en vertu de la solidarité liant Dieu à Sa création, l'unicité métaphysique devient diversité et pluralisme dans le monde physique ». La question « Islam et islamisme : différence de nature ou de degré ? » nous égare dans un monde de la confusion. Pour éviter le piège de la confusion des genres, nous devons constamment veiller à garder présent à l'esprit qu'avec le terme « islam », nous sommes dans le registre de la religion, tandis qu'avec le concept d'« islamisme », nous sommes dans le registre de la politique. Comme on le constate, c'est le dévoiement du terme « islamisme » qui prête à confusion et sème, *in fine*, le doute dans les esprits.

Si l'islam est une religion à l'instar du christianisme et du judaïsme, dans sa réalité pratique, à l'image du monde chrétien et du monde juif, le monde islamique est pluriel. Pour ce qui est de l'islam : d'après les classements en vigueur, nous avons d'abord trois branches dont deux majoritaires, à savoir : sunnite, chiite et khârijite. Au sein de ces deux branches majoritaires que sont le chiisme et le sunnisme, nous avons différentes sectes. Sabrina Mervin[1] distingue quatre écoles juridiques au sein de la branche sunnite : mâlikites, hanéfites, hanbalites et châfi'ites. La branche chiite comporte : les zaydites, les ismaéliens et les duodécimains. Chacune de ces sous-branches se subdivise en différents courants. Ce qui semble aujourd'hui poser problème pour les pays occidentaux, c'est le terrorisme islamique qui est essentiellement le fait d'un certain courant du wahhabisme (le salafisme djihadiste) [2] qu'il faut oser nommer et dénoncer.

---

[1] Sabrina Mervin, *Histoire de l'Islam - Fondements et doctrines*, Paris, Flammarion - Champs Histoire, 2010.
[2] Encore plus rigoriste que le salafisme, le wahhabisme est une doctrine développée par le prédicateur Muhammed Ibn Abdel Wahhab (1703-1792). Comme il a été signalé précédemment, aujourd'hui, le salafisme se conjugue au pluriel : salafisme quiétiste, politique et djihadiste. C'est le Saoudien Ben Laden qui, avec l'aide des Etats-

Wahhabisme ou salafisme, la position d'Olivier Roy[1], d'un point de vue terminologique, est nuancée. De facto, il vise le wahhabisme mais préfère opter pour le terme néo-fondamentalisme qu'il considère comme équivalent à la tendance conservatrice du salafisme[2]. Il rappelle que : « le terme de « salafisme » date de la fin du XIXe siècle, avec Jamaluddin al-Afghani. L'idée est de contourner une tradition religieuse sclérosée et une histoire politique où les musulmans se sont aliénés dans le colonialisme, en revenant aux textes originaux et au modèle de société du temps du Prophète. Il faut donc ouvrir les portes de l'interprétation (*ijtihad*) ». Précisons que qui dit revenir au modèle de société du temps du Prophète en ouvrant les portes de l'interprétation, dit respecter l'« esprit » et non pas la « forme » du modèle de société du temps du Prophète. C'est-à-dire être avant-gardiste comme celui-ci l'était à son époque. « Mais aujourd'hui, précise Olivier Roy, ceux qui se réclament du salafisme incarnent surtout une tendance conservatrice, proche du wahhabisme saoudien (et souvent identique à lui) ».

Or, précise Olivier Roy, « les Saoudiens ont joué un rôle clé dans l'expansion du néo-fondamentalisme. Afin de couper l'herbe sous le pied tant du nationalisme arabe que du chiisme iranien ou du communisme, ils ont encouragé un sunnisme très conservateur sur le plan doctrinal, mais aussi très hostile à l'Occident sur le plan strictement religieux »... « Grâce aux pétrodollars, l'Arabie saoudite a pu jouer un rôle de plus en plus déterminant dans les

---

Unis et de l'Arabie saoudite, a promu et banalisé le terrorisme djihadiste (Cf. supra, chapitre 3).
[1] Olivier Roy, *Islam mondialisé*, Paris, Seuil, 2004, chapitre 6.
[2] D'après Olivier Roy, « le néo-fondamentalisme représente une vision très stricte et littéraliste du message coranique, dans la tradition hanbalite (la plus littéraliste des écoles de l'islam) », op. cit., p.146.

réseaux d'éducation religieuse du monde musulman », précise-t-il. « Les wahhabis saoudiens se sont bien gardés de diffuser le wahhabisme en tant que tel, se contentant d'insuffler leur doctrine dans l'enseignement des autres écoles, en dénonçant les formes nationales ou populaires de l'islam, en marginalisant tout ce qui s'articule sur les grandes cultures du monde musulman (littérature, philosophie) et en soulignant tout ce qui va dans le sens du hanbalisme. Le contenu pédagogique a été allégé au profit de manuels plus courts, axés avant tout sur du *fiqh* (droit appliqué) et les *ibadat* (dévotion). De toute façon, l'anti-intellectualisme propre au néo-fondamentalisme n'encourage ni à écrire ni à lire des ouvrages longs et complexes. »

Gilles Kepel[1] est non moins catégorique, selon lui, le choc pétrolier de 1973 consécutif à la guerre du Kippour a été une aubaine pour l'Arabie saoudite qui a pu, grâce à ses pétrodollars, concrétiser son ancienne ambition de « wahhabisation de l'islam mondial ». Avant la manne pétrolière, « la doctrine wahhabite ne jouissait de prestige, en dehors de la péninsule, que parmi les milieux rigoristes (ou salafistes) ». Pour avoir un panorama clair de la situation de l'islam dans le monde à cette époque, il est important de rappeler que « Les traditions islamiques nationales et locales enracinées dans la piété populaire, les clercs se réclamant des diverses écoles juridiques du sunnisme implantées dans les grandes régions du monde musulman (hanéfite dans les zones turques et l'Asie du Sud, malékite en Afrique, chaféite en Asie du Sud-Est) ou du chi'isme, conservaient encore partout avant 1973 une position dominante. Ils tenaient en suspicion le puritanisme

---

[1] Gilles Kepel, « La victoire du pétro-islam et l'expansion wahhbite : 1973 » in *Jihad – Expansion et déclin de l'Islamisme*, Paris, Gallimard, 2000, chapitre 2.

d'inspiration saoudienne, incriminant son caractère sectaire. »

Gilles Kepel dépeint comme suit le processus de wahhabisation de l'islam dans le monde après 1973 : « après cette date, précise-t-il, les institutions wahhabites changent de dimension et se livrent au prosélytisme à grande échelle dans l'univers sunnite (les chi'ites, tenus pour hérétiques, restent en dehors du mouvement). Leur objectif sera à la fois de faire de l'islam un acteur de premier plan sur la scène internationale, le substituant aux nationalismes défaits, et de réduire les modes d'expression pluriels de cette religion au credo des maîtres de la Mecque. Leur zèle embrasse le monde entier, par-delà les frontières traditionnelles de l'islam et jusqu'en Occident, où les populations immigrées musulmanes constituent une cible de prédilection du prosélytisme saoudien ».

Les propos du ministre belge Madrane résumés dans l'article « Le péché originel, en Belgique, a été de confier les clés de l'islam à l'Arabie saoudite en 1973 » publié par le quotidien *La Libre Belgique* le jeudi 19 novembre 2015, sont très éclairants quant à la descente en enfer actuelle. Selon le ministre de la Jeunesse belge, en 1973, en échange de l'approvisionnement énergétique, les clés de l'islam ont été confiées à l'Arabie saoudite, c'est ainsi que l'islam malakite (islam apaisé) a été balayé au profit du wahhabisme et du salafisme.

Ce qu'il importe de souligner ici, c'est que, si la propagation du wahhabisme dans les sociétés occidentales ne s'est pas faite à l'insu des gouvernements des pays concernés, elle ne s'est pas faite non plus pour des raisons d'ordre exclusivement économique d'approvisionnement énergétique. En effet, comme il a été signalé précédemment, l'instrumentalisation de l'islam n'est pas le seul fait des pays musulmans, l'Occident a excellé en la matière. En effet, selon Olivier Roy[1], « la propagande saoudienne a aussi bénéficié de l'approbation tacite des grands pays occidentaux ou musulmans, car elle était vue dans les années 1980 comme un utile contre-feu aux radicalismes de l'époque (l'islamisme iranien, le nationalisme arabe ou le communisme). Enfin, étant donné l'excellence des relations entre la monarchie saoudienne et les gouvernements occidentaux, on pensait que cette prédication resterait sous contrôle politique. Comment, de plus, refuser un visa demandé par une ambassade saoudienne ? Les pays ont donc, par impuissance ou indifférence plus que par calcul, laissé l'influence saoudienne se développer parmi leurs populations musulmanes ».

En réalité, les gouvernements occidentaux n'étaient pas aussi naïfs que cette analyse s'évertue à le penser. La politique étant ce qu'elle est dans une démocratie qui ne peut faire fi des agendas électoraux, les gouvernements occidentaux de l'époque ont évalué la situation à l'aune de leurs intérêts à court terme, comme ils continuent à le faire aujourd'hui ; ce qui primait avant tout, dans un environnement économique concurrentiel, c'était d'assurer un approvisionnement énergétique au meilleur coût possible. Par ailleurs et ce pour ce qui est du long terme, les axiologies stratégiques des acteurs dominants étant ce qu'elles sont, la propagande saoudienne a aussi bénéficié de l'approbation tacite des grands pays occidentaux parce qu'elle était vue dans les années 1980 comme un utile contre-feu aux mouvements progressistes et/ou anti-occidentaux de l'époque : l'islam politique iranien (anti-impérialiste), le nationalisme arabe (tiers-mondiste) ou le communisme (anticapitaliste).

---

[1] Olivier Roy, *Islam mondialisé*, op.cit.

Mais Olivier Roy a parfaitement raison de déclarer que « l'explication par l'argent saoudien ne suffit pas ». En effet, comme nous avons eu l'occasion de le souligner précédemment, le contexte des années 70 a été particulièrement favorable à la montée du fait religieux en raison, entre autres, de la crise des idéologies. Et le wahhabisme a été considéré par les puissances occidentales comme le courant de pensée islamique le mieux à même de défendre ses intérêts évalués à l'aune d'un arsenal axiologique périmé au regard de la configuration mondiale actuelle.

Il s'agit maintenant de voir comment l'on peut, en évitant la question « Islam et islamisme : différence de nature ou de degré ? » qui nous égare plus qu'elle nous éclaire, résoudre le problème qui se pose à nous. Pour ce faire, il faut commencer par identifier le ou les problème(s). Quel est le problème de fond ? Le retour des religions ? Le retour du religieux ? L'islam ? L'islam politique ? La montée en force ou en visibilité de l'islam ? L'instrumentalisation de l'islam à des fins de politique interne et/ou de géopolitique ? Le terrorisme des jeunes Occidentaux de confession musulmane (convertis inclus) ? Le terrorisme de Daech ? La montée des mouvements terroristes instrumentalisant l'islam ? La déterritorialisation des actes terroristes commis au nom de l'islam ?

La situation étant complexe, il existe, en réalité, un amoncellement de problèmes bien plus nombreux que ceux énumérés ci-dessus, problèmes avec des chaînes de causalité multiples. Mais, pour les néo-réactionnaires islamophobes, la réponse est simple, le vrai problème c'est l'islam. En effet, si jadis tous les chemins menaient à Rome, aujourd'hui tous les problèmes les mènent à l'islam et tous les moyens (essais, articles, journaux écrits et audiovisuels, œuvres littéraires, cinématographiques et télévisuelles, séries policières, jeux vidéo, etc.) sont permis pour arriver à inculquer cette idée au commun des mortels, jusqu'à ce qu'elle soit bien ancrée dans les esprits. Mais, comme la propagande dite djihadiste est fondée sur cette même idée, on en vient à se demander si fiction et réalité n'en font pas qu'une et s'il ne s'agit finalement pas d'une seule et même réalité.

Prenons le cas de la violence de guerre. Il n'est pas certain que les photos des corps déchiquetés par les bombes du camp des « civilisés » soient plus soutenables que celles du camp des « barbares » qui, pour intimider et faire peur, se plaît à diffuser des photos en couleur de macabres mises en scène de décapitations. Edgar Morin et Michel Onfray, qui ne peuvent être accusés d'être du même bord, le reconnaissent tous deux. En effet, Edgar Morin écrit à ce sujet : « l'Occident dénonce avec horreur le terrorisme aveugle qui tue civils, femmes et enfants, sans se soucier que, dans le monde arabo-musulman, on dénonce avec horreur les bombardements aveugles qui tuent civils, femmes et enfants, les assassinats ciblés par drones ou autres »[1].

Michel Onfray[2], qui, de son côté, donne l'impression de prendre le parti de Daech, proteste à tort au regard du droit international en vigueur -- pour les raisons que nous avons exposées précédemment -- que leur soit « dénié le droit de dire qu'ils sont un Etat islamique » mais s'insurge à juste raison qu'ils soient seuls à être qualifiés de « barbares (alors qu'ils font à la disqueuse et au marteau piqueur ce que l'Occident fait avec des avions furtifs – je vous rappelle, précise-t-il, qu'une partie des sites mésopotamiens ont été détruits par les bombardements américains sans émotion

---

[1] Edgar Morin, « Essayons de comprendre », in Eric Fottorino (dir.), *Qui est Daech ? Comprendre le nouveau terrorisme*, Paris, Ed. Le1, 2015, p.9.
[2] Michel Onfray, « La France doit cesser sa politique islamophobe » in *Le Point* N° 2254, du 19 novembre 2015.

internationale...), les qualifier de terroristes (alors que, certes, ils tuent des victimes innocentes avec des kalachnikovs ou des couteaux mais que l'Occident fait de même à plus grande échelle avec des bombes qui tuent femmes et enfants, vieillards et hommes qui n'ont rien à se reprocher, sinon d'habiter le pays associé à l'« axe du mal », tout ça fait que nous sous-estimons en tout point leur nature véritable qui n'est pas à mépriser ».

Il est important de souligner ici que Michel Onfray a parfaitement raison de ne pas vouloir essentialiser les actes de violence, car, ainsi que nous avons déjà eu l'occasion de le signaler ailleurs dans un essai consacré à la question du « progrès du genre humain »[1], ce ne sont pas les êtres humains qui doivent être qualifiés de « barbares »[2] mais les actes de violence qui sont par définition des actes de « barbarie » compris au sens d'inhumains puisqu'ils ne peuvent tous être qualifiés de criminels du fait que certains sont estimés légitimes au regard du « droit international » et des rapports de force en vigueur.

La question se pose maintenant de savoir pourquoi Daech ne défend pas sa cause en diffusant les photos des images insoutenables des horreurs des bombardements du camp des alliés plutôt que celles de ses décapitations macabrement mises en scène. La raison paraît évidente :

---

[1] Ninou Garabaghi, « Les organisation internationales et régionales et le progrès du genre humain : Quel avenir pour la culture de la paix et l'éthique de la non-violence ? », Revue *Géostratégiques* N° 44, avril 2015.

[2] Les jeunes qui ont commis des meurtres au nom de l'islam en France n'étaient pas des barbares (étrangers, au sens originel du terme grec *barbaros*) mais des Français. Etant donné que l'être humain est capable du meilleur comme du pire, quelles que soient sa religion et sa culture, si l'on tient à utiliser cette référence, il conviendrait de limiter l'usage du terme barbarie pour qualifier le caractère de l'acte commis et non l'être humain dans son essence.

Daech ne cherche pas à convaincre ni à défendre sa cause auprès de l'Occident qu'il estime mécréant et quasi imperméable à sa cause mais, étant faible, il a besoin de faire peur pour s'imposer dans les esprits à défaut de pouvoir s'imposer sur le terrain militaire. Si les images chocs de Daech sont destinées à combler le vaste écart existant entre leur intention de nuire et leur capacité à le faire, elles font également partie de la stratégie de Daech, qui œuvre à la montée de l'islamophobie au sein des sociétés occidentales. Reste à comprendre pourquoi le chroniqueur d'une chaîne d'information publique accepte d'enfreindre la *common decency* au point de faire appel aux photos de propagande de Daech pour manipuler par les images, à défaut d'avoir pu manipuler par la rhétorique, les auditeurs indécis ou réfractaires à ses idées.

En effet, la chronique de Brice Couturier du 16 décembre 2015, intitulée « L'ensauvagement, c'est maintenant », où les idées du chroniqueur ont été comme à l'accoutumée savamment agencées de façon à susciter la haine de l'islam, est mise en ligne sur le site d'information de France Culture le jour même. Pour toucher les esprits anti-islamophobes, est mise en exergue une photo insoutenable de propagande de Daech destinée à briser tout risque de résistance mentale. Ironie du sort, le 16 novembre, accusée d'incitation à la haine et d'utilisation des photos de propagande à des fins d'intérêt politique, Marine Le Pen est poursuivie par la justice et sommée de retirer les photos immondes de Daech de son compte twitter. Scandale ou pas, Brice Couturier n'estimera pas pour autant nécessaire de s'excuser d'avoir heurté les esprits policés peu accoutumés aux images de violence extrême.

La question qui se pose maintenant consiste à savoir pourquoi un journaliste, dont la chronique est consacrée du label « les idées claires », a besoin de recourir à de tels procédés obscurs. Pourquoi certains intellectuels médiatiques œuvrent-ils à la montée de l'islamophobie ? La raison serait-elle esclave des passions au point de ne pouvoir y échapper, à moins que, contrairement aux scientifiques, le journaliste, à l'instar du politique, ne puisse opérer que dans le registre de l'émotion. L'éthique de la conviction l'emporterait-elle sur celle de la responsabilité ? A moins que nous ne nous trouvions dans la configuration où c'est l'idée même d'éthique qui est lestée sur l'autel de la « fin justifie les moyens » ?

Par exemple, le binôme « Alain Finkielkraut et Brice Couturier » semble davantage animé par la haine de l'islam que par le mépris arabo-musulman qui paraît être le propre du binôme « Pascal Bruckner et Caroline Fourest ». Mais, avec Alain Finkielkraut et Pascal Bruckner qui ont l'avantage de la franchise et de la sincérité, nous sommes, semble-t-il, en présence d'une haine par conviction pour le premier et d'un mépris par intérêt pour le second. Pour Brice Couturier et Caroline Fourest, la fin semble justifier les moyens, de sorte que ceux-ci n'ont apparemment aucun scrupule à faire appel si nécessaire à l'arme de la désinformation pour influencer et manipuler l'opinion publique.

En 1941, dans « littérature et totalitarisme », George Orwell écrivait : « le totalitarisme n'exerce pas son contrôle sur la pensée de manière seulement négative mais aussi positive. Il ne vous interdit pas seulement d'exprimer et même de penser certaines pensées, il vous dicte ce que vous devez penser. Il crée une idéologie qui devient la vôtre ». Brice Couturier, à l'instar de Caroline Fourest, a pour habitude, non seulement de réorienter le discours des invités dans le sens qu'il souhaite donner à la discussion, mais, dans les émissions de la matinale de France Culture où il faisait office de chroniqueur, il avait pour habitude d'empêcher les invités d'exprimer une idée qui pouvait contrer « son » idéologie (pour ne pas dire sa propagande anti-islamique). Si, en situation de stress extrême, il n'avait aucun scrupule à interrompre l'invité pour couper court à un raisonnement qui ne lui convenait pas, il avait davantage pour habitude de s'adonner à une sorte de « propagande par glissement » qui consiste à diffuser subrepticement des réalités tronquées et à propager par ce biais « sa vérité » sous couvert de liberté d'expression.

Prenons un autre exemple : Invitée des matins de France Culture le 15 janvier 2016, Delphine Horvilleur, rabbin au Mouvement Juif Libéral de France (MJLF), est interrogée par le journaliste responsable de l'émission, sur la question du port de la kippa. Faisant référence au débat qualifié d'ancien en France sur les signes religieux considérés ostentatoires, le journaliste demande pourquoi la kippa serait considérée comme licite et pas le voile. C'est alors que Delphine Horvilleur déclare : « Tous les rites ne disent pas la même chose : le port de la kippa signifie, "moi homme" je suis tout petit au regard de Dieu » tandis que « le port du voile signifie "moi femme" je suis toute petite par rapport à l'homme, ce qui n'est pas la même chose »[1]. Pour éviter tout malentendu, la rabbine tient à préciser que la religion juive n'est pas prosélyte.

Brice Couturier, après avoir pris soin dans sa chronique de flatter l'égo du responsable de l'émission[2] pour le dissuader de réagir, souligne : « les Juifs français, qui

---

[1] Il convient de préciser à cet égard que le responsable du programme, le journaliste Guillaume Erner, a eu la délicatesse de rappeler que « la plupart des religions sont assez misogynes. Dans la religion juive, a-t-il précisé, il est de coutume pour la femme de cacher ses cheveux, de porter une perruque, de porter un foulard également. Il y a une forme de discrimination entre homme et femme. »
[2] Pour flatter l'égo du responsable de l'émission, Brice Couturier a pris soin de citer sa thèse qui a été publiée sous forme d'essai.

constatent, avec tristesse, qu'il semble plus facile, dans un pays comme le nôtre, de faire disparaître la kippa que la burqa... Alors que la seconde, contrairement à la première, constitue un geste d'hostilité envers notre pays » (difficiles à déchiffrer, les points de suspension après la « burqa » sont de l'auteur de la chronique). Comme on le constate, dans le discours de Delphine Horvilleur, la femme musulmane est humiliée mais il lui reste une porte de sortie par laquelle elle peut se prévaloir de bénéficier du statut de victime de discrimination[1]. Mais c'est sans compter avec la perspicacité de notre chroniqueur des "idées claires" pour qui le port du voile est signe d'« hostilité envers son pays ». Depuis la tragédie du 13 novembre, journée de la gentillesse profanée, la France est en guerre et Brice Couturier, qui aime à se faire passer pour le Zoro des matins de France Culture, a trouvé l'ennemie de la patrie, elle était cachée sous la burqa.

Etre victime de discrimination ou coupable de complot contre la patrie, ce n'est évidemment pas la même chose. Il est vrai que Brice Couturier n'a rien inventé mais, convaincu probablement d'avoir raison, il donne la désagréable impression de vouloir jouer le rôle de porte-parole d'un courant de pensée qui a pour tâche de diaboliser l'islam. Si, avant la tragédie de Charlie Hebdo, les islamophobes opéraient par déni du problème comme nous avons pu le constater ci-dessus, depuis les attentats de novembre 2015, ils « revendiquent un droit à l'islamophobie »[2].

---

[1] Si comparaison n'est pas raison, on ne peut pour autant ignorer que la condition des femmes de la communauté harédim en Israël n'est pas tellement meilleure que celle des femmes de la communauté wahhabite en Arabie saoudite.
[2] Pascal Boniface : « L'image de la France dans le monde musulman », Tribune du quotidien *La Croix* du 18/01/2016.

Les causes de la montée de l'islamophobie sont multiples. Il y a d'abord et surtout la montée de l'islamophobie consécutive à la persistance et à l'extension du terrorisme orchestré d'abord par al-Qaïda, ensuite par Daech et d'autres organisations terroristes qui instrumentalisent l'islam, il y a ensuite, et tout aussi efficace, la propagande anti-islamique d'une partie des élites : dirigeants, intellectuels, journalistes. Non moins dramatique est la montée de l'islamophobie consécutive à l'instrumentalisation de l'islam à des fins bassement électorales. Certains, pour ne pas dire la plupart des membres de l'élite islamophobe, de notoriété ou pas, sont indéniablement voués à terme à l'oubli mais leur pensée, leurs discours et leurs actes n'auront pas été sans conséquences néfastes pour leur pays et le reste du monde.

L'islamo-bashing de ces élites néoconservatrices et/ou néo-réactionnaires contribue à la montée de la violence islamophobe des couches populaires fragilisées d'un point de vue économique et/ou identitaire, ce qui contribue à la montée de la haine et du ressentiment des populations violentées dans leur chair et/ou leur âme (dignité). Effet de boomerang, *in fine*, l'islamophobie caractérisée par la montée des partis d'extrême droite en Europe fait le lit de Daech et d'autres entités terroristes salafo-djihadistes. C'est ainsi que deux fractions minoritaires de populations : les mouvements populistes islamophobes et les mouvements terroristes djihadistes sont en voie d'imposer leur loi pour la plus grande joie du complexe militaro-industriel.

CHAPITRE 7

Le salafisme djihadiste avatar de la politique de wahhabisation de l'islam mondial

Pour contrer le panarabisme nassérien, le prince Fayçal d'Arabie saoudite décide au début des années soixante de promouvoir le panislamisme, pour la plus grande joie des grandes puissances et plus spécifiquement des Etats-Unis. Il est utile d'ouvrir ici une parenthèse, pour noter que cette volonté manifeste d'instrumentalisation de l'islam n'a pas heurté à l'époque les consciences au point de faire ressentir à l'intelligentsia occidentale le besoin de dévoyer le terme islamisme.

Pour mener à bien sa politique panislamiste, l'Arabie saoudite prend l'initiative de créer en 1962 la Ligue islamique mondiale qui va, à coups de milliards de dollars, financer sa politique de wahhabisation de l'islam mondial. La tendance à la « wahhabisation de l'islam mondial » en tant que réalité factuelle ne peut pas ne pas soulever des questions quant à la légitimité d'une hégémonie religieuse fondée exclusivement sur le pouvoir de l'argent. En effet, une politique expansionniste, quelle qu'elle soit, soulève des questions au regard du respect du principe de non-ingérence inscrit dans la Charte des Nations Unies. Et la prétention à la suprématie d'une branche de pensée islamique à l'encontre de toutes les autres pose d'autant plus de problèmes qu'elle équivaut à l'imposition d'une culture exogène, c'est-à-dire d'une coutume, d'un mode de vie, d'une conception du monde et de la vie, etc., à l'encontre d'autres, ceux de la culture endogène. Lorsqu'on refuse à l'autre sa vérité (le mot de vérité étant compris au sens générique du terme) et que l'on veut lui imposer sa vérité à soi, alors l'on doit être à même de justifier la prétention à la supériorité de cette vérité, il s'agit, en d'autres termes, de légitimer cette vérité en apportant les preuves de son bien-fondé.

En tout état de cause, nous devons admettre que, nonobstant toute politique expansionniste, même en l'absence d'une quelconque volonté de puissance de la part d'une culture à l'encontre d'une autre, à l'ère de la mondialisation et d'une interdépendance de fait, nous sommes nécessairement confrontés à des problèmes de cohérence et d'harmonisation des systèmes de rationalité éthique et esthétique. En effet, les processus de mondialisation des différentes sphères de la vie des sociétés et plus spécifiquement le fait que nous vivons de fait dans un monde rétréci à l'état de village planétaire, expression d'un cosmopolitisme *de facto*, ont pour conséquence de scotomiser la complexe et quasi insoluble question de la conciliation des valeurs consécutive à l'insurmontable problème d'une harmonisation simultanée des différences de valeur et de temporalité entre communautés au sein des sociétés, d'une part, et des différences de culture et de temporalité entre sociétés, d'autre part, ce qui *de facto* génère des situations explosives et à bien des égards immaîtrisables.

Pour ce qui concerne l'Arabie saoudite, foyer du wahhabisme, qui nous intéresse plus spécifiquement ici, il faut admettre que les pratiques sociales, culturelles et politiques en vigueur dans ce pays posent problème quant à leur validité, à leur justesse et à leur légitimité : validité des pratiques sociales au regard des droits de l'homme estimés définir les contours de la morale de la communauté internationale ; justesse du wahhabisme au regard de l'esprit de l'islam ; légitimité de la politique de wahhabisation de l'islam mondial au regard du principe de non-ingérence, corollaire du droit à l'autodétermination des populations. Ce qui est le plus malsain avec la politique de wahhabisation de l'islam mondial de l'Arabie saoudite, c'est que son arsenal de mécanismes de socialisation des esprits au wahhabisme opère, le plus souvent, envers et contre la volonté des dirigeants des pays concernés, et, dans tous les cas, à l'insu de leur population.

En effet, d'après Olivier Roy[1], « Forte de la légitimité que lui donnait le contrôle des lieux saints, mais handicapée par son attachement au wahhabisme, doctrine récusée par une grande partie des oulémas sunnites, l'Arabie saoudite a tenté, grâce aux pétrodollars, de développer une propagande religieuse strictement fondamentaliste », évitant de poser explicitement « la question du pouvoir politique. C'est le sens de la création en 1962, de la *Rabita*, la Ligue islamique mondiale ; celle-ci imprime corans et livres de piété et subventionne dans le monde entier mosquées et instituts islamiques, payant par exemple le salaire de nombreux imams de mosquées d'Europe ». Il va de soi que celui qui rémunère est maître du discours professé. Si, avant la révolution iranienne de 1979, l'Arabie saoudite se contente, « par l'édition, les bourses et l'organisation de colloques et séminaires, d'infléchir la production islamiste dans un sens conservateur », à partir de 1980, elle étend son action « en encourageant, de manière plus ou moins empirique, la formation de réseaux fondamentalistes sunnites essentiellement tournés vers les pays non arabes ou vers l'émigration en Europe. Pour s'attribuer « une légitimité religieuse que les islamistes lui refusent, le roi Fahd prend en 1986 le titre de « Gardien des lieux saints » (la Mecque et Médine) ».

---

[1] Olivier Roy, *Echec de l'islam politique*, Paris, Seuil, 2015, pp 170-171.

Pour contrer l'Iran qui est considéré comme l'adversaire principal dans sa volonté de puissance, l'antagonisme est traduit en termes religieux : islam véritable contre hérésie chiite ».

Pour s'arroger le rôle de leadership du monde musulman, « l'Arabie saoudite s'est ainsi trouvée financer soit des réseaux fondamentalistes, certes conservateurs, mais violemment antioccidentaux, soit des groupes islamistes beaucoup plus radicaux qui, eux, revendiquaient le pouvoir politique, mais paraissaient plus aptes à empêcher l'Iran d'occuper le terrain de la contestation »[1]. Pour vaincre toute résistance auprès des sunnites récalcitrants à la doctrine wahhabite, l'Arabie saoudite est contrainte d'avancer masquée sous le drapeau du « salafisme ».

L'instrumentalisation des mouvements religieux islamiques par les dirigeants politiques saoudiens n'est pas un fait nouveau. Si la dynastie Saoud a été instaurée grâce à son alliance avec Mohammad Ben Abdel Wahhab, fondateur du courant religieux wahhabite, elle n'a par la suite cessé d'établir des alliances de circonstance avec différents mouvements religieux islamiques y compris des courants d'inspiration shiite.

C'est ainsi que, contrairement aux dires, fin 2017, du prince héritier Mohammad Ben Salmane, nouvel homme fort du royaume saoudien, bien avant l'avènement de l'islam politique en Iran, l'Arabie saoudite a pendant la décennie 60 soutenu l'islam politique et servi de terre d'asile pour ses représentants attitrés, les Frères musulmans, ce pour promouvoir le panislamisme à l'encontre du panarabisme de Nasser comme il a été signalé précédemment. Et, durant la décennie 1979-1989, allié des

---

[1] Ibid.

Etats-Unis, elle a soutenu, financé et armé les talibans en Afghanistan[1].

En 2015, non contente de fournir des contingents à Daech, elle en devient le principal bailleur de fonds. D'après Pierre Conesa[2], ancien haut fonctionnaire au ministère français de la Défense, « près de 2.500 combattants saoudiens sont dans les rangs de l'EI. Parmi eux, des rejetons de riches familles qui financent le combat de leur progéniture ». Selon un rapport du Congrès américain[3], rien que sur la période 2013-2014, Daech a bénéficié de transferts d'argent en provenance d'Arabie saoudite, du Qatar et du Koweït, pour un montant équivalent à plus de 40 milliards de dollars. Et on ne peut ignorer, comme l'historien Nabil Mouline le souligne, que parmi tant d'autres, c'est « le royaume des Saoud qui est le principal propagateur de la doctrine qui justifie religieusement les atrocités d'al-Qaïda et de Daech ».

Comme par hasard, l'Arabie saoudite, pas plus que le Qatar ou le Koweït, ne figure sur la liste noire des pays à majorité musulmane dont les ressortissants ont été interdits d'entrée aux Etats-Unis par le décret anti-immigration du président Trump[4]. On a beau être à la tête de la plus grande puissance du monde, on ne peut décemment pas déclarer « persona non grata », les ressortissants d'un pays disposé à vous concéder des « méga-contrats excédant 380 milliards

---

[1] Cf. supra, chapitre 3.
[2] Vincent Monnier, « L'Arabie saoudite, principal bailleur de fonds de Daech ? », *L'Obs* du 21 janvier 2016.
[3] Rapport paru fin avril 2015.
[4] La liste des pays faisant partie de la liste noire a changé, au gré des protestations des gouvernements des pays concernés et le décret et sa version révisée ont été suspendus par des juges fédéraux des Etats-Unis, cf. entre autres : Feriel Alouti, « Décret anti-immigration suspendu : « Une épreuve symbolique pour Trump » », le quotidien *Le Monde* du 16.03.2017.

de dollars, dont 110 pour des ventes d'armements »[1] destinés en principe à être utilisés contre d'autres pays musulmans.

La politique de wahhabisation de l'islam mondial conduit à des changements de mode de vie des populations de confession musulmane via l'adoption de pratiques et de normes sociales spécifiques à l'Arabie saoudite, conséquence néfaste de l'acculturation desdites populations concernées, ce sans garantie préalable d'une vie meilleure ici-bas et/ou au-delà après la mort. Par soucis de rigueur et d'ouverture d'esprit, nous nous devons d'examiner, ne serait-ce que brièvement, la question de la validité des pratiques et des normes sociales (formelles et informelles) en vigueur dans ce pays, à la lumière des principaux courants de pensée politiques en vogue, d'une part, et de différents points de vue moraux et théologiques, d'autre part. Nous essayerons de voir pour terminer s'il est possible d'en tirer des conclusions quant à la légitimité de la politique de wahhabisation de l'islam mondial, compte dûment tenu des principes du droit international.

Pour ce faire, nous allons commencer par le courant de pensée conservateur américain qui a pour pendant européen les identitaristes intransigeants. Du point de vue ethnosociologique, ce courant de pensée politique relativement influent aux Etats-Unis est affilié au modèle culturaliste[2]. Pour les tenants occidentaux du « relativisme

---

[1] Selon les informations communiquées par FranceInfo le 21 mai 2017, à l'occasion de son premier voyage en Arabie saoudite, « le président des Etats-Unis, Donald Trump, a bénéficié d'un accueil royal ponctué par l'annonce de méga-contrats excédant 380 milliards de dollars, dont 110 pour des ventes d'armements à Riyad, visant en particulier à contrer les *"menaces iraniennes"* ».
[2] Voir, entre autres, Ralph Linton (1945), *Le Fondement culturel de la personnalité*, Paris, Bordas, 1977.

culturel », les systèmes sociaux et les valeurs idéologiques des autres cultures sont parfaitement valables[1]. Il est vrai que la « reconnaissance de la validité morale d'autres civilisations » ne les a pas empêchés, par le passé ni maintenant, d'assigner la plus haute valeur à leur propre civilisation. Car, comme Henry Kissinger le souligne à juste titre dans son constat sur l'« ordre mondial » : « Tout au long de l'histoire, chaque civilisation, se considérant comme le centre du monde et regardant ses principes comme universellement pertinents, a défini sa propre conception de l'ordre ». Pour les wahhabites, l'argument du relativisme culturel ne peut opérer puisque, pour les salafistes d'inspiration wahhabite, la vie sociale et privée doit être conforme à la loi islamique telle que le wahhabisme la perçoit, c'est-à-dire de façon à la fois fixiste et holistique, il n'y a donc pas de place pour la culture à leurs yeux. Et, comme leur doctrine est censée représenter le vrai islam, il semble donc exclu, de leur point de vue, d'envisager une quelconque modernisation des institutions sociales.

Les conservateurs étant respectueux du principe de non-ingérence pour les raisons exposées, il serait bon de voir si les néo-conservateurs, qui ont pour pendant européen les néo-réactionnaires[2], peuvent nous être d'une quelconque utilité.

Dans son célèbre livre *Le choc des civilisations*, Samuel Huntington[3], éminent représentant du courant de pensée néo-conservateur, écrit : « Les musulmans en grand

---

[1] Robert Cresswell, *Eléments d'ethnologie*, Paris, Armand Colin, 1975, tome 1.
[2] Pascal Durand & Sarah Sindaco (dir.), *Le discours "néo-réactionnaire"*, Ed. CNRS, 2015.
[3] Samuel Huntington, *Le choc des civilisations*, Paris, Odile Jacob, 1997, pp. 117-118.

nombre, se tournent vers l'islam comme source d'identité, de sens, de stabilité, de légitimité, de développement, de puissance et d'espoir, espoir symbolisé par le slogan « l'islam est la solution ». Cette Résurgence de l'islam, par son ampleur et sa profondeur, est la dernière phase du réajustement de la civilisation musulmane par rapport à l'Occident. C'est un effort pour trouver la « solution » non plus dans les idéologies occidentales mais dans l'islam. Elle se traduit par l'acceptation de la modernité, le rejet de la culture occidentale et le réengagement dans l'islam comme guide de vie dans le monde moderne ». Et le meilleur est à venir, il précise à l'appui de son argumentaire : « Comme l'expliquait un haut fonctionnaire saoudien en 1994, les "importations de l'étranger" sont sympathiques quand il s'agit de "choses" belles ou sophistiquées, mais des institutions sociales et politiques intangibles venues d'ailleurs peuvent être mortelles –demandez au shah d'Iran (...). Pour nous, l'islam n'est pas seulement une religion, c'est un mode de vie. Nous autres Saoudiens voulons nous moderniser, mais pas nécessairement nous occidentaliser ». A croire ce haut fonctionnaire, la modernisation est une bonne chose mais elle devrait se limiter aux « importations de "choses" belles et sophistiquées ».

Etrange conception de la religion en général et de l'islam en l'occurrence. Car, ainsi que nous serons amenés à le constater plus loin, si critique de la société occidentale il y a, cette critique a d'abord et surtout été une critique de la civilisation matérialiste. Le problème de fond, avec la remarque susmentionnée du haut fonctionnaire saoudien, est que demeurer fidèle à l'islam tout en se modernisant ne consiste justement pas à consommer de belles choses importées tout en continuant à voiler les femmes, à les confiner dans la sphère domestique, à leur assigner le statut de mineure à vie ; à décapiter au sabre les condamnés à mort ; à lapider les femmes infidèles ; à infliger des châtiments corporels (mutilation, flagellation, etc.).

Nous reviendrons plus loin sur le problème fondamental de la condition féminine. Limitons-nous pour l'heure à la question des biens de consommation. Lorsqu'on parle de biens de consommation licites ou illicites (halal), il s'agit de savoir avant toute chose si ces biens ont été produits dans des conditions respectueuses de la personne humaine, qu'il s'agisse du consommateur ou du producteur. Demeurer fidèle à l'esprit de l'islam, c'est éviter que la société soit prise au piège des mécanismes d'aliénation inhérents à la société de consommation, pourvoir à la couverture des coûts de l'homme en général et des coûts humains du travail en particulier, s'assurer que les biens consommés sont produits dans des conditions respectueuses des normes sociales islamiques, à savoir : la juste rétribution du travailleur, le refus de la consommation d'êtres humains pour la production de « "choses" belles et sophistiquées »[1], etc.

Et le plus grave au regard des pratiques susmentionnées, estimées surannées et inhumaines, est que, le monde s'étant rétréci à l'état de village planétaire et que, du fait notamment des NTIC, les frontières étant de moins en moins étanches nonobstant les murs dressés, le problème de l'autre a vite fait de devenir le mien. Ceci d'autant plus que les promoteurs de l'idéologie wahhabite ne se limitent pas à préserver un mode de vie moyenâgeux mais ambitionnent d'universaliser des pratiques et normes sociales perçues aujourd'hui comme barbares au monde musulman d'abord, ensuite aux impies après les avoir convertis au préalable. Or, ainsi que Tareq Oubrou[2] le rappelle à juste titre,

---

[1] Selon François Perroux, la non-couverture des coûts humains du travail équivaut à la consommation d'êtres humains pour la production de choses.
[2] Tareq Oubrou, *Un Imâm en colère*, Paris, Albin Michel, 2015, p.67.

« L'Arabie saoudite n'est pas le Vatican des musulmans. Ses options politico-religieuses n'incombent qu'à ceux qui la dirigent ». Il nous faut reconnaître à cet égard que, sans l'apport des civilisations perse, gyptienne, ottomane et de bien d'autres, l'Islam n'aurait probablement pas connu la gloire qui fut la sienne au cours des siècles passés.

La question consiste maintenant à savoir si les pratiques et normes sociales en vigueur en Arabie saoudite, foyer du wahhabisme mondial, sont propres à l'islam ou inhérentes aux coutumes ancestrales de ce pays, susceptibles d'évoluer sans enfreindre l'aqîda (fondement de la foi religieuse) et la sharia (fondement de la loi religieuse).

En effet, la première question qui se pose au regard desdites pratiques et normes sociales consiste à savoir dans quelle mesure les pratiques et normes en vigueur dans ce pays sont imputables à l'islam à proprement parler. Cette question en sous-entend une autre qui est celle de savoir : si le Coran avait été révélé dans un autre contexte (un autre lieu et/ou époque), aurait-il été parfaitement identique à ce qu'il est ? Car ce qu'il importe de souligner ici, c'est que le Coran est composé de passages principiels qui énoncent des vérités constantes/universelles et des passages circonstanciels.

Pragmatiques, les versets circonstanciels étaient destinés à répondre aux impératifs de gestion de la communauté bédouine au Moyen-Age. Et ce qu'il convient de préciser, quitte à se répéter, au sujet des versets circonstanciels, c'est que les pratiques prescrites à l'époque du Prophète Mohammad étaient pragmatiques, certes, mais d'esprit avant-gardiste.

La question importante, qui se pose au regard des normes, règles et pratiques sociales en vigueur en Arabie saoudite, consiste à savoir si, au nom d'un antiuniversalisme occidentalo-centré, on serait en droit de refuser tout jugement de valeur en la matière, surtout lorsqu'on est en présence d'institutions sociales censées être fondées sur une doctrine religieuse à visée hégémonique. Evidemment que non, mais juger est une chose et agir en fonction du verdict du jugement porté en est une autre ; pour ce qui est de l'action politique, la panoplie des mesures à disposition des gouvernements est longue : arrêt des relations diplomatiques, sanctions économiques, etc.

Pour ce qui est du cas de non-respect des droits de l'homme par les tenants du wahhabisme qui nous intéresse plus spécifiquement ici, si nous nous plaçons dans la posture de défenseur du relativisme culturel, on est alors en droit de penser que tant que le wahhabisme est confiné au sein du royaume wahhabite d'Arabie saoudite, le principe de non-ingérence cher au paradigme du droit international westphalien peut être invoqué et l'on se trouve dans l'obligation de dire que ce sont leurs affaires après tout.

Mais, étant donné que l'Arabie saoudite a opté pour une politique hégémoniste de wahhabisation de l'islam mondial par la force de l'argent, dans ce cas, le wahhabisme devient un problème mondial et le principe de non-ingérence ne peut plus être invoqué, même si l'on s'astreint au cadre étriqué du paradigme du droit international westphalien.

L'on ne peut ignorer à cet égard que Daech n'a pas le monopole des actes qualifiés de barbares, qu'il s'agisse de décapitations, de châtiments corporels, de destruction de patrimoines culturels. Prenons le cas spécifique du vandalisme du patrimoine culturel de l'humanité non conforme aux croyances religieuses néo-fondamentalistes (patrimoine préislamique, chiite, etc.). Le patrimoine culturel et intellectuel de l'Iran préislamique a fait l'objet de destructions massives au cours de l'invasion arabe à l'époque de l'empire Sassanide. Et aujourd'hui, « au-delà des destructions liées aux opérations militaires, "certains experts soupçonnent l'Arabie saoudite de cibler précisément le patrimoine yéménite" »[1].

Le pire est que les destructions perpétrées par l'Arabie saoudite ne se limitent pas aux patrimoines situés hors de ses territoires mais elles concernent également ceux situés sur son propre territoire. Destruction du patrimoine préislamique bien sûr mais islamique également. En effet, comme Gilles Kepel[2] le rappelle, « A peine maîtres des villes saintes, les wahhabites saccagèrent les tombeaux des imams et de Fatima, la fille du Prophète, dont la vénération par les chi'ites constituait à leurs yeux une idolâtrie inacceptable ». Il en a été de même pour les constructions de l'époque ottomane qui ont été complètement rasées. Mais, comme un chroniqueur de l'hebdomadaire *Economie Matin*, à l'instar de bien d'autres, le rappelle avec amertume, « l'Arabie saoudite est un allié du monde occidental, alors évidemment il ne faut pas se fâcher avec lui, même si le comportement de ce pays, au niveau du patrimoine comme de celui des droits humains, ne diffère pas énormément de l'Etat islamique »[3].

Kamel Daoud[4], journaliste et écrivain algérien, déclare à cet égard que l'Arabie saoudite est « un Daech qui a réussi ». Et le plus inquiétant, car non perçu à sa juste

---

[1] Selon le magazine *Le Journal des Arts* (article « Le patrimoine yéménite ravagé » 8-21 janvier) cité par Philippe Herlin, « Destruction du patrimoine culturel : l'Arabie saoudite ne vaut guère mieux que l'Etat islamique », *Economie Matin* du 15.01.2016.
[2] Gilles Kepel, *Jihad – Expansion et déclin de l'Islamisme*, op.cit., p75.
[3] Philippe Herlin, « Destruction du patrimoine culturel : l'Arabie saoudite ne vaut guère mieux que l'Etat islamique », *Economie Matin* du 15.01.2016.
[4] « L'Arabie saoudite, un Daech qui a réussi », Tribune de Kamel Daoud publiée le 20 novembre 2015 par le *New York Times*.

gravité, est que ces deux entités œuvrent à la faveur du même problème de fond qui est celui de la « wahhabisation des esprits ». Or, si les tentatives de familiarisation avec la doctrine wahhabite à l'instar, de la connaissance d'autres doctrines peuvent s'avérer utiles, la politique de wahhabisation des esprits peut en revanche être considérée comme incompatible avec l'esprit de coexistence pacifique et la politique de wahhabisation de l'islam mondial comme illégitime au regard du principe de non-ingérence inscrite dans la Charte des Nations Unies. Face à la gravité de la situation, la question se pose évidemment de savoir ce qu'il faudrait faire. Et que peut-on faire ?

Pour le courant de pensée néo-réactionnaire fervent défenseur d'un universalisme totalitaire qui a pour pendant américain le néo-conservatisme interventionniste, ce sont les valeurs et institutions occidentales qui doivent en toute circonstance servir de base de référence, de sorte que toute pratique et/ou mode de comportement qui y dérogent seront critiqués et taxés de rétrogrades. Etant entendu que ce courant de pensée politique dominant, nonobstant le principe d'indivisibilité et d'interdépendance des droits humains, aura, dans sa mise en pratique, tendance, sous couvert d'universalisme des droits de l'homme, à privilégier le corpus des droits civils et politiques. Ce qui, dans les contrées non occidentales, a eu et continue à avoir pour conséquence (pour ne pas dire inconséquence), avec ou sans ingérence, la prise de pouvoir par des mouvements fondamentalistes. Tandis que, au sein des sociétés occidentales, l'instrumentalisation, par les néo-réactionnaires, des principes de la « laïcité » a pour conséquence la stigmatisation de la population de culte musulman wahhabisée, ce qui conduit à son rejet social : discrimination à l'emploi, à l'accès au logement, etc. Ce qui, a contrario, du fait du ressentiment généré, a pour effet le rejet desdits principes par la partie la plus vulnérable de ladite couche de population stigmatisée et marginalisée.

A l'affût de la nouveauté, les néo-progressistes sont évidemment à l'avant-garde du progrès mais, tout aussi convaincus qu'ils sont d'être les seuls détenteurs de la vérité, ils ne sont malheureusement pas exempts de toute inconséquence. Prenons le cas de la défense de la cause des homosexuels par ce courant de pensée. Dans les critiques adressées par le courant de pensée « néo-progessiste » à l'encontre des condamnations à mort pour homosexualité dans les pays non occidentaux, ce n'est pas tant la condamnation à mort des homosexuels qui est remise en cause que la criminalisation de l'homosexualité. Et pour preuve, la peine de mort n'a toujours pas été abolie aux Etats-Unis. Comme on le constate, on est confronté à une incohérence en matière de rationalité d'éthique qui se traduit par une inversion de la hiérarchie des valeurs. Dans le même registre, d'aucuns auraient pu croire que la dépénalisation de l'homosexualité allait progressivement sonner le glas du « mariage ». Paradoxalement il n'en est rien, tout à la fois progressistes et conservateurs, les homosexuels revendiquent le droit au mariage. Se pose alors la question de savoir comment on peut être tout à la fois progressiste et conservateur. A quoi tient la permanence dans une société de certaines valeurs que l'on pourrait considérer comme dépassées au regard de la validation par ladite société de certaines pratiques jadis considérées comme moralement inacceptables au point de pouvoir faire l'objet de sanctions pénales ?

La permanence de la valeur « famille » dans les sociétés postmodernes s'explique par le besoin de sécurité au sens large du terme qui, loin de baisser, a augmenté dans les sociétés occidentales, il n'en va pas de même de la valeur « mariage ». Le maintien dans la société postmoderne de la valeur « mariage » et sa généralisation à des couples homosexuels ne s'expliquent pas par le besoin de sécurité pour les raisons qui suivent. Si l'on considère le cas de la France, on constate qu'il existe pour l'heure trois modalités d'union : le concubinage, le Pacs (pacte civil de solidarité) et le mariage. Si l'on fait abstraction du concubinage, on constate que les obligations et avantages du mariage et du Pacs sont identiques, à ceci près que les pacsés ne sont pas astreints au devoir de secours mutuel et les formalités de la séparation les concernant sont fort simples puisqu'elles se limitent à une simple déclaration au greffe. La différence majeure entre les deux modalités d'union se situe au niveau de l'adoption, d'après les règlements en vigueur, seul le mariage donne droit à une adoption conjointe, les deux autres modalités d'union doivent se contenter d'une adoption individuelle. Comme on le constate, le Pacs aurait pu être amendé, de façon à permettre une adoption conjointe sans avoir besoin de généraliser le mariage aux couples homosexuels.

Ce n'est donc pas du point de vue légal qu'il faut chercher à comprendre l'attachement à la valeur « mariage » des différents protagonistes : opposants et partisans du mariage homosexuel. Opérant dans le registre du symbolique, les partisans du mariage homosexuel misent sur les changements de mentalité ; ils cherchent à agir sur les consciences espérant acquérir à terme une légitimité morale au niveau relationnel.

La question se pose alors de savoir pourquoi les associations de défense des couples homosexuels ont opté pour la légalisation du mariage homosexuel plutôt que de miser et d'œuvrer à la faveur du dépassement de la valeur « mariage ». Il n'est pas inutile de rappeler à cet égard que le mariage n'est pas une institution universelle, les NA de Chine, par exemple, ne connaissent ni la paternité, ni le mariage. Alors même qu'en toute cohérence on était en droit de s'attendre qu'au nom de la défense de l'homosexualité, le mariage soit considéré comme une institution dépassée ; c'est à l'aune de la postmodernité, que le courant de pensée néo-progressiste va œuvrer au renforcement de cette institution au prix d'un oxymore : le « mariage homosexuel ».

Qu'à cela ne tienne, l'oxymore a vite fait d'être résorbé. Il suffisait, pour ce faire, de changer la définition du mot « mariage » jadis limité à l'union de personnes de sexe différent[1] et le reste, c'est-à-dire le changement de « notre » perception/conception du « mariage », ce qui n'est pas des moindres, n'est qu'affaire de temps. Sinon pourquoi avoir cherché à masquer la réalité des faits en ayant recours à l'expression « mariage pour tous » ? A croire que, étant consacrée par toutes les formes d'union, plus personne n'aurait le droit de s'opposer à la légitimité des relations homosexuelles, quitte à désacraliser l'institution « mariage ». L'histoire du concept d'« islamisme » est là pour convaincre les sceptiques en matière de changement de mentalité et/ou de perception. En entendant le mot « islamisme », qui pourrait croire aujourd'hui que ce mot -- qui dans le meilleur des cas est compris comme « islam politique »-- il y a moins d'une génération, était utilisé pour désigner une religion et/ou croyance au même titre que le judaïsme, le christianisme, le mazdéisme, le taoïsme, le

---

[1] Dans le dictionnaire Robert de 1970, le terme « mariage » est défini comme suit : « Union légitime d'un homme et d'une femme ». Dans le dictionnaire en ligne Larousse de 2017, le terme mariage est défini comme suit : « Acte solennel par lequel un homme et une femme (ou, dans certains pays, deux personnes de même sexe) établissent entre eux une union dont les conditions, les effets et la dissolution sont régis par le Code civil *(mariage civil)* ou par les lois religieuses *(mariage religieux)* ; union ainsi établie ».

bouddhisme, l'hindouisme, le shintoïsme, l'animisme, le chamanisme, etc ?

Sauf mauvaise foi, il nous faut admettre qu'il n'est de perception parfaitement « neutre », chaque personne est déterminée par un héritage culturel et une histoire personnelle qui lui sont propres. De sorte qu'en percevant les modes de vie et les comportements d'autrui et plus spécifiquement des cultures qui lui sont étrangères, chaque personne porte consciemment ou inconsciemment un jugement conditonné par ses déterminismes. Si aujourd'hui, à l'ère de la mondialisation, nous sommes, tout un chacun, en mesure de réceptionner la même information et/ou image, grâce aux NTIC, nous ne sommes en revanche pas en mesure de les percevoir ni de les interpréter d'une façon identique. Sans parler qu'au niveau de nos actions, nous sommes tous (individus et pays), plus ou moins subtilement mais nécessairement conditionnés et guidés par notre histoire, nos valeurs, nos intérêts, nos rêves/utopies et nos projets. Et, si les rêves peuvent être de couleur d'azur, la réalité est toujours nécessairement de couleur grise.

Il est à noter à cet égard que l'intervention du gouvernement néo-conservateur de George W. Bush en Irak en 2003, contrairement à ce qui fut annoncé, n'a pas été une croisade en faveur de la défense des droits humains universels en général et plus spécifiquement des valeurs de liberté et de démocratie. Et pour preuve, républicains ou démocrates, les gouvernements des Etats-Unis ont non seulement permis à l'Arabie saoudite de bafouer les droits humains sur son propre territoire mais de mener une politique de wahhabisation de l'islam mondial en toute impunité.
Si les démocrates d'abord et les républicains ensuite ont soutenu l'Irak contre l'Iran dans la guerre meurtrière de la

décennie 80, ce n'est pas parce qu'ils étaient contre un régime islamique déclarant la suprématie du droit divin sur les droits humains, mais parce que, contre toute attente, ce régime, qui avant son accession au pouvoir avait accepté le soutien financier des Etats-Unis, s'est retourné contre eux en instrumentalisant la religion à des fins de lutte anti-impérialiste.

Qu'ils soient progressiste ou conservateurs, le pragmatisme des intérêts géostratégiques semble, dans tous les cas, l'emporter sur la cause des droits humains, ce qui donne raison à Noam Chomsky[1] lorsqu'il déclare : « Un refrain souvent entendu est que le risque posé par l'islamisme radical oblige à s'opposer (à contrecœur) à la démocratie et ce pour des raisons pragmatiques. Si la formule mérite considération, elle est néanmoins trompeuse. La vraie menace est l'indépendance. Les Etats-Unis et leurs alliés ont régulièrement soutenu des islamistes radicaux, parfois pour éliminer la menace d'un nationalisme laïque. Un exemple connu est celui de l'Arabie saoudite, le centre idéologique de l'Islam radical (et du terrorisme islamiste). Un autre sur la longue liste est Zia ul-Haq, le plus brutal des dictateurs pakistanais et le préféré du Président Reagan, qui a mené un programme d'islamisation radical (financé par les Saoudiens). »

La réalité des relations internationales semble donner raison à la Russie qui rappelle, à intervalles réguliers, que les Etats, surtout depuis une vingtaine d'années, n'ont ni amis ni ennemis mais des intérêts nationaux. Sinon, comment expliquer que certains Etats, ne pouvant se défaire de la conception schmittienne de la politique, et à défaut de

---

[1] Noam Chomsky, « Ce n'est pas l'Islam radical qui préoccupe les États-Unis, mais l'indépendance », *The Guardian*, 4 février 2011 (article traduit en français par et pour le quotidien *Le Grand Soir*).

véritable ennemi, s'attellent à diaboliser l'adversaire par la construction de la figure du mal ?

Un nombre considérable de chercheurs se sont penchés sur cette question fondamentale de la construction de l'ennemi sans qu'aucun n'ait pu apporter, à défaut d'un remède efficace, une réponse, tout au moins satisfaisante à la problématique sous-jacente.

Reste à savoir dans quelle mesure les Etats sont à même de concilier la défense des valeurs dites universelles et leurs intérêts nationaux. Et le plus important ici consiste à savoir : quelles sont ces valeurs universelles ? Qui a décidé de leur caractère universel ? Les valeurs dites progressistes sont-elles nécessairement supérieures aux valeurs dites traditionnelles ? Les droits de l'homme des sociétés dites « civilisées » sont-ils plus importants que les droits de la nature des sociétés dites « primitives » ?

Selon Noam Chomsky[1] : « Dans le futur, les historiens, s'il y en a, regarderont le spectacle curieux de ce début du XXIe siècle où, pour la première fois dans l'histoire humaine, les hommes font face au désastre résultant de leur propre action, qui menace notre propre survie. Ces historiens verront que le pays le plus riche et le plus puissant dans l'histoire, qui jouit d'avantages sans pareil, est à la tête des efforts conduisant au désastre pourtant prévisible. A la tête des efforts pour préserver les conditions qui donneront une vie décente à nos descendants se trouvent en revanche les sociétés dites « primitives » : les premières nations, les tribus indigènes et aborigènes. Les pays où les populations indigènes sont importantes mènent l'effort pour la préservation de la planète, tandis que les pays qui ont détruit ces populations ou les ont rendues marginales courent vers leur propre destruction. »

---

[1] Noam Chomsky, « La civilisation peut-elle survivre au capitalisme ? », publié le 28 juillet 2014, par Noam-Chomsky.fr.

Si la sphère de la réalité empirique des affaires humaines, en général, et du discours et de l'action politiques, en particulier, n'a cesse de nous décevoir à bien des égards, celle des débats théoriques n'est pas des plus satisfaisantes pour autant. La rationalité étant toujours située, il est aujourd'hui communément admis qu'elle ne peut assurer une objectivité absolue, même en sciences exactes où Bachelard[1] invite à penser en termes d'« objectivation d'une pensée en quête du réel.. ». Il semble donc assez naturel que, dans les sciences sociales et humaines, la question de l'objectivation soit d'autant plus problématique que nous devons tenir compte des causes non matérielles pour décrire le réel[2], donnant ainsi l'impression désagréable que chacun, privilégiant son point de vue, tente d'imposer sa perception de la réalité telle qu'elle est et/ou telle qu'il souhaiterait qu'elle soit. Reste à savoir qui va, dans ces circonstances, décider du bien et du mal, des principes et des valeurs auxquels tout un chacun doit se soumettre ? La communauté internationale ? Oui, mais quelle communauté internationale ? Peut-on identifier en bonne et due forme cette entité supranationale ? Le FMI, l'OMC ou l'ONU ? L'Assemblée générale de l'ONU ou son Conseil de sécurité ? De qui ces entités tiennent-elles leur légitimité ? Des Etats ou des nations ?

---

[1] Gaston Bachelard, *Epistémologie*, Paris, PUF, 1974.
[2] Selon le sociologue Raymond Boudon, « Si l'objectif de toute science est de décrire le réel tel qu'il est, il faut admettre une priorité du réalisme sur le matérialisme. Cette priorité est sans conséquence dans le cas des sciences de la nature, car ici réalisme et matérialisme vont de pair. Mais le matérialisme rentre en collision avec le réalisme dans le cas des sciences de l'homme, dès lors que l'on ne s'intéresse pas à l'être humain en tant qu'organisme biologique, car le second suppose qu'on ignore les motivations et les raisons des hommes ou qu'on les traite comme dépourvues de toute valeur explicative ».

Comme l'histoire nous l'enseigne, l'universalité des valeurs ne se décrète pas ; elle ne s'impose que par la force des "choses", lorsque les valeurs acquièrent à force d'habitude le statut de « vérités acceptées ». Quand le respect des « droits de la nature » défendus par les sociétés dites « primitives » devient une condition pré-requise aux « droits de l'homme » défendus par les sociétés dites « civilisées », on comprend que l'universalisme ne peut être le fait d'une civilisation décrétée supérieure aux autres mais une œuvre commune de l'humanité[1]. Il est intéressant de noter à cet égard que la première partie de la "Charte africaine des droits de l'homme et des peuples" est composée de deux chapitres, le premier est consacré aux droits de l'homme et des peuples, le second aux devoirs. A titre d'exemple, selon l'article 27 alinéa 1, « Chaque individu a des devoirs envers la famille et la société, envers l'Etat et les autres collectivités légalement reconnues et envers la Communauté internationale. » D'après l'alinéa 2, « Les droits et les libertés de chaque personne s'exercent dans le respect du droit d'autrui, de la sécurité collective, de la morale et de l'intérêt commun ». Pour mettre en exergue la nécessité du juste équilibre entre les droits et les devoirs, il serait bon que le terme « devoir » puisse figurer dans l'intitulé même de ladite Charte africaine.

Par ailleurs, pour décrisper les débats autour des droits de l'homme, critiqués par certains de ses détracteurs en raison de leur caractère prétendument occidental, il y a lieu de rappeler que, bien avant la Magna Carta (1215), la Constitution des Etats-Unis (1787) et la Déclaration des droits de l'homme et du citoyen (1789), en 539 avant J.-C., après avoir conquis la ville de Babylone, Cyrus le Grand, le roi de Perse, « libéra les esclaves, déclara que toutes les personnes avaient le droit de choisir leur propre religion et

---

[1] Cf. Ninou Garabaghi, *Les espaces de la diversité culturelle*, Paris, Karthala, 2010.

établit l'égalité raciale. Ces décrets et bien d'autres furent enregistrés sur un cylindre d'argile rédigé en akkadien et en caractères cunéiformes. Connu aujourd'hui sous le nom de cylindre de Cyrus, ce document antique est maintenant identifié comme la première Déclaration des droits de l'Homme dans le monde. Il est traduit en chacune des six langues officielles de l'ONU et ses clauses sont analogues aux quatre premiers articles de la Déclaration universelle des droits de l'homme.»[1] Une réplique du cylindre de Cyrus, qui est conservé au British Museum, est exposée au siège des Nations Unies à New York, au second étage, entre les chambres du Conseil de sécurité et du Conseil économique et social de l'ONU.

Il s'agit de voir maintenant si, en renvoyant dos à dos les deux positions extrêmes exposées ci-dessus (partisans du relativisme culturel versus partisans de l'universalisme ethno-centré) il est possible, ne serait-ce que d'un point de vue conceptuel, d'arriver à dégager un *consensus situé* en matière des droits humains. Pour éviter tout malentendu, il convient de préciser que le qualificatif de « situé » a été utilisé pour éviter l'écueil de la critique postmoderniste des grands récits obtenus à coups de consensus forcés, fondés sur l'illusion quant à la possibilité de convenir d'un consensus rationnel universel.

Ainsi que nous avons déjà eu l'occasion d'en débattre longuement ailleurs[2], il s'agirait, *in fine*, d'œuvrer à la faveur d'un universalisme ouvert. Les travaux de Mireille Delmas-Marty[3], dans le domaine du droit, en sont un

---

[1] Cf. « Un regard sur le passé des droits de l'homme » sur le site « www.humanrights.com ».
[2] Voir, entre autres, l'ouvrage *Les espaces de la diversité culturelle*, op.cit.
[3] Voir, entre autres, les quatre volumes de son cours professé au Collège de France sous le titre *Les forces imaginantes du droit* publiés aux éditions du Seuil.

exemple. En quête d'une communauté de droit et de valeurs, Mireille Delmas-Marty[1] déclare : « Comment oser parler de communauté de droit à l'échelle d'une planète livrée aux affrontements, à la violence et à l'intolérance ? Et comment seulement concevoir les contours d'une communauté de valeurs par-delà la diversité des cultures et l'opposition des intérêts ? S'il est vrai que le droit international connaît un développement sans précédent et que les juridictions internationales se multiplient, la réalité quotidienne démontre davantage le grand désordre du monde que l'émergence d'un ordre juridique mondial, légitime et efficace ». Membre de l'Académie des sciences morales et politiques, professeure au Collège de France, l'intéressée a le courage de citer des exemples d'entorses portées en toute impunité par les plus puissants au modèle universaliste qui est censé privilégier la justice sur la politique et soumettre au droit l'usage de la force.

La question essentielle qui se pose à nous ici consiste à savoir s'il existe une incompatibilité insurmontable entre l'islam et les droits de l'homme ainsi que certains islamophobes néo-conservateurs de droite et néo-progressistes de gauche veulent le faire croire. La réponse est évidemment non et pour preuve l'Organisation de la coopération islamique (OCI), la Ligue arabe et l'Union africaine (UA) ont toutes trois adopté des instruments régionaux relatifs aux droits de l'homme. Il est vrai, par contre, qu'il existe une incompatibilité entre les droits de l'homme et la charia wahhabite en vigueur en Arabie saoudite et la version encore plus rigoriste de Daech. Si, ainsi qu'il a été signalé précédemment, selon la conception wahhabo-salafiste de l'islam, religion et culture sont identiques, c'est que, pour ses concepteurs, l'islam a cessé d'être la soumission libre et

---

[1] Mireille Delmas-Marty, *Vers une communauté de valeurs ?*, Vol.4 de la série *Les forces imaginantes du droit*, Paris, Seuil, 2011, p.7.

consentie à Dieu[1] pour devenir la soumission contrainte et forcée aux mœurs et traditions de l'Arabie du VIIe siècle et d'aujourd'hui.

Prenons la question brûlante des Droits des femmes, les instruments régionaux adoptés par l'OCI, la Ligue arabe, l'UA reconnaissent toutes trois l'égalité entre hommes et femmes. L'article 6, alinéa 1 de la Déclaration du Caire sur les Droits de l'homme en Islam stipule que : « La femme est l'égale de l'homme dans la dignité humaine ; ses droits sont équivalents à ses devoirs. Elle a une personnalité civile, une responsabilité financière indépendante, et le droit de conserver son nom patronyme et ses liens de famille. ». Selon l'al.2, « Le mari a la charge de l'entretien de la famille et la responsabilité de sa protection ». D'après l'art.1, al.2 : « Personne n'est supérieur à personne, sauf par la piété et les bonnes œuvres ».

Plus catégorique, l'article 3, alinéa 3 de la Charte islamique des droits de l'homme stipule que : « L'homme et la femme sont égaux sur le plan de la dignité humaine, des droits et des devoirs dans le cadre de la discrimination positive instituée au profit de la femme par la charia islamique et les autres lois divines et par les législations et les instruments internationaux. En conséquence, chaque État partie à la présente Charte s'engage à prendre toutes les mesures nécessaires pour garantir la parité des chances et l'égalité effective entre l'homme et la femme dans l'exercice de tous les droits énoncés dans la présente Charte »[2]. Selon l'art.33, al.1 de ladite Charte : « Il ne peut

---

[1] Selon le Coran, « Point de contrainte en religion » (2 : 256).
[2] Il convient de préciser que l'Arabie saoudite, qui fait partie des dix Etats membres de la Ligue arabe qui ont à ce jour ratifié la Charte islamique des droits de l'homme entrée en vigueur le 15 mars 2008, continue à faire subir aux femmes des discriminations surannées.

y avoir de mariage sans le plein et libre consentement des deux parties », l'al.2 fait état de : « l'interdiction de toutes les formes de violence ou de mauvais traitements dans les relations entre ses membres, en particulier à l'égard de la femme et de l'enfant ». Encore plus catégorique, l'art.34 al.4, stipule : « Il est interdit de faire une distinction entre l'homme et la femme dans l'exercice du droit de bénéficier de manière effective d'une formation, d'un emploi, de la protection du travail et d'un salaire égal pour un travail de valeur et de qualité égales ».

D'après l'article 2 de la Charte africaine des droits de l'homme et des peuples[1], « Toute personne a droit à la jouissance des droits et libertés reconnus et garantis dans la présente Charte sans distinction aucune, notamment de race, d'ethnie, de couleur, de sexe, de langue, de religion, d'opinion politique ou de toute autre opinion, d'origine nationale ou sociale, de fortune, de naissance ou de toute autre situation. ». Selon l'art.18, al.3 : « L'Etat a le devoir de veiller à l'élimination de toute discrimination contre la femme et d'assurer la protection des droits de la femme et de l'enfant tels que stipulés dans les déclarations et conventions internationales. »

Selon l'historien S. D. Goitein[2], « Du point de vue religieux, selon l'islam ancien, la femme est l'égale de l'homme ; lui incombe l'obligation d'observer les prières quotidiennes, voire d'étudier la loi religieuse (en fait, elle est soumise à toutes les obligations religieuses). Cependant,

---

[1] Il nous a paru important de citer la Charte africaine des droits de l'homme pour la très bonne raison que nombre de pays arabes de culte musulman font partie des Etats signataires de la Charte.
[2] S.D. Goitein, *Studies in Islamic History and Institutions*, Leyde, 1966, p.122 note 1, cité par Ahmad Hasnâwi, « L'Islam, la conquête, le pouvoir », in François Châtelet (dir.), *Histoire des idéologies*, Paris, Hachette, 1978, tome 1, p.334.

elle n'appartient pas au corps politique, dans lequel ne sont membres que les hommes libres, porteurs d'armes ».

D'après le théologien Tareq Oubrou[1], « Tout en énonçant l'égalité de principe, le Coran a formulé des dispositions différentes selon les sexes. Cet aspect relatif de l'égalité hommes / femmes correspond à un contexte patriarcal où l'économie et le pouvoir politique étaient essentiellement liés à la force physique, donc masculine[2]. Aussi les femmes ne pouvaient-elles bénéficier d'aucune véritable indépendance par rapport à l'homme. Cette situation ne pouvait évoluer que peu à peu » précise Tareq Oubrou.

En effet, il est à noter à cet égard qu'en Arabie à l'époque de Mohamet[3], il n'était pas rare que les filles soient enterrées vivantes pour parer au déséquilibre hommes/femmes résultant de la surmortalité masculine consécutive aux razzias. La pratique de la polygamie était une autre réponse moins barbare destinée à parer au déséquilibre démographique. L'infanticide des filles a été interdit par le Prophète mais la polygamie qui, à l'époque, jouait le rôle d'instrument d'inclusion, avait été maintenue comme un moindre mal.

Il paraît évident que cette pratique désuète et injuste n'a plus raison d'être, ce d'autant plus que, même pour les plus rigoristes, il n'est pratiquement pas possible de se conformer aux prescriptions qui exigent un traitement rigoureusement égalitaire des épouses. Car, s'il est possible

---

[1] Tareq Oubrou, *Un Imâm en colère*, op. cit.
[2] Etant donné qu'aujourd'hui ce sont les capacités intellectuelles qui jouent un rôle déterminant, on comprend pourquoi les groupes terroristes tels que les talibans, Boko Haram, etc. font tout ce qui est dans leur pouvoir pour empêcher les filles d'avoir accès à l'éducation.
[3] L'orthographe du nom du Prophète dépend des usages locaux : Mohammad en Iran, Mehmet en Turquie, Muhammad au Moyen-Orient arabe, Mohamed au Maghreb, Mohamet en France (selon la tradition orientaliste).

d'assurer un traitement matériel égalitaire entre ses épouses, il est impossible de les aimer de façon identique pour la simple raison que « le cœur a ses raisons que la raison ignore » comme le dit si bien dit l'adage philosophique emprunté aux *Pensées* de Blaise Pascal[1].

Il y a lieu de rappeler qu'avant l'islam, c'est la "polygamie sans limite ni condition" qui était pratiquée dans les contrées du Prophète (nombre illimité de femmes et sans prescription). Il est vrai que les plus forts se sont, de tout temps, octroyé des droits qu'ils ont ensuite érigés en morale sociale pour légitimer leur pratique. Il est d'ailleurs fort probable que si les femmes avaient été plus fortes physiquement que les hommes, c'est la polyandrie qui aurait à juste titre fait l'objet de procès aujourd'hui. Comme il a été précisé, si l'on se situe d'un point de vue religieux, l'impossibilité pratique de la "polygamie équitable" (respect impératif du principe d'équité prescrit dans le Coran) équivaut à son interdiction de droit pour les raisons exposées ci-dessus (selon le verset 4 : 129, il est impossible d'être équitable).

Admettons maintenant que la polygamie soit considérée comme un fait culturel, dans ce cas, on doit admettre que, si la morale sociale accepte la polygamie, pour être juste, elle devrait accepter tout autant la polyandrie. Comme on le constate clairement avec l'exemple de la pratique de la polygamie, refuser la distinction entre culture et religion, ce serait vouloir justifier la morale sociale par la volonté divine, ce serait accepter ainsi de sombrer soit dans l'ignorance soit dans l'hypocrisie et le mensonge.

Car, comme Tareq Oubrou[2] le souligne à juste titre, « dès qu'on passe à l'aspect horizontal des pratiques de

---

[1] La citation exacte est la suivante « Le cœur a ses raisons que la raison ne connaît point », *Pensées* 273/432.
[2] Tareq Oubrou (entretiens avec Michael Privot et Cédric Baylocq), *Profession Imâm*, Paris, Albin Michel, 2015.

l'islam, à savoir le droit et la morale, les variables sociologiques entrent en jeu parce qu'il n'y a pas de pratiques morales ou juridiques sans le substrat culturel ». Le plus grand danger avec la politique de wahhabisation de l'islam mondial menée par certaines monarchies pétrolières réside dans cette volonté d'universalisation des pratiques culturelles propres à l'Arabie du début du VIIè siècle, ce nonobstant la ratification par ces pays d'instruments juridiques destinés à assurer, autant que faire se peut, le respect des droits humains.

Le plus important, pour ne pas dire le plus grave, est que ces pratiques injustes légitimées par des interprétations sélectives et machistes de certains passages du Coran ont réussi à scotomiser les réalités théologiques qui instaurent d'emblée « une égalité morale, spirituelle et ontologique entre l'homme et la femme »[1].

Un nombre non négligeable de religieux, en terre d'islam et en Occident, dénoncent les interprétations patriarcales et misogynes des textes coraniques. D'après le président de la République islamique d'Iran, Hassan Rohani[2] : « Selon les règles islamiques, l'homme n'est pas le premier sexe ni la femme le second sexe (...) les femmes sont aux côtés des hommes et les deux sont égaux. ». D'après le théologien, Tareq Oubrou[3], « Le Dieu du Coran ne s'est pas fait homme... Il n'est donc ni du côté de l'homme, ni du côté de la femme. Le Coran ne connaît pas non plus le récit de la Genèse qui trace une chronologie commençant par la création d'Adam et se poursuit avec celle d'Eve, venue lui tenir compagnie. Selon le Coran,

---

[1] Tareq Oubrou, *Un imam en colère, op.cit.*, p.143.
[2] Pauline Verduzier, « Iran : pressions sur le magazine féministe qui réinterprète l'islam », Le magazine *Le Figaro Madame* du 08 septembre 2014.
[3] Tareq Oubrou, *Un imam en colère, op.cit.*, pp.142-143.

Adam et Eve furent créés par/pour Dieu, issus d'une même nature, d'un même être (nafs wâhida), lequel a donné deux entités sexuées, habitées par une âme (rouh) commune, ni féminine ni masculine, à l'image de l'Ange qui l'a déposée. Quant au péché originel, la femme, Eve, n'y est pour rien. Tous deux, nous dit le Coran, furent simultanément tentés par Satan, péchèrent et se repentirent. Aucune description d'une Eve tentatrice et alliée à Satan n'existe dans les textes scripturaires de l'islam. Ce point théologique est important, car il instaure d'emblée une égalité morale, spirituelle et ontologique entre l'homme et la femme ».

Comme on le constate, il serait faux de vouloir prétendre que le Coran est un texte plus misogyne que les autres textes religieux, au contraire. En fait, la misogynie est une réalité propre à toutes les cultures patriarcales[1]. Contrairement aux autres religions monothéistes, l'islam ne culpabilise pas davantage la femme que l'homme. En effet, si l'islam ne déculpabilise par entièrement l'être humain en général, la femme n'est tout au moins pas à l'origine du péché originel. Si danger d'islamisation du monde il y a, c'est bien de ce côté qu'il faut regarder mais le danger aurait pu exister si tous les pays du monde étaient régis par les trois religions monothéistes. Ce qui est loin d'être le cas.

Par ailleurs, outre la sécularisation et la montée de l'athéisme, il y a le fait que certaines croyances ont l'avantage d'être exemptes du péché originel, il est vrai aussi que d'autres, pires encore que la croyance dans le péché originel, tendent à légitimer des situations injustes sur terre par la croyance dans le karma des individus.

*

---

[1] Voir, entre autres, Sherin Khankan, *La femme est l'avenir de l'islam*, Paris, Stock, 2017.

Reste maintenant à savoir pourquoi d'aucuns ont pu s'inquiéter de l'islamisation du monde.

Hanté par la crainte de l'islamisation de la France, Michel Houellebecq fantasme *Soumission*[1]. En mal de notoriété planétaire ou marqué à jamais par l'épouvantable décennie de terrorisme en Algérie, Boualem Sansal s'inscrit dans la filiation d'Orwell pour fantasmer *La fin du monde en 2084*[2]. La question se pose alors de savoir si les statistiques laisseraient entrevoir un quelconque risque d'explosion démographique des populations de confession musulmane. A moins qu'il ne s'agisse de conversion massive des populations de confession non musulmane à l'islam ? Toute conversion de force étant à exclure au regard des rapports de force existant, reste le scénario d'une massive conversion de gré. Dans ce cas, la question se pose de savoir comment l'islam pourrait se prévaloir comme une alternative plausible à la civilisation capitaliste. Car on ne peut ignorer que c'est l'hypothèse de l'islam comme solution alternative qui a fasciné et servi de justification au dévoiement du terme « islamisme » par l'intelligentsia occidentale en quête d'alternative à l'échec de la pensée communiste.

---

[1] Titre d'un roman publié en 2015 par l'éditeur Flammarion.
[2] Roman intitulé *2084 : la fin du monde* publié en 2015 chez Gallimard.

CHAPITRE 8

Du mythe de l'islamisation du monde : montée en force et/ou en visibilité de l'islam

La montée en visibilité de l'islam est une réalité incontestable comme l'est celle du retour du religieux comme phénomène mondial. Reste à savoir si nous avons également affaire à une montée en force de l'Islam ainsi que certains islamophobes le craignent ou s'il ne s'agit là que d'un simple fantasme qu'il y aurait lieu de balayer en vue d'apaiser le climat anxiogène créé. La question que nous allons commencer à étudier consiste à savoir si, au-delà de la montée en visibilité de l'islam, nous avons affaire à une véritable expansion de cette religion. Dans ce cas, nous devrons examiner de près les statistiques pour voir s'il existe une croissance en nombre et en pourcentage des populations de confession musulmane telle qu'on puisse parler d'une tendance à l'islamisation du monde. A moins que le "risque" ne soit circonscrit à l'Europe puisque, selon l'orientaliste Bernard Lewis : « L'Europe sera musulmane d'ici à la fin du siècle »[1].

Comme il ressort du premier tableau présenté ci-après[2], de moins de 1,6 milliard en 2010, la population de confession musulmane a été estimée croître de plus d'un milliard pour s'établir à 2,76 milliards d'individus en

---
[1] Alain Gresh, « Bernard Lewis et le gène de l'islam », *Le Monde diplomatique*, août 2005.
[2] Les tableaux sont regroupés à la fin du présent chapitre 8.

l'an 2050. De près de 2,17 milliards en 2010, la population de confession chrétienne devrait s'élever à 2,92 milliards en 2050. Révisé à la hausse, le taux de fécondité de la population de confession musulmane a été estimé à 3,1 enfants par femme, contre un taux de l'ordre de 2,7 pour les femmes de confession chrétienne, 2,4 pour les femmes de confession hindouiste et 2,3 pour les femmes de confession juive. Estimé à 1,7, le taux de fécondité du groupe des personnes non-affiliées (les athées, les agnostiques et les personnes qui ne s'identifient à aucune religion et/ou croyance spirituelle) ne devrait pas permettre à ce groupe de maintenir son rang de troisième dans le classement des groupes par ordre décroissant des effectifs en 2050.

Il est à noter que la forte croissance de la population de confession musulmane est essentiellement le fait de l'Afrique subsaharienne où le taux de fécondité, toutes religions confondues, a été estimé à 4,8 et celui des femmes de confession musulmane à 5,6. Il importe de préciser que les différences de taux de fécondité s'expliquent davantage par les disparités de niveaux de vie et d'éducation des femmes que par la religion des groupes concernés. En effet, comme les données longitudinales l'ont révélé par le passé, il existe une forte corrélation entre le niveau de développement des pays et la baisse du taux de fécondité des femmes. De sorte que, si le taux de fécondité des femmes de confession musulmane au Niger a été estimé à 6,9, ce taux n'est que de l'ordre de 2 chez la même catégorie de femmes en Indonésie, 1,6 en Iran et 1,5 en Roumanie.

De 16,4% de la population mondiale en 2010, le groupe des non-affiliés devrait chuter à 13,2% en 2050, tandis que, de 23,2% en 2010, le groupe des personnes de confession musulmane devrait s'élever à 29,7% en 2050. Le pourcentage des juifs et des chrétiens devrait demeurer constant. Majoritaire en 2010, la population de confession chrétienne devrait le demeurer en 2050. En déclin et/ou perte d'influence en Europe[1], le christianisme devrait continuer à être la première religion du monde. Comme on le constate, la population islamophobe peut dormir sur ses deux oreilles, d'après les statistiques, il n'y a apparemment aucun risque d'islamisation du monde.

Il convient de préciser que les statistiques relatives aux effectifs de la population de confession musulmane surestiment la réalité du fait religieux pour deux raisons tout au moins. La première raison tient au fait que, même non pratiquants, les individus nés de parents musulmans se déclarent de confession musulmane par hérédité, pour ne pas dire par habitude, et ce a fortiori s'ils habitent dans un pays à majorité musulmane. Il y a par ailleurs le biais « risque de takfir ». Il s'agit pour les personnes athées ou agnostiques de la crainte d'être déclarées apostats ce qui, dans certains pays musulmans, peut avoir des conséquences pénales graves pouvant aller jusqu'à la peine de mort. Il est à noter à cet égard que, à l'instar de la peine de mort par lapidation, la peine de mort pour apostasie a été abolie dans la République islamique d'Iran.

Comme on a pu le constater, les statistiques établies au niveau mondial contredisent le discours de certains intellectuels médiatiques que d'aucuns qualifient d'islamophobes, quant au "risque" d'islamisation du monde en général et de l'Europe en particulier. Il est vrai que les théories du complot ne sont pas propres à un groupe spécifique à l'exclusion de tous les autres et pour preuve, la pseudo-thèse politique connue sous le nom d'Eurabia. Qu'en est-il alors du discours de certains dirigeants populistes européens quant à l'islamisation rampante de leur pays ? La réalité de la situation présente et à venir est habilement masquée au moyen de phrases ambiguës et/ou savamment

---

[1] Frédéric Lenoir, « La chrétienté est morte. Vive l'Evangile ! », *Le Monde des religions*, septembre-octobre 2010.

construites pour faire peur, telles que : « l'islamisation de l'Europe »[1], visible à travers le port du foulard ; la maison du maître d'école « vendue par la mairie et transformée en mosquée »[2] ; etc. Qu'en est-il de la réalité ?

Comme nous avons vu, données statistiques à l'appui, il n'y a pas de risque d'islamisation du monde. Y aurait-il un quelconque risque d'islamisation de l'Europe ou des Etats-Unis du fait des migrations ?

Comme il ressort clairement des données factuelles établies par région et présentées ci-après (voir tableau 2), avec ou sans migration, il n'y a pas le moindre risque d'islamisation de l'Europe ou de l'Amérique du Nord. De 74,5% en 2010, le pourcentage de la population européenne de confession chrétienne devrait en 2050 baisser à 66,7% en l'absence d'émigration, et à 65,2% si l'on tient compte des flux migratoires. En réalité, cette baisse du pourcentage des chrétiens ne s'opère pas au bénéfice du groupe des personnes de confession musulmane mais du groupe des non-affiliés, c'est-à-dire des personnes qui se déclarent « athées », « agnostiques » ou « sans religion ». Détaillées au niveau des pays, les données du tableau 4 sont assez éloquentes à cet égard. Pour ce qui concerne la France, par exemple, de 22% en 2010, le pourcentage des non-affiliés pourrait grimper à 44,1% en 2050. Bien que le taux de fécondité de ce groupe ait été estimé comme relativement faible, le groupe des non-affiliés, qui représente 18,8% de la population européenne en 2010, devrait s'élever à près de 24% en 2050.

---

[1] Expression utilisée dans le programme du FN lors de l'élection présidentielle de 2012.
[2] Propos d'Alain Finkielkraut sur l'antenne d'Europe 1. Cf. l'article daté du 11 avril 2011 du journaliste, Guillaume Stoll, dans le magazine *Le Nouvel Observateur*.

La population européenne de confession musulmane, estimée à 5,9% en 2010, devrait en 2050 s'élever à 8,4% sans migration et à 10,2% en tenant compte des flux migratoires (cf. tableau 2). Il importe de noter à cet égard que les musulmans en Europe ne sont pas tous des immigrés ou des descendants d'immigrés. Une partie de la population européenne de "souche" est de confession musulmane. Et il paraît d'autant plus risqué de nier l'hétérogénéité des peuples que, comme en témoignent les pays de l'Europe de l'Est, la population européenne de souche de confession musulmane n'est pas exclusivement constituée par des convertis de fraîche date, loin s'en faut (cf. tableau 4).

Estimée à 1% en 2010, la population de confession musulmane en Amérique du Nord devrait s'élever en 2050 à 1,4% seulement, en l'absence de tout flux migratoire et à 2,4% avec la prise en compte du flux des migrants. Pour ce qui est des Etats-Unis, comme il ressort du tableau 4, estimée à 0,9% en 2010, la population de confession musulmane, compte dûment tenu des flux migratoires, devrait s'élever à 2,1% en 2050. Comme on le constate, les Etats-Unis sont à l'abri de tout risque d'islamisation mais qu'en est-il de l'Europe ? Longtemps terre d'émigration, l'Europe serait-elle devenue une terre d'immigration ? La récente crise migratoire laisserait-elle entrevoir un quelconque risque d'islamisation de l'Europe par immigration ? Selon les données factuelles, la réponse est non. Lorsqu'on examine les statistiques relatives au flux des migrants sur la période 2010-2015, on constate que la population de confession chrétienne constitue 46% des 19,22 millions de migrants dans le monde contre seulement 30% pour la population de confession musulmane.

Le tableau 3 permet d'avoir une idée exacte de l'impact des migrations de populations de confession musulmane sur la population des pays européens. Comme on le constate, c'est la Suède qui doit accueillir le plus fort contingent de migrants au regard de sa population, +5,6% en 2050, vient ensuite le Danemark qui devrait enregistrer une augmentation de 3,8% durant la même période, après l'Espagne +3,4% puis le Royaume-Uni +3,0%. Avec +1,8 % en 2050, la France se situe au 9e rang ex aequo avec l'Irlande. Comme il ressort clairement du tableau 4, en 2050 la population allemande devrait baisser de plus de 12 millions ; suite à sa nouvelle politique d'accueil massif des migrants en 2015, la position de l'Allemagne (15e rang d'après les données du tableau 3) devra être révisée en conséquence. En effet, en date d'avril 2015, les statistiques présentées dans le tableau 3 ne tiennent pas compte des migrants clandestins et des réfugiés en provenance des pays en guerre depuis l'établissement de ces statistiques.

Les réfugiés se succèdent, les tragédies avec eux. Après le drame des boat people vietnamiens, vint celui d'autres damnés de la terre. Nous sommes aujourd'hui confrontés au drame des réfugiés syriens, érythréens, afghans, etc. D'après les chiffres du gouvernement allemand, 1,09 million de réfugiés ont été enregistrés entre janvier et décembre 2015 en Allemagne[1]. Tous ces réfugiés ne sont évidemment pas de confession musulmane et, quel que soit leur effectif, ils ne risquent pas d'islamiser l'Europe et encore moins l'Occident.

En résumé, l'impact des migrations de populations de confession musulmane sur la population européenne a dans

---

[1] Source : AFP, publié le 30/12/2015 par le magazine en ligne *Le Point.fr*. Il est à noter qu'en 2016, l'Allemagne a accueilli plus de 280 000 nouveaux demandeurs d'asile. Le parti bavarois CSU, allié conservateur d'Angela Merkel, a exigé, à l'époque, que le gouvernement se fixe un quota de 200 000 demandeurs d'asile. Ce que la chancelière avait refusé et qui lui a valu des déboires politiques fin 2017, début 2018. Cf. le quotidien *Le Monde* du 11.01.2017.

l'ensemble été évalué à 1,8% pour l'Europe et 1% pour l'Amérique du Nord. Il nous faut donc écarter tout risque d'islamisation du monde occidental par migration. Mais, comme les temps sont durs et qu'il n'y a aucune cause ou idéologie nouvelle à défendre, la question se pose de savoir si la "paranoïa antimusulmane" n'est pas fondée sur l'attractivité de l'islam comme idéologie alternative. Les données statistiques laisseraient-elles entrevoir un quelconque risque d'islamisation par conversion ?

Comme on le constate à partir des données présentées dans le tableau 5, les changements de religion ne devraient pas avoir d'impact sur les pourcentages de la population de confession musulmane dans la région Asie et Pacifique ainsi qu'en Amérique latine et dans les Caraïbes. Les données relatives au groupe d'ensemble constitué par les populations du Moyen-Orient et de l'Afrique du Nord posent problème du fait que les personnes interrogées, par crainte d'être mal perçues et/ou pour éviter tout risque de sanction pénale pour apostasie, n'osent pas se prononcer en toute sincérité sur les questions afférentes à leurs convictions religieuses. Les données relatives à l'Afrique subsaharienne doivent également souffrir du biais « risque de takfir ».

Les données relatives aux deux groupes Europe et Amérique du Nord, où la liberté de conscience est garantie, sont plus parlantes. D'après ces données, les changements de religion devraient avoir un impact négatif sur le pourcentage de la population de confession musulmane. En Europe, bien que positif, l'impact est quasi négligeable. En conclusion, il est évident, statistiques à l'appui, que, contrairement aux prévisions attribuées à Bernard Lewis, il n'y a aucun risque d'islamisation de l'Europe par conversion des populations.

Comme on a pu le constater à partir de l'évaluation de la force démographique des principales religions du monde, ce n'est pas du côté du poids démographique de la population de confession musulmane qu'il y aurait lieu de s'inquiéter. D'où vient donc cette gêne, ce malaise, cette peur diffuse de l'islam ?[1] Qui dit religion, pense à l'indicible. Il n'est évidemment pas question d'apporter ici une réponse simple, nous rendant apte à élucider toute la complexité d'une situation devenue quasi ingérable à maints égards. Pour mieux comprendre la situation, disons que nous sommes en présence tout au moins de deux phénomènes qui nous interrogent. Le premier est celui de la montée de la religiosité. Le second concerne l'instrumentalisation du « djihad mineur » par des mouvements radicaux.

Commençons par la première question qui a trait au retour du religeux. Nous reviendrons plus loin sur le second phénomène qui a, d'ores et déjà, été longuement examiné.

Ainsi que nous avons déjà eu l'occasion de le préciser, ce phénomène n'est pas exclusif à l'islam. Il n'est d'ailleurs pas certain que la religion n'ait jamais disparu comme il a pu sembler en apparence, car il ne faut pas confondre ce que d'aucuns (Marcel Gauchet et d'autres) ont qualifié de « sortie de la religion », et la « fin du religieux ». Plus catégorique que nous, le sociologue américain Peter L. Berger estime que : « L'idée selon laquelle nous vivons dans un monde sécularisé est fausse. Le monde

---

[1] Moins par souci de dépassionner le débat que de relativiser la gravité de la question, il est à noter à cet égard que quatre émissions radiophoniques du service religieux organisé par la Fédération protestante de France ont été consacrées à la thématique « Qui a peur des religions ? », émissions en date de 2010, rediffusées en 2011. Un livre, coécrit par Tariq Ramadan, Elie Barnavi et Mgr di Falco, avait précédemment été publié en 2008 sous le titre de *Faut-il avoir peur des religions ?*

d'aujourd'hui est aussi furieusement religieux qu'il l'a toujours été ».

Il conviendrait, en effet, de faire une distinction entre la sécularisation de fait de la société et la sécularisation du politique, car il est à croire que le vacarme du combat des deux idéologies politiques dominantes durant la guerre froide a fortement contribué à étouffer le son des cloches et le chant du muezzin en provenance des églises et des mosquées. Mais, s'il est vrai que les statistiques établies au niveau mondial (cf. tableau 1) confirment les dires de Peter L. Berger, les données relatives aux pays de l'Europe de l'Ouest et des Etats-Unis laisseraient penser, en revanche, que la sécularisation de ces sociétés s'accompagne d'une diminution de l'engagement religieux des populations de confession chrétienne. Mais, ici encore, c'est sans compter avec les évangélistes dont le rôle, dans l'instrumentalisation de la religion en général et de l'islam, en particulier, n'est pas des moindres (cf. supra chapitre 4).

Dissertant sur les résultats d'une récente enquête sociologique portant sur des jeunes lycéens (enquête que nous allons examiner ci-après), le journaliste[1] d'un quotidien réputé pour son sérieux, sous-titre son article par l'annonce de la « montée de la religiosité chez les jeunes musulmans ». L'enquête ne fournissant aucune donnée longitudinale, la question se pose de savoir comment, à partir de données ponctuelles, on peut conclure à une dynamique à la hausse. L'histoire nous apprend que les données ethniques et religieuses sont par définition des données sensibles qu'il convient de manier avec beaucoup de précautions. La première condition étant évidemment l'honnêteté intellectuelle, il convient avant toute chose de se méfier des

---

[1] Pierre Wolf-Mandroux, « Les adolescents les plus pratiquants affichent leur défiance envers l'école », le quotidien *La Croix* du 04/02/2016.

risques de biais dans l'interprétation des données tenant à différents facteurs. Ceci dit, il est vrai que l'examen des données temporelles sur l'évolution de l'engagement religieux permet de croire en la montée de la religiosité dans le monde, certes, mais pas exclusivement chez les musulmans.

En tout état de cause, en l'absence de données factuelles sur l'évolution de la religiosité dans l'enquête susmentionnée, l'on est en droit de se demander s'il ne s'agit pas plutôt d'un « sentiment » du journaliste plutôt que d'une réalité sûre et certaine ; tant il est vrai qu'il n'est pas permis à partir de données comparatives certes, mais transversales et ponctuelles sur la croyance de jeunes lycéens d'un département, de tirer des conclusions sur la montée de la religiosité d'une partie spécifique de la population d'un pays multiconfessionnel.

Ceci dit, il est vrai que l'enquête sociologique intitulée « Les adolescents et la loi », qui a été menée par le CNRS et Sciences-Po Grenoble auprès de 9000 collégiens des Bouches-du-Rhône entre avril et juin 2015, met en exergue une plus forte religiosité chez les adolescents de confession musulmane de cette zone géographique de la France. Il est à noter que Jean-Paul Willaime, sociologue des religions et Directeur d'études à l'École pratique des hautes études, se demande si le choix de ce département, plus marqué par la pluralité religieuse, ne conduit pas à un biais. En effet, d'après ce chercheur, la part des musulmans dans ce département est nettement supérieure à celle enregistrée en 2008 à l'échelle nationale (25,5 % contre 13 %) et celle du groupe des « sans religion » apparaît relativement faible (38,8 %).

Un autre risque de biais tient au fait que cette enquête, qui commence à identifier huit catégories différentes à partir des réponses à sa question « As-tu une religion, si oui laquelle », recentre les réponses aux questions destinées à mesurer « la religiosité, le fondamentalisme et la confiance dans les services publics, dont l'enseignement », sur trois à quatre catégories selon les besoins de l'analyse. Les huit groupes identifiés par ordre d'importance démographique sont les suivants : Pas de religion (38,4%), chrétiens catholiques (30,1%), musulmans (25,3%), chrétiens protestants (1,7%), juifs (1,6%), autres chrétiens (1,1%), autres religions (0,8%), pas de réponse (0,9%). Les quatre groupes retenus aux fins d'analyse sont soit les quatre catégories : athées, catholiques, musulmans et autres religions, soit les trois catégories : catholiques, musulmans et autres.

La prise en compte des autres groupes minoritaires (protestants, juifs, etc.) dans l'analyse des réponses aurait probablement eu pour effet de tempérer les résultats qui semblent mettre en exergue la plus forte religiosité des adolescents de confession musulmane sans qu'on puisse mesurer le poids de la religion comme marqueur identitaire, du fait de la non-prise en compte des autres minorités. C'est ainsi qu'à la question, « la religion est-elle importante pour toi ? » : 61,4% des lycéens de confession musulmane ont répondu « très », contre 6,2% de ceux de confession catholique et 22,5% du groupe constitué par les autres religions. Etant donné qu'en période de crise, le sentiment de religiosité augmente chez les minorités, il aurait été intéressant de connaître la réponse des autres minorités et plus spécifiquement des juifs. Il est à noter à cet égard que le pourcentage des lycéens qui considèrent la religion comme « importante » est de l'ordre de 15,8% pour les catholiques, 20,6% pour les musulmans et 25% pour les autres religions.

Ainsi que l'enquête "Banlieue de la République" de 2011[1] l'a révélé, plus l'individu se sent inséré d'un point de vue socio-économique, plus la gestion qu'il fait de ses identités territoriale, culturelle et religieuse est harmonieuse et pacifiée. C'est pourquoi il est important de garder présente à l'esprit l'influence du milieu socio-économique des lycéens qui ont fait l'objet de l'enquête. Il y a lieu de préciser à cet égard que le pourcentage des lycéens interrogés ayant spécifié habiter dans une HLM est de l'ordre de 47,7% pour les lycéens de confession musulmane contre 13,9% pour les lycéens de confession catholique et le pourcentage de ceux qui ne savent pas s'ils habitent une HLM ou pas est de l'ordre de 20,3% pour les lycéens de confession musulmane contre 11,3% pour les lycéens de confession catholique. Par ailleurs, le pourcentage des lycéens bénéficiant d'un enseignement privé est de l'ordre de 36,6% pour les lycéens de confession catholique, contre 7,7% des lycéens de confession musulmane. Comme on le constate, les données relatives au sentiment de religiosité des lycéens enquêtés ne peuvent être interprétées dans l'absolu, sans tenir compte de tous les facteurs déterminants.

Il est à noter à cet égard que, le 31 mars 2010, Soheib Bencheikh, Grand mufti de la mosquée de Marseille, est invité à intervenir sur la thématique « Islam : foi et valeurs » dans le cadre d'un séminaire « Psychiatrie, psychothérapie, cultures » ; avant de lui céder la parole, le Dr. Bertrand Piret, Président de Parole Sans Frontières, prend soin d'exposer les raisons de son invitation. La raison centrale invoquée par l'intéressé, qui nous paraît des plus éloquentes au regard de nos préoccupations, est d'ordre clinique. Le

---

[1] Enquête de 2011 de l'Institut Montaigne auprès des habitants de l'agglomération de Clichy-sous-Bois et Montfermeil en Seine-Saint-Denis, épicentre des émeutes de 2005.

Président de Parole Sans Frontières précise : « nous rencontrons dans nos consultations – et c'est ce qui fait le socle sur lequel s'appuie ce séminaire depuis dix ans – des patients d'origine étrangère, des patients immigrés qui sont en lutte, pour résumer, contre deux facteurs de difficultés. Le premier facteur est le passage d'un monde de traditions, d'un monde souvent rural à un monde urbain, dit moderne. Le deuxième facteur est le passage d'un monde sacré à un monde dit laïc ».

Shlomo Sand[1], qui fait état de son désaccord concernant la « loi sur le voile » à l'appui de son plaidoyer en faveur d'une laïcité bien comprise, rapporte que ses « deux aïeules du ghetto de Lodz ont été gazées en 1941, alors qu'elles portaient des coiffes religieuses juives. Toutes deux étaient des croyantes traditionnelles, comme la plupart des femmes juives de leur âge, issues des milieux populaires ». Dans le cadre de la campagne présidentielle américaine de 2016, ce que Donald Trump[2] avec ses propos insultants, reproche en réalité aux migrants mexicains c'est d'appartenir aux couches défavorisées et/ou populaires de la population mexicaine. Si l'on fait abstraction du cas des réfugiés politiques, la majorité des migrants de confession musulmane dans les pays européens appartiennent à des classes populaires de leurs pays d'origine.

Si nous nous en tenons au cas français, en référence à l'enquête auprès des lycéens susmentionnés, ce qu'on exige de ces enfants d'immigrés des classes défavorisées c'est d'accomplir un triple exploit. Ces enfants doivent pour la plupart surmonter trois catégories de difficultés tenant à

---

[1] Shlomo Sand, *La fin de l'intellectuel français ?*, Paris, La Découverte, 2016.
[2] Voir, entre autres, AFP, « Propos de Donald Trump "absurdes" (Mexique) », le quotidien *Le Figaro* du 17.06.2015.

leur statut d'enfants d'immigrés de classes populaires en situation de précarité vivant dans un environnement de plus en plus hostile aux étrangers en général et à ceux de confession musulmane en particulier. Juridiquement « Français » mais de fait « enfant d'immigrés », donc « immigré » en raison de ton milieu et « étranger » en raison de ta religion, tu resteras. Et, ironie du sort, ce sont souvent des immigrés et enfants d'immigrés[1] qui reprochent à leurs congénères de ne pas avoir réussi l'exploit qu'eux, dans des circonstances différentes, ont eu la chance de pouvoir accomplir.

Vu la situation, il serait sage, de tenir compte du contexte actuel caractérisé par la montée du populisme, du fanatisme religieux et de l'islamophobie et, de garder présent à l'esprit que nombre de jeunes Occidentaux de confession musulmane ont depuis leur naissance assimilé l'idée selon laquelle « islamisme » équivaut à « islam politique ». Comme aujourd'hui d'aucuns ne se lassent pas de leur faire comprendre qu'islam et islamisme sont une seule et même chose, certains d'entre eux finissent par penser qu'à moins de renier leur religion, ils n'ont d'autre choix que d'assumer leur "identité d'islamiste".

*

Après examen des données d'enquête nous en venons à la question de fond concernant la problématique de la mesure de la montée de la religiosité, qui est celle de savoir ce qu'on entend par « sentiment de religiosité ».

Si le but ultime de la religion est le salut dans l'au-delà et la béatitude ici-bas, on n'a pas besoin de lire Simone

---

[1] Alain Finkielkraut, par exemple, ne comprend pas que, si certains jeunes en sont là où ils en sont, c'est justement parce qu'ils n'ont pas pu s'intégrer comme il le fallait.

Weil, Mowlavi[1], Hafez, Sohravardi, Hallaj, Attar ou tant d'autres, pour savoir qu'il s'agit là d'un tout autre exploit, gratifiant certes, mais exploit dans tous les cas. Pour ce qui est du sentiment religieux du commun des mortels, il serait, *in fine*, indiqué de se fier au bon sens des sociologues, selon qui la vitalité des religiosités ne doit pas faire oublier que l'humanité se partage entre deux petites minorités opposées : la première est formée par le groupe restreint mais en progression constante des athées et la seconde par ceux qui ont une foi indéfectible en Dieu. Et le reste, c'est-à-dire la très grande majorité des individus, est constitué par toute la panoplie des personnes qui, ni incroyantes ni détentrices de la vérité, sont plus ou moins fortement engagées dans une religion donnée. « Cette majorité oscille entre foi et agnosticisme, se conformant au scepticisme de Montaigne - croire, mais sans certitude. » Statistiquement vérifiable, cette réalité mériterait d'être étayée par des données chiffrées récoltées à partir des différentes modalités d'enquête[2] et schématisée en bonne et due forme par la courbe y afférente qui devrait être de forme Gauss selon toute vraisemblance.

*

Reste la question de la plus grande visibilité de l'islam, avec ou sans « i » majuscule[3].

Plus grande visibilité au niveau mondial, en raison du fait du terrorisme islamique qui est devenu le trou noir de la

---

[1] Mowlânâ Djalâl Od-Dîn Rûmî en persan, la transcription du nom en français diffère d'une traduction à l'autre.
[2] Pour ce qui concerne les techniques d'enquête, voir, entre autres, Christophe Monnot, « Mesurer la pratique religieuse », in *Archives de sciences sociales des religions*, 2012/2 (N° 158), pp 137-156.
[3] Le terme islam sans majuscule est conventionnellement réservé pour désigner la religion et Islam avec majuscule se réfère à la civilisation islamique.

géopolitique mondiale. Plus grande visibilité au sein des pays à majorité musulmane, en raison de la montée du fondamentalisme et de la résurgence de régimes autoritaires d'obédience islamique. Plus grande visibilité au sein des pays occidentaux, en raison de la vogue du voile et de l'édification de mosquées. Il ne s'agit pas d'ouvrir ici un nouveau débat sur le voile, débat qui nous paraît dépassé à maints égards, disons seulement qu'il existe toute une panoplie de situations et de motivations quant à son port, qui vont de la servitude volontaire, à la mise sous tutelle forcée, de la femme. Nous savons aujourd'hui que le port du voile peut être autoritaire (pression de la police des mœurs), prescrit (pression de la famille et de la société) ou volontaire, dans le meilleur des cas. Volontaire, "dans le meilleur des mondes", qu'il soit dans ce cas un voile de conviction ou un voile militant de contestation, comme l'histoire des femmes nous l'a appris, dans tous les cas, nous sommes, du point de vue de la réalité pratique, en présence d'un indicateur d'asservissement de la femme. Et, d'un point de vue théologique, le voile comme norme vestimentaire imposée aux femmes ne peut que faire du tort à l'image de l'islam en tant que religion.

Après la polémique sur « le voile à l'école » des années 90, puis celle sur « le niqab dans l'espace public » des années 2006-2010, vint, courant été 2016, le temps de la polémique sur « le burkini dans l'espace public ». Exemple typique de récupération des faits divers par le monde politique, la polémique sur le burkini, qui a occupé l'essentiel de l'actualité politique française au cœur du mois d'août 2016 et qui est censée être close par la décision du Conseil d'Etat d'invalider les arrêtés « anti-burkini », a de fortes chances de rebondir dans les années à venir pour se clore courant 2020 par une nouvelle loi anti-…, les pointillés laissés au gré de l'imagination débordante du monde politique. Ce qui est intéressant, au regard de la polémique sur la tenue vestimentaire des femmes musulmanes, c'est que, tant que celles-ci se baignaient vêtues de robes et de pantalons, leur mode vestimentaire étant assimilé à de l'archaïsme, il ne posait apparemment pas de problème majeur. Ce n'est que lorsque les femmes ont commencé à se vêtir de « burkini », que la tenue a été considérée comme scandaleuse. Il y a cinquante ans, c'est le bikini qui faisait scandale, maintenant c'est le burkini ! Entre le string qui fait figure de bikini et le burkini qui s'est substitué aux robes et pantalons, on est en droit de se demander lequel est le plus scandaleux au regard de la norme, c'est-à-dire indécent au regard de l'opinion majoritaire qui fait la norme : le string ou le burkini ? Au regard de la « norme », certains diront les deux. Mais pourquoi alors le string est-il toléré et le burkini rejeté ?

Commençons par le string. La nudité, marqueur de la « sauvagerie », aujourd'hui identifie le « civilisé », mais si le « civilisé », à l'instar du « sauvage », porte un string, ceux-ci diffèrent en ceci que le sauvage est nu tandis que le civilisé s'est dénudé. Le string, effet pervers de la postmodernité, fait un étalage ostentatoire d'une libération des mœurs. La question se pose maintenant de savoir pourquoi le « burkini » dérange tant les consciences ; de sorte que, même parmi tous ceux qui tolèrent le voile, certains commencent à se poser des questions. Si l'on réfléchit de plus près au cas du burkini, on constate que la raison tient au fait qu'avec cette tenue vestimentaire, nous sommes en présence d'un cas flagrant d'islamisation de la modernité. Tant que les femmes se baignaient vêtues de robes et de pantalons, on pouvait penser qu'il y avait là un semblant de modernisation des mœurs puisque ces femmes pouvaient enfin se baigner, et d'ici peu elles allaient pouvoir se baigner « normalement », vêtues de maillots. Le problème le plus révoltant au regard de l'évolution des mœurs vestimentaires dans le monde musulman, aujourd'hui encore plus qu'hier, réside dans l'inégalité des genres. En effet, le fait le plus aberrant, et qui n'a d'ailleurs malheureusement pas été relevé, est la discrimination positive flagrante en faveur des hommes au regard de l'évolution vestimentaire qui s'est traduite par le fait que : les hommes ont droit à la modernisation des mœurs vestimentaires et les femmes, à l'islamisation desdits mœurs.

Pour ce qui concerne la question de la plus grande visibilité de l'islam consécutive à l'édification de mosquées, il est nécessaire, comme Olivier Roy l'illustre pour le cas de la France, de faire la part des choses. D'après Olivier Roy[1], « le surgissement brutal de centaines de mosquées dans les années 1980 a pu donner l'impression d'une expansion rapide de l'islam. Or, le nombre de moquées était très faible autour de 1970. Lorsque les travailleurs immigrés font venir leur famille et quittent donc les foyers de travailleurs (où une pièce servait de mosquée), les plus pratiquants d'entre eux éprouvent le besoin d'une mosquée de quartier comme on a une église de quartier. L'apparition de mosquées correspond ici à un besoin d'insertion sociale et à une recherche de notabilité de la part d'immigrés qui réalisent qu'ils ne retourneront plus au pays. Mais le nombre et la taille des mosquées, ramenés à celui de la population musulmane, font que les musulmans ne disposent sans doute pas encore d'une superficie de lieux de culte en rapport avec leur nombre. Plus qu'une réislamisation, nous assistons à l'apparition au grand jour d'une religion quasiment clandestine jusqu'en 1980 ».

---

[1] Olivier Roy, *Généalogie de l'islamisme*, *op.cit.*, pp 101-102.

En France comme dans les autres pays de l'Europe de l'Ouest qui ont fait appel à de la main-d'œuvre de confession musulmane, le nombre de mosquées augmente, en raison de l'accroissemnt de la population concernée certes, mais le problème ne tient pas seulement au nombre mais à la spécificité du lieu de culte. En raison du fait même qu'il s'agit de mosquées et non d'églises, ces lieux de culte deviennent plus visibles. Sinon, que dire des cinq dômes dorés de l'Eglise orthodoxe inaugurée le 15 mars 2016 à Paris. Si l'inégalité des moyens est manifeste, dans tous les cas, il est certain que nous nous situons dans un contexte caractérisé par le retour du religieux et ce retour ne se limite évidemment pas à l'islam, mais concerne toutes les religions comme nous avons déjà eu l'occasion de le préciser et de l'analyser.

*

Reste maintenant à examiner la question de la plus grande visibilité de l'islam due au phénomène du terrorisme wahhabo-salafiste consécutif à la montée de l'intégrisme religieux.

Il nous faut rappeler ici que la montée de l'intégrisme religieux n'est pas un phénomène limité aux religions monothéistes, loin s'en faut. Il y a également les cas non moins préoccupants, des pays de confession hindouiste ou bouddhiste, comme l'Inde ou la Birmanie, qui sont également frappés par la montée de l'intégrisme religieux. Pour ce qui concerne le christianisme, nous avons affaire à l'essor des évangélistes, des pentecôtistes et au renouveau de la religion orthodoxe suite à l'effondrement du communisme ; pour ce qui est du judaïsme, il y a le mouvement techouva. Et concernant ce mouvement, on ne peut ignorer que la condition des femmes de la communauté harédim en

Israël n'est pas tellement meilleure que celle des femmes de la communauté wahhabite en Arabie saoudite. Le problème avec la montée de l'intégrisme religieux, en général, et du terrorisme wahhabo-salafiste, en particulier, est que l'idée de progrès est sacrifiée sur l'autel de la sécurité qui a fini par faire office de projet de société. Avec l'ennemi pour seul horizon d'avenir, les sociétés occidentales conjurent leur peur de l'« islamisme » par la hantise des catastrophes écologiques et technologiques (risque nucléaire inclus). C'est dans cette perspective que nous devons examiner la question de l'instrumentalisation du djihad mineur par des mouvements radicaux. D'après les enquêtes[1] portant sur la psychologie et les motivations des djihadistes candidats aux attentats suicides, il ressort que, si l'appartenance religieuse joue par définition un rôle important, ceux-ci ne naissent et ne vivent pas nécessairement dans un milieu plus religieux que les autres.

Selon Jacques Julliard[2], « le goût de la violence et la haine du christianisme expliquent la fascination pour l'islam radical » des intellectuels de gauche, « pour la plupart agnostiques et libertaires ». D'après cet historien français, que d'aucuns qualifient d'anticlérical, ces intellectuels se seraient « brusquement pris de passion pour la religion la plus fermée, la plus identitaire, et, dans sa version islamiste, la plus guerrière et la plus violente à la surface du globe ». La violence de la haine de Jacques Julliard envers l'islam laisse perplexe mais plus encore c'est l'impressionnante amnésie des faits historiques qui surprend. Nous avons pourtant affaire à un historien qui est

---

[1] Andrew Silke, "Holy Warriors : Exploring the Psychological Processes of Jihadi Radicalization", *European Journal of Criminology*, 01/2008, vol. 5 (1), pp 99-121.
[2] Jacques Julliard, « Aux sources de l'islamo-gauchisme », le quotidien *Le Figaro* du 02.05.2016.

censé être au fait de la violence des guerres de religion. Et, pour ce qui est de l'histoire immédiate, la violence du 21ᵉ siècle serait-elle le seul fait des terroristes djihadistes ?

Car, s'il est vrai que toute l'histoire de l'humanité est jalonnée de violence, contrairement à ce que Jacques Julliard semble laisser entendre, la violence n'a pas été le seul fait du fascisme, du communisme ou du djihadisme, mais elle a tout autant été le fait des croisades, du colonialisme et de l'impérialisme, comme Alain Gresh[1] le rappelle à juste titre. Il est vrai que mue davantage par la volonté de puissance que par la recherche d'intérêts économiques, raciste, la conception messianique de l'impérialisme colonial de Jules Ferry était teintée de bonne conscience tandis que l'impérialisme économique qui lui a succédé a eu "l'intelligence" de se délester du pesant fardeau de la bonne conscience.

« Fascination pour l'islam radical », c'est là où Jacques Julliard a vu juste. En effet, l'ampleur et la virulence des débats, que Jacques Julliard contribue d'ailleurs à entretenir, témoignent de cette fascination et Rémi Brague, historien des religions, explique le pourquoi de cette fascination. Invité de France culture[2], Rémi Brague déclarait que l'Occident envie ces gens qui ont des « idéaux » pour lesquels ils sont prêts à mourir. Serions-

---

[1] Dans sa réplique à Jacques Julliard intitulée « Fantasmes et approximations de Jacques Julliard » en date du lundi 2 mai 2016, Alain Gresh précise : « On pourrait longuement parler de la violence à travers l'histoire. Il est étonnant d'évoquer la violence fasciste et communiste, sans dire un mot de la violence coloniale, des millions de gens exterminés, du Congo à l'Algérie. Ni de la violence actuelle, des bombardements, des drones, des assassinats ciblés menés par les troupes américaines, russes ou françaises. La destruction de l'Irak par les Etats-Unis, rappelons-le, a permis la création de l'Organisation de l'Etat islamique (OEI). Je souhaiterais savoir qui justifie la violence d'Al-Qaida ou de l'OEI ? Pour Julliard, seule la violence des groupes non-étatiques est condamnable, jamais celle des Etats. ».
[2] Les matins de France Culture du 21 novembre 2015.

nous en présence d'une guerre des subconscients, nostalgie de cette ferveur de la foi religieuse du temps de la splendeur de l'empire byzantin, qui fut troquée en vain contre la foi dans le progrès comme faisant pendant à l'enthousiasme suscité par le mythe du Califat brandi par Daech ? Il est à noter que la remarque de Rémi Brague relève davantage d'un lapsus que d'un aveu, tant il est mal vu d'essayer de comprendre la situation au risque de la justifier, et tant les intellectuels sont sommés de la désavouer sans plus, c'est-à-dire sans oser l'expliquer en quoi que ce soit, sous peine d'être attaqués et voués à l'opprobre par ceux-là mêmes qui prétendent œuvrer à la libération de la parole[1]. Et pour preuve, la vive polémique autour de certaines des paroles de bon sens émises par Michel Onfray.

Dans un article intitulé : « Faut-il brûler Onfray ? », le magazine *L'Express*, le cite déclarant, dans un entretien publié par le *Corriere della Sera* : « L'islam manifeste ce que Nietzsche appelle "une grande santé" : il dispose de jeunes soldats prêts à mourir pour lui. [...] Quel Occidental est prêt à mourir pour les valeurs de notre civilisation : le supermarché et la vente en ligne, le consumérisme trivial et le narcissisme égotiste, l'hédonisme trivial et la trottinette pour adultes ? » et conclut avec à-propos : « ni adorer ni brûler, mais comprendre » le citoyen et philosophe Onfray[2].

En effet, les intellectuels devraient non seulement avoir le droit de chercher à comprendre la situation pour y remédier mais il est également de leur devoir de le faire le

---

[1] Dans un éditorial au titre tapageur, « La parole libérée », du magazine *Marianne* en date du 28 novembre 2015, Jacques Julliard déclare : « Les attentats ont fait exploser cette orthodoxie de la pensée et du langage, ont libéré la parole et ont posé au grand jour des questions qu'il était déjà jugé « courageux » de murmurer en douce ».
[2] Le magazine *L'Express* N° 3374, semaine du 2 au 8 mars 2016.

plus honnêtement possible compte tenu de la mondialisation et de l'interdépendance de fait qui font qu'il n'est pas d'action sans conséquence, de sorte que même les plus forts ne peuvent échapper à la loi du "karma". Pour une grande partie, produit des blessures mal cicatrisées, l'histoire nous révèle que, si la mémoire ici a tendance à refouler ce qui dérange, la mémoire là-bas a vite fait de resurgir pour faire état du ressentiment de cet ailleurs.

Chercher à comprendre l'islam radical, c'est commencer par définir ce qu'on entend par « islam radical ». Après avoir circonscrit ce qu'on entend par islam radical, il s'agit de comprendre la ou les raison(s) d'être des mouvements djihadistes en général. Ensuite, chercher à identifier et comprendre les motivations des kamikazes djihadistes. Et enfin, tenter d'évaluer l'impact de la violence djihadiste sur l'Occident : réflexions et engagements des intellectuels, réponses des classes dirigeantes, réaction de la société civile, etc.

*

Pour ce qui est des motivations des kamikazes djihadistes, nous nous limiterons ici au cas des djihadistes d'origine occidentale pour la très bonne et simple raison que le changement majeur entre les kamikazes djihadistes de la génération al-Qaïda et celle de Daech tient à l'origine de ces kamikazes. Aujourd'hui, on voit moins de djihadistes venant du Moyen-Orient frapper l'Europe que de djihadistes produits par l'Europe frapper les sociétés du Moyen-Orient. Contrairement à al-Qaïda, c'est au sein même des sociétés occidentales que Daech recrute ses candidats au djihad terroriste. Plus radical sur le plan pratique et moins rigoureux sur le plan théologique, Daech a réussi au moyen de divers stratagèmes à brouiller tous les repères et à créer le chaos pour un certain temps, le triste débat en France sur la déchéance de nationalité en a été un exemple. Moins obnubilé par les questions religieuses qu'Al-Qaïda, Daech a recruté ses candidats parmi les jeunes Occidentaux davantage en fonction de leur haine et de leur capacité de violence que de leur foi religieuse et de leur idéal humanitaire.

C'est en ce sens que le politologue spécialiste de l'islam Olivier Roy[1] et l'anthropologue Alain Bertho[2] ont permis de mieux appréhender le cas spécifique d'une grande partie de la jeunesse radicalisée embrigadée par Daech. Contrairement à ce que les critiques formulées par Scott Atran[3] peuvent laisser entendre, Olivier Roy a raison de parler d'« islamisation de la radicalité ». Mais là où Scott Atran a en revanche raison, c'est de vouloir mettre en garde contre l'erreur et le piège qui consisteraient à penser que toutes les personnes enrôlées par Daech auraient un antécédent de délinquance ou que l'on n'aurait affaire qu'à des analphabètes en matière religieuse, ceci au risque de dédiaboliser des actes barbares commis et légitimés au nom d'une certaine interprétation des textes sacrés. Gilles Kepel, pour sa part, « considère que le passage à l'acte terroriste découle d'une interprétation radicale de l'islam relayée notamment par les mosquées salafistes ». En vérité, ce qu'il y a lieu de mettre à jour et de dénigrer fermement, c'est l'instrumentalisation de l'islam via sa radicalisation par des entités étatiques et des mouvements terroristes.

Ainsi qu'Olivier Roy, Alain Bertho et Rik Coolsaet le soulignent, le changement central qui s'est opéré en matière

---

[1] Olivier Roy, « Le djihadisme est une révolte générationnelle et nihiliste », le quotidien *Le Monde* du 24 novembre 2015.
[2] Catherine Tricot (entretien) « Alain Bertho : « Une islamisation de la révolte radicale », *Regard.fr*, 11 mai 2015.
[3] Scott Atran, *L'État islamique est une révolution*, Paris, Ed. Les liens qui libèrent, 2016, p.12.

de politique de recrutement des djihadistes tient aux connaissances théologiques en matière d'islam. La réalité, et c'est ce qui a fait le succès de Daech, c'est que contrairement à al-Qaïda, le djihadisme version Daech est très peu exigeant en termes religieux, sur tous les plans, qu'il s'agisse du mode de vie, des pratiques ou des connaissances théologiques et le pire est que « l'organisation fournit une licence pour le crime au nom d'un but supérieur ». Cette politique de recrutement a permis d'élargir considérablement l'« armée de réserve » des personnes parmi lesquelles Daech était susceptible et continue de trouver des recrues djihadistes en permettant à des individus ayant des profils socio-économiques et identitaires variés et mus par tout un éventail de motivations distinctes de se porter candidat.

De sorte que les candidats au djihad ne sont pas exclusivement constitués par les couches défavorisées de la population issue de l'immigration mais comportent également l'élite intellectuelle pour qui la religion n'est pas une fin mais un moyen. La motivation première de ces derniers n'est pas la foi et la religion mais la frustration relative (écart entre ce à quoi je crois avoir droit et ce à quoi j'ai effectivement droit), le ressentiment consécutif à la guerre des mémoires et à la montée de l'islamophobie, et la soif de vengeance.

Un autre changement important a trait à l'origine des ascendants des candidats au djihad qui, loin de se limiter à des populations issues de l'immigration, comportent un nombre croissant de convertis occidentaux appartenant à la population de « souche » européenne (Français, Anglais, Allemand, Danois, Espagnol, etc.).

Il importe de garder présent à l'esprit que la socialisation à l'Idée des droits de l'homme suscite des attentes chez la jeunesse occidentale. C'est pourquoi parmi les candidats au djihad, se trouvent aussi et surtout des déçus du système, en quête d'utopie politique et des idéalistes en quête de cause humanitaire et non pas exclusivement des personnes qui ont de l'appétence pour la violence et pour les discours anti-systèmes. D'après Rik Coolsaet[1], Daech offre un catalogue de solutions aux différentes frustrations et manques de perspectives que les hommes et femmes semblent ressentir dans leur propre pays, « en leur proposant pouvoir, sentiments d'appartenance, camaraderie, respect, reconnaissance, aventure, héroïsme, et martyr. En plus de la richesse matérielle, l'organisation fournit une licence pour le crime au nom d'un but supérieur ».

En ce qui concerne ces derniers et leur goût pour la lutte, le danger et la mort, Scott Atran[2] cite George Orwell qui, dans sa critique de Mein Kampf, dépeint ainsi l'essence du problème : « Hitler sait que les êtres humains ne désirent pas seulement le confort, la sécurité, moins d'heures de travail et une meilleure santé… et en général du bon sens ; ils veulent aussi, du moins de façon intermittente, du combat et du sacrifice ». Il convient de préciser que, dans les groupes activistes, que l'on ait affaire à des manifestants, des supporters ou des insurgés, le noyau dur des plus violents (les casseurs, les hooligans, ou les terroristes) est toujours formé par des radicaux ultra-minoritaires mais à très forte capacité de nuisance. Qu'il s'agisse des hooligans ou des terroristes, drogués à l'adrénaline de leurs expériences exaltantes, ils ont le sentiment de former une famille unie par des valeurs communes. Impuissants et/ou désœuvrés, les groupes humains (entités étatiques compris) peuvent devenir violents.

---

[1] Rik Coolsaet, « Faire face à la quatrième vague de combattants étrangers. Qu'est-ce qui conduit les Européens en Syrie et vers Daech ? Aperçu du cas belge », intervention dans le cadre du Colloque international « Le djihadisme transnational, entre l'Orient et l'Occident » co-organisés par l'Institut Montaigne et la Maison des sciences de l'homme (31 mai – 2 juin 2016).
[2] *op. cit.*, p. 59.

*

Au terme de notre étude il est plausible de conclure que source de confusions dans les mondes de la science, de la politique et dans la vie courante des sociétés, le dévoiement du terme islamisme a généré plus de problèmes qu'il n'en a résolu. Et, *in fine*, le concept d'islamisme peut être considéré comme le trou noir de la géopolitique du XXIe siècle. Avec une mondialisation tous azimuts qu'aucune « gouvernance mondiale » ne parvient à maîtriser, nous sommes entrés dans l'ère de la complexité propice à l'essor d'entités illégales qui échappent à tout contrôle. Aujourd'hui le vocable « islamisme » est utilisé pour désigner les mouvements « wahhabo-djihadistes » tels qu'al-Qaïda et Daech. Nous avons déjà examiné la raison d'être de ces mouvements terroristes : al-Qaïda est le produit de l'instrumentalisation de l'islam aux fins de la guerre froide par les Etats-Unis et Daech la conséquence de la destruction de l'Irak par les Etats-Unis de G.W. Bush.

Il est certain qu'aujourd'hui le terrorisme wahhabo-salafiste ne peut passer inaperçu. Mais, lorsqu'on analyse la situation dans toute sa complexité pour faire le bilan, on constate que, si l'on raisonne en termes de nombre de morts, ce sont des musulmans qui font d'abord et surtout les frais du problème du terrorisme wahhabo-salafiste. Si les Rohingyas de Birmanie n'ont pas eu besoin d'être des « islamistes » pour être persécutés et massacrés, ce sont pour l'essentiel des musulmans « non-islamistes » qui ont perdu et continuent de perdre la vie sous les bombes de la coalition occidentale en Afghanistan, en Iran, en Irak, en Libye, en Syrie et ailleurs. Sans parler du fait qu'avec le problème de l'« islamisme », le commun des musulmans a acquis le statut de souffre-douleur des sociétés occidentales en mal de bouc émissaire et de nouvel ennemi.

Le fait marquant du terrorisme wahhabo-salafiste est de nous rappeler la réalité de l'interdépendance. Ainsi que nous avons déjà eu l'occasion de le signaler[1], il est de plus en plus difficile de penser en termes de « nous » et de « eux » ; l'autre n'est pas ailleurs, il est ici. On ne peut plus exercer une ingérence en toute immunité et toute impunité, ni réagir à l'ingérence par un terrorisme sans conséquence. Avec les nouvelles technologies de l'information et de la communication, l'intensification des migrations, avec les nouvelles guerres (guerre dans les cyberespaces), avec les nouvelles technologies de guerre (drones et autres engins sophistiqués), la frontière entre le dedans et le dehors s'estompe, ici comme là-bas, il devient de plus en plus difficile d'avoir des ennemis.

Le problème est que cette réalité est tout aussi valable pour l'Occident que pour l'Orient. Barak Obama a pourtant essayé mais, malgré toute sa bonne volonté, il n'a pas réussi à inverser la situation et, tout aussi inconséquent que cela puisse paraître ici comme là-bas, c'est la posture hobbesienne qui semble être privilégiée. Et les diatribes et mesures anti-musulmans de Donald Trump, qui ont pour effet d'attiser la haine des islamophobes et de renforcer le malaise du commun des musulmans, n'augurent rien de bon.

---

[1] Voir, entre autres, l'essai « Les Organisations internationales et régionales et le progrès du genre humain : quel avenir pour la culture de la paix et l'éthique de la non-violence », op.cit.

Tableau 1 : Evolution du poids démographique absolu et relatif des principales religions du monde

	Population 2010	% de la population mondiale en 2010	Prévision de la population 2050	% de la population mondiale en 2050	Croissance de la population 2010-2050
Chrétiens	2 168 330 000	31.4%	2 918 070 000	31.4%	749 740 000
Musulmans	1 599 700 000	23.2	2 761 480 000	29.7	1 161 780 000
Non-affiliés	1 131 150 000	16.4	1 230 340 000	13.2	99 190 000
Hindouistes	1 032 210 000	15.0	1 384 360 000	14.9	352 140 000
Bouddhistes	487 760 000	7.1	486 270 000	5.2	-1 490 000
Religions autochtones	404 690 000	5.9	449 140 000	4.8	44 450 000
Autres religions	58 150 000	0.8	61 450 000	0.7	3 300 000
Juifs	13 860 000	0.2	16 090 000	0.2	2 230 000
Population mondiale	6 895 850 000	100.0	9 307 190 000	100.0	2 411 340 000

Source : Tableau construit à partir des données extraites de : Pew Research Center, April 2, 2015, "The Future of World Religions: Population Growth Projections, 2010-2050".

Tableau 2. Changement du poids démographique des différentes religions par région en 2050, avec migrations et sans migration

Régions	Années	Chrétiens	Musulmans	Non-affiliés	Hindouistes	Bouddhistes	Juifs	Autres religions
Afrique sub-saharienne	2010, % de la population de la région	62,9%	30,2%	3,2%	0,2%	0,0%	0,0%	3,5%
	2050 avec migration	58,5%	35,2%	2,7%	0,1%	0,0%	0,0%	3,4%
	2050 avec zéro migration	58,5%	35,3%	2,6%	0,1%	0,0%	0,0%	3,4%
Amérique du Nord	2010, % de la population de la région	77,4%	1,0%	17,1%	0,7%	1,1%	1,8%	0,9%
	2050 avec migration	65,8%	2,4%	25,6%	1,3%	1,4%	1,4%	2,1%
	2050 avec zéro migration	66,0%	1,4%	27,1%	0,8%	1,2%	1,5%	2,1%
Amérique latine & Caraïbes	2010, % de la population de la région	90,0%	0,1%	7,7%	0,1%	0,1%	0,1%	1,9%
	2050 avec migration	88,9%	0,1%	8,7%	0,1%	0,1%	0,1%	2,1%
	2050 avec zéro migration	89,0%	0,1%	8,7%	0,1%	0,1%	0,1%	2,1%
Asie & Pacifique	2010, % de la population de la région	7,1%	24,3%	21,2%	25,3%	11,9%	0,0%	10,3%
	2050 avec migration	7,7%	29,5%	17,0%	27,7%	9,6%	0,0%	8,4%
	2050 avec zéro migration	7,8%	29,6%	16,9%	27,7%	9,6%	0,0%	8,4%
Europe	2010, % de la population de la région	74,5%	5,9%	18,8%	0,2%	0,2%	0,2%	0,2%
	2050 avec migration	65,2%	10,2%	23,3%	0,4%	0,4%	0,2%	0,4%
	2050 avec zéro migration	66,7%	8,4%	24,0%	0,2%	0,2%	0,2%	0,2%
Moyen-Orient & Afrique du Nord	2010, % de la population de la région	3,7%	93,0%	0,6%	0,5%	0,1%	1,6%	0,4%
	2050 avec migration	3,1%	93,7%	0,6%	0,6%	0,2%	1,4%	0,4%
	2050 avec zéro migration	2,9%	94,3%	0,6%	0,3%	0,1%	1,4%	0,4%

Source : Cf. tableau 1.

Tableau 3. Poids démographique des musulmans dans les pays de l'Europe de l'Ouest, en 2010 et 2050

	% Musulmans en 2010	% Musulmans en 2050 avec migration anticipée	% Musulmans en 2050 sans nouvelle migration	Différence en 2050 avec/sans migration
Suède	4,6 %	12,4 %	6,8 %	+ 5,6 %
Norvège	3.7	8.9	5.2	+ 3.8
Espagne	2.1	7.5	4.1	+ 3.4
Royaume-Uni	4.8	11.3	8.3	+ 3.0
Italie	3.7	9.5	7.2	+ 2.2
Finlande	0.8	3.4	1.4	+ 2.0
Belgique	5.9	11.8	9.9	+ 1.9
Danemark	4.1	8.5	6.7	+ 1.9
France	7.5	10.9	9.0	+ 1.8
Irlande	1.1	3.0	1.2	+ 1.8
Grèce	5.3	7.8	6.1	+ 1.6
Pays-Bas	6.0	9.4	8.1	+ 1.3
Portugal	0.3	1.3	0.3	+ 1.0
Autriche	5.4	8.9	8.1	+ 0.8
Allemagne	5.8	10.0	9.4	+ 0.6
Suisse	4.9	7.6	7.4	+ 0.3
Luxembourg	2.3	2.3	2.3	+ 0.0

Source : Cf. tableau 1.

Tableau 4. Changements démographiques et évolution de l'importance relative des populations selon leur confession dans les principaux pays occidentaux (2010-2050)

	Années	Population	Chrétiens	Musulmans	Non-affiliés	Hindouistes	Bouddhistes	Juifs	Autres religions
Albanie	2010	3 200 000	18,0	80,3	1,4	<0,1	<0,1	<0,1	0,2
	2050	2 810 000	13,4	85,9	0,4	<0,1	<0,1	<0,1	0,2
Allemagne	2010	82 300 000	68,7	5,8	24,7	<0,1	0,3	0,3	0,1
	2050	70 220 000	59,3	10,0	29,8	0,1	0,4	0,3	0,1
Autriche	2010	8 390 000	80,4	5,4	15,5	<0,1	0,2	0,2	0,2
	2050	8 460 000	72,6	8,9	17,2	0,2	0,5	0,3	0,4
Belgique	2010	10 710 000	64,2	5,9	29,0	<0,1	0,2	0,3	0,2
	2050	11 120 000	52,8	11,8	33,6	0,3	0,6	0,3	0,6
Bulgarie	2010	7 490 000	82,1	13,7	4,2	<0,1	<0,1	<0,1	<0,1
	2050	5 180 000	78,7	15,0	6,2	<0,1	<0,1	<0,1	<0,1
Bosnie-Herzégovine	2010	3 760 000	52,3	45,2	2,5	<0,1	<0,1	<0,1	<0,1
	2050	2 620 000	48,5	49,4	2,1	<0,1	<0,1	<0,1	<0,1
Danemark	2010	5 550 000	83,5	4,1	11,8	0,4	0,2	<0,1	0,1
	2050	5 820 000	78,6	8,5	11,0	0,9	0,8	<0,1	0,2
Espagne	2010	46 080 000	78,6	2,1	19,0	<0,1	<0,1	0,1	0,1
	2050	52 310 000	65,2	7,5	26,5	0,2	0,1	0,1	0,3
Etats-Unis	2010	310 380 000	78,3	0,9	16,4	0,6	1,2	1,8	0,8
	2050	394 350 000	66,4	2,1	25,6	1,2	1,4	1,4	2,0
Finlande	2010	5 360 000	80,1	0,8	19,1	<0,1	<0,1	<0,1	0,1
	2050	5 570 000	72,5	3,4	23,2	0,2	0,4	<0,1	0,1
France	2010	62 790 000	63,0	7,5	28,0	<0,1	0,5	0,5	0,5
	2050	69 300 000	43,1	10,9	44,1	0,1	0,6	0,5	0,7

Source : Cf. tableau 1.

Tableau 4 (suite). Changements démographiques et évolution de l'importance relative des populations selon leur confession dans les principaux pays occidentaux (2010-2050)

	Années	Population	Chrétiens	Musulmans	Non-affiliés	Hindouistes	Bouddhistes	Juifs	Autres religions
Grèce	2010	11 360 000	88,1	5,3	6,1	0,1	<0,1	<0,1	0,1
	2050	10 120 000	86,1	7,8	5,5	0,3	<0,1	<0,1	0,2
Irlande	2010	4 470 000	92,0	1,1	6,2	0,2	0,2	<0,1	0,3
	2050	6 640 000	83,5	3,0	12,1	0,6	0,3	<0,1	0,4
Italie	2010	60 550 000	83,3	3,7	12,4	0,1	0,2	<0,1	0,2
	2050	56 080 000	72,8	9,5	16,3	0,5	0,4	<0,1	0,4
Kosovo	2010	1 770 000	6,1	93,8	0,1	<0,1	<0,1	<0,1	<0,1
	2050	1 970 000	4,7	95,2	0,1	<0,1	<0,1	<0,1	<0,1
Norvège	2010	4 880 000	84,7	3,7	10,1	0,5	0,6	<0,1	0,3
	2050	5 850 000	73,7	8,9	14,5	0,9	1,5	<0,1	0,4
Pays-Bas	2010	16 610 000	50,6	6,0	42,1	0,5	0,2	0,2	0,4
	2050	17 050 000	39,6	9,4	49,1	0,7	0,5	0,2	0,5
Portugal	2010	10 680 000	91,9	0,3	7,5	<0,1	0,2	<0,1	<0,2
	2050	8 690 000	86,3	1,3	11,3	0,1	0,3	<0,1	0,7
Royaume-Uni	2010	62 040 000	64,3	4,8	27,8	1,4	0,4	0,5	0,8
	2050	68 610 000	45,4	11,3	38,9	2,0	0,9	0,3	1,1
Russie	2010	142 960 000	73,3	10,0	16,2	<0,1	0,1	0,2	0,2
	2050	123 960 000	71,3	16,8	11,3	0,1	0,1	<0,1	0,2
Suède	2010	9 380 000	67,2	4,6	27,0	0,2	0,4	0,1	0,4
	2050	10 960 000	52,3	12,4	32,8	0,5	1,2	0,2	0,8
Suisse	2010	7 660 000	72,7	4,9	20,9	0,5	0,5	0,3	0,3
	2050	7 860 000	61,9	7,6	28,4	0,6	0,8	0,2	0,4

Tableau 5. Impact des conversions sur le poids démographique de la population de confession musulmane en 2050

	% Musulmans, avec les changements de religion en 2050	% Musulmans, sans les changements de religion en 2050	Impact des changements de religion
Afrique subsaharienne	35,2%	34,9%	+ 0,3
Amérique du Nord	2,4%	2,6%	- 0,2
Europe	10,2%	10,1%	+ 0,1
Asie & Pacifique	29,5%	29,5%	0,0
Moyen-Orient & Afrique du Nord	93,7%	93,7%	0,0
Amérique latine & Caraïbes	0,1%	0,1%	0,0
Monde	29,7%	29,6%	- 0,1

Source : Tableau construit à partir des données extraites de : Pew Research Center, April 2, 2015, "The Future of World Religions: Population Growth Projections, 2010-2050"

CHAPITRE 9

De la crise des idéologies politiques à l'échec de l'« islamisme » comme idéologie alternative

La civilisation matérialiste que le capitalisme a réussi à mondialiser offre des avantages indéniables qui ont permis son succès et continuent à assurer sa pérennité.

Le problème est que, après l'échec du communisme d'abord et l'absence de réponses idoines de la part du maoïsme maintenant, la pérennité du+ système ne semble être remise en cause que par ses propres contradictions internes qui, d'après les pronostics d'André Gorz, ne sont pas des moindres. En 2007, un an avant la désastreuse crise bancaire de 2008, A. Gorz[1] écrit : « La question de la sortie du capitalisme n'a jamais été plus actuelle. Elle se pose en des termes et avec une urgence d'une radicale nouveauté. Par son développement même, le capitalisme a atteint une limite tant interne qu'externe qu'il est incapable de dépasser et qui en fait un système qui survit par des subterfuges à la crise de ses catégories fondamentales : le travail, la valeur, le capital ».

D'après ce théoricien de l'éco-socialisme : « La "restructuration écologique" ne peut qu'aggraver la crise du système. Il est impossible d'éviter une catastrophe climatique sans rompre radicalement avec les méthodes et la logique économique qui y mènent depuis 150 ans... La décroissance est donc un impératif de survie. Mais elle

---

[1] André Gorz, « La sortie du capitalisme a déjà commencé », *El Correo*, le 20 novembre 2012.

suppose une autre économie, un autre style de vie, une autre civilisation, d'autres rapports sociaux. En leur absence, l'effondrement ne pourrait être évité qu'à force de restrictions, rationnements, allocations autoritaires de ressources caractéristiques d'une économie de guerre. La sortie du capitalisme aura donc lieu d'une façon ou d'une autre, civilisée ou barbare. ».

Si ces propos peuvent heurter les consciences, c'est en partie parce qu'André Gorz parle vrai. Si l'univers est infini, il n'en va pas de même pour la planète terre, de plus en plus prégnante, cette réalité s'impose à l'espèce humaine qui doit admettre aujourd'hui qu'à plus ou moins brève échéance elle est vouée à disparaître en tant qu'espèce vivante, à moins que grâce aux progrès fulgurants dans les domaines de la biotechnologie, de la nanotechnologie et de l'intelligence artificielle, l'*Homo sapiens* hissé au rang d'*Homo deus*[1] puisse être sauvé par ses robots armés d'une sagesse supérieure.

En ce qui concerne la question des changements climatiques, il est à noter que Barak Obama, dans l'interview qui a été considérée comme son « Testament diplomatique »[2], déclare *"isis is not an existential threat to the United States"* en revanche *"Climate change is a potential existential threat to the entire world if we don't do something about it."*

Après la Cop20, nous avons eu droit à l'édition 2015 de la Conférence des Nations Unies sur le climat (Cop21). Lasse des guerres idéologiques, l'humanité s'est résignée à des « Cops » qui se succèdent, celles-ci sont destinées à colmater les fissures du réacteur mais n'apportent aucune solution viable pouvant assurer la pérennité du système. Ce

---

[1] Cf. entre autres, Yuval Noah Harari, *Homo deus : Une brève histoire du futur*, Paris, Albin Michel, 2017.
[2] Jeffrey Goldberg, "The Obama Doctrine ", *The Atlantic*, April 2016.

d'autant plus que le succès du texte adopté lors de la Cop21 a été battu en brèche par Donald Trump, le nouveau président des Etats-Unis, qui, fidèle à sa promesse électorale d'« annuler l'accord de Paris sur le climat »[1], a fini par annoncer le retrait de l'accord de son pays. Pour les écolos anticapitalistes, il s'agissait de toute façon d'une cause perdue d'avance, car c'est sans compter les capacités de résistance du système. En effet, d'après une certaine pensée critique, dans la lignée de la société du Mont Pèlerin chère à Hayek, le mentor attitré du « libéralisme sans frontières ni limites », et du World Economic Forum (Davos) qui en a pris le relais en 1982, « le World Business Council for Sustainable Development, créé en 1995, poursuit l'objectif *d'habiller en vert* les stratégies d'expansion du capital des monopoles, et, par ce moyen, de rallier les opinions écologistes qui ont le vent en poupe »[2].

La mondialisation tous azimuts et la financiarisation du capitalisme ont donné lieu à des changements majeurs. Le monde change, les concepts et les pratiques aussi. "Planification économique", "projet de société", tous ces concepts et pratiques qui font étalage de la présence de la main visible de l'Etat dans l'économie sont tombés en désuétude pour le monde politique. Lorsque "le politique" est inopérant au niveau local, "la politique" qui demeure locale devient inconsistante. Car si, concrètement, le politique se manifeste au niveau de l'intervention de l'Etat dans l'économie, avec la montée des entités supra-

---

[1] Vittorio De Filippis, Isabelle Hanne et Frédéric Autran, « Sept chantiers du nouveau mandat », le quotidien *Libération* du 8 novembre 2017.
[2] Alberto Rabilotta et Michel Agnaïeff, *D'une civilisation capitaliste industrielle vers une barbarie ploutocratique*, 13.04.2015, Essai mis en ligne par l'Agence latino-américaine d'information (ALAI).

nationales et une mondialisation gouvernée pour l'essentiel par les rapports de force, il nous faut admettre, comme un fait, la perte de la souveraineté économique des entités étatiques et le caractère inopérant du politique au niveau local[1]. Dans ces conditions, demeurant au niveau local, la politique devient inconsistante, la lutte pour le pouvoir dans un climat d'incertitude et d'insécurité crée un terreau favorable à la montée des populismes.

« Une autre fin du monde est possible », éloquent, ce nouveau slogan des mouvements « anti-système » est l'expression de la peur panique des catastrophes climatiques. Grave et problématique, la thématique environnementale peine à mobiliser. Car, après avoir, grâce à l'idéologie matérialiste et progressiste, intégré l'idée d'infini et de « *no limit* », on demande maintenant aux nouvelles générations de réintégrer l'idée de finitude, de rareté et de catastrophe imminente alors que, via la modernité, l'individu en Occident s'est affranchi de l'autorité de tous ceux qui jouaient le rôle de filet protecteur et/ou qui permettaient de faire sens (Dieu, le Pape, le Roi). Concrètement, le problème est que même l'idée du Progrès a malencontreusement été tronquée dans sa mise en pratique. Pris au piège des progrès matériels, scientifiques et techniques, plus angoissé que jamais, l'individu athée doit maintenant, orphelin de l'idéologie du progrès, admettre qu'il vit délesté de tout filet de protection dans un monde fini et voué à se fondre dans un univers infini et ouvert de toutes parts. Désespéré et démuni, myope et/ou imperméable à la question du progrès moral, éthique et spirituel, sa réponse pour l'heure consiste à courir encore plus vite vers le précipice.

---

[1] Cf. Ninou Garabaghi, « L'Union européenne en quête de sens ? », revue de l'Académie de géopolitique de Paris, *Géostratégiques*, N° 35, 1er trimestre 2012.

Polysémique donc, ô combien plus pertinent au regard de la complexité de la situation, serait le slogan « d'autres fins sont possibles ». Peur des pandémies, peur des catastrophes nucléaires, peur des catastrophes environnementales, peur des crises financières, peur des attentats terroristes, peur du chômage, etc. Si, aujourd'hui, les moins jeunes ont l'obsession de la sécurité, c'est que les jeunes comme les moins jeunes plus que jamais veulent vivre et vivre pleinement leur vie : d'aucuns à l'affût de nouveauté recherchent une vie d'aventure, d'autres optent pour le retour à la terre et au "naturel", d'autres plus nombreux encore parmi les jeunes sont en quête de sens, d'authenticité, de repères moraux et éthiques, d'identité[1]. La nouvelle génération a besoin d'idéal, de projets de société qui fassent sens, d'idéologies mobilisatrices qui fassent rêver. Marx et Mao étant morts et bien enterrés, les déçus de la civilisation matérialiste comme les laissés pour compte de l'économie capitaliste financiarisée, qui a de plus en plus recours à une automatisation à marche forcée, sont en quête de nouveaux prophètes. Le fait dramatique de la civilisation du capitalisme financier est que les laissés pour compte font davantage figure d'inutiles rebuts, à l'image du « monstrueux insecte » dans la *métamorphose* de Kafka[2], que de chômeurs rejoignant l'armée industrielle de réserve de l'époque de Marx et d'Engels.

---

[1] Se référant aux résultats de *l'Enquête sur la diversité des populations en France*, parue fin 2015, Jean-Louis Schlegel note que « la religiosité des jeunes musulmans (et juifs) est de 10 % supérieure à celle de leurs aînés », Cf. Jean-Louis Schlegel, « La capacité d'intégration de la France », Revue *Esprit*, mars-avril 2016. Comme nous avons pu le constater dans les chapitres précédents, les défauts d'intégration d'abord et la montée de l'islamophobie ensuite expliquent pour partie « le renforcement du religieux identitaire chez les jeunes générations ».
[2] Kafka, *La métamorphose*, Paris, Gallimard, 1955.

Une partie de la jeunesse est en quête d'alternatives qui fassent rêver les uns, booster les autres, mais qui dans tous les cas leur permettent d'évacuer le trop plein d'énergie qui caractérise cette période, la plus belle de leur vie selon d'aucuns, par un acte d'engagement. Ces jeunes ne veulent pas se contenter d'avoir à empêcher le monde de se défaire, ils voudraient pouvoir refaire le monde. Refaire un monde où tout un chacun a sa place. Plus le monde se complexifie, plus les jeunes ont besoin de points de repère, de perspectives d'avenir à espérer. Paradoxalement, pour ce qui est des mœurs par exemple, dans les sociétés occidentales, le champ des interdits régresse comme peau de chagrin. Jadis dissimulées, les défaillances éthiques de la sphère politico-économique sont mises à nu et étalées au grand jour par les nouveaux moyens de communication instantanée.

Les jeunes ne sont pas tous également armés pour se forger une morale individuelle qui tienne le coup face aux vicissitudes de la route truffée d'embûches, qu'est la vie. Il semble communément admis aujourd'hui que l'idée du progrès est morte et enterrée à Hiroshima. Mais, en réalité, ce ne sont pas la science et la technique qui ont failli à Hiroshima, mais la morale et le politique. Ce n'est pas parce que, épaulé par la science et la technique, le complexe militaro-industriel peut être mis au service du pire que les jeunes n'ont plus besoin de continuer à espérer en la possibilité d'un avenir meilleur.

Tandis qu'en Chine, fidèle à l'idéologie matérialiste sous couvert d'un confucianisme de façade, la population, à l'image de la besogneuse fourmi, s'acharne au travail et les dirigeants tissent patiemment leur toile et posent leurs pions un peu partout à travers le monde dans l'espoir de hisser la Chine au rang de la future première superpuissance mondiale, l'Occident, en général, et la France, en particulier, semblent enlisés dans la guerre contre le terrorisme. En témoigne le discours historique du président chinois Xi Jinping au sommet annuel des élites économiques mondiales à Davos le 17 janvier 2017. « A Davos, Le rendez-vous d'un monde à l'envers » titrait le quotidien Le Monde[1]. Effectivement, qui aurait cru, il y a à peine quelques années, que la Chine, « dernier empire communiste de la planète » selon le quotidien Libération[2], allait, à l'encontre des « positions protectionnistes et isolationnistes » du nouveau président des Etats-Unis Donald Trump, se poser comme « champion d'un monde ouvert et connecté », pourfendeur du protectionnisme de la première superpuissance économique mondiale ? En parfait super-businessman, Donald Trump a bien flairé le danger, lui reste à prouver qu'en ravissant le terrain aux altermondialistes, il saura apporter la bonne réponse aux problèmes.

Oui problèmes il y a, mais avec les populismes, nous avons droit à un immense silence dans un vacarme assourdissant.

Incapables de répondre aux nouveaux défis de la mondialisation et d'une Europe en quête de sens[3], les élites dirigeantes semblent en panne d'idées. En France, par exemple, alors que, traumatisée par les attentats terroristes commis en 2015 et 2016 par et/ou au nom de Daech, la population s'efforce de faire preuve de résilience, un nombre non négligeable de responsables politiques davantage préoccupés par leur carrière que par le destin du pays attisent la peur et la haine suscitées par les tragiques événements au lieu de les apaiser. Le chef du gouvernement

---

[1] Sylvie Kauffmann, « A Davos, le rendez-vous d'un monde à l'envers », le quotidien *Le Monde* du 17 janvier 2017.
[2] Raphaël Balenieri, « Anti-Trump – A Davos, le président chinois en apôtre du libre-échange », le quotidien *Libération* du 18 janvier 2017.
[3] Ninou Garabaghi, « L'Union européenne en quête de sens ? », revue de l'Académie de géopolitique de Paris, *Géostratégiques*, N° 35, 1er trimestre 2012.

allant jusqu'à se poser la question de savoir si « l'islam est compatible avec la république »[1], ce qui revient pour ainsi dire à se demander si « la laïcité est compatible avec la république ».

La bonne volonté ne manque pas chez les chercheurs, soucieux d'outiller au mieux les dirigeants et les médias, mais malheureusement, aux moments critiques, l'action de certains dirigeants se détache de la réflexion. Entre-temps, censés éclairer les acteurs clés, dont l'opinion publique, nombre d'intellectuels parmi ceux qui ont voix au chapitre, en panne d'idées et de solutions d'avenir, se querellent entre eux : les néo-réactionnaires contre les néo-progressistes, à défaut d'innovations idéologiques universalisables, se défoulent sur la question en vogue qu'est devenu l'« islamisme ». Il est vrai que les circonstances de la géopolitique mondiale aidant, l'islam politique est la seule nouvelle idéologie alternative présente sur le marché des idéologies.

En effet, la Déclaration islamique des droits de l'homme stipule : « Réaffirmant le rôle civilisateur et historique de la communauté islamique (oummah), la meilleure communauté que Dieu ait créée et qui a donné à l'humanité une civilisation universelle équilibrée, alliant la vie présente à l'au-delà, et la connaissance à la foi, et réaffirmant le rôle espéré que cette communauté devrait jouer aujourd'hui pour guider l'humanité plongée dans la confusion à cause de croyances et d'idéologies différentes et antagonistes, et pour apporter des solutions aux problèmes chroniques de cette civilisation matérialiste ». Comme on le constate, nous sommes en présence d'une idéologie totalisante pour ne pas

---

[1] Propos attribué à Manuel Valls, le Premier ministre de la France, par Michel Tubiana, président d'honneur de la Ligue des droits de l'homme, dans l'émission les matins d'été de France Culture « Comment dépasser l'Etat d'urgence ? », diffusée le 21 juillet 2016.

dire totalitaire, il est vrai que l'islam, à l'instar du christianisme et du judaïsme, est une religion à vocation universelle.

A ne pas confondre avec la Charte arabe des droits de l'homme à laquelle il a déjà été fait référence, la Déclaration du Caire sur les droits de l'homme en Islam a été adoptée en 1990 par l'Organisation de la coopération islamique (OCI). Celle-ci se base sur une interprétation spécifique des droits et des libertés selon la Charia, qui est considérée comme « l'unique référence pour l'explication ou l'interprétation de l'un quelconque des articles » de ladite Déclaration (art. 25). "Contrairement à la Charte arabe, la Déclaration du Caire n'a pas de portée juridique mais une signification symbolique et une importance indirecte dans la politique des droits de l'homme. Sur ce plan, elle s'érige en contre-projet islamique à la Déclaration universelle des droits de l'homme".

Il est certain que, pour éviter tout risque d'anachronisme, il faut garder présent à l'esprit que le Coran n'a pas pour vocation de servir de manuel d'économie. Mais, comme Bruno Etienne[1] le rappelle avec à-propos : « les musulmans ont plutôt plus de facilités que les croyants d'autres religions à intégrer l'économie de marché dans la mesure où l'islam est né à la croisée de deux modes de production, le commerce caravanier et l'agro-pastoralisme ». Lorsqu'on se réfère au Qatar ou à l'Arabie saoudite, les deux principaux promoteurs du fondamentalisme salafiste et wahhabite, on constate que l'économie de ces deux pays participe de la financiarisation du capitalisme mondial, à croire que les élites dirigeantes de ces pays musulmans ne voient pas de contradiction entre le fondamentalisme wahhabite et le fondamentalisme du marché des « Chicago Boys ».

---

[1] Bruno Etienne, *Islam, les questions qui fâchent*, Paris, Bayard, 2003.

Le problème est que nous sommes là en présence d'une adhésion flagrante à la civilisation matérialiste. La question se pose alors de savoir si nous sommes en présence d'une contradiction propre à l'idéologie wahhabite, à moins qu'il ne s'agisse d'une affaire de corruption des élites ?

Dans un de ses articles, Régis Debray[1] écrit : « Voltaire était schizophrène comme tout le XVIIIe siècle : on prêche la liberté mais il y a des esclaves dans nos îles. Je ne dis pas que Voltaire était esclavagiste, simplement, il ne faisait aucun rapport entre les revenus de ses investissements et l'esclavage. C'était un autre monde. Le XVIIIe siècle pouvait fonctionner à deux niveaux. Un armateur à Nantes avait baptisé son plus beau navire négrier *Contrat social*. La moitié mourrait en chemin mais ils étaient dans le *Contrat social* ! On peut très bien vivre à deux étages. Nous vivons très bien à côté de la Libye. Nous sommes allés y mettre une pagaille noire, un chaos atroce, mais au nom de la protection des populations. Nous aussi, nous sommes un peu schizophrènes. Voltaire est le penseur de la France d'aujourd'hui ».

Avec l'exemple susmentionné concernant la participation du Qatar et de l'Arabie saoudite à la financiarisation du capitalisme mondial, on pourrait conclure qu'il n'y pas que la France qui est schizophrène au niveau de sa politique étrangère. D'ailleurs, comme Nietzsche l'a si bien dit : « Et ceux qui se nomment les bons et les justes il ne leur manquait que le pouvoir pour devenir des pharisiens ». La question qui se pose maintenant consiste à savoir ce qui advient dans la pratique de la critique de la civilisation matérialiste mise en exergue dans la Déclaration islamique des droits de l'homme adoptée à Riyad car nous devons

---

[1] Régis Debray, « Contre les fanatismes », entretien réalisé par Franz-Olivier Giesbert et Valérie Toranian, Revue des Deux Mondes, avril 2015.

garder présent à l'esprit qu'avec le salafisme et/ou le wahhabisme nous avons affaire à une doctrine ultraconservatrice.

En fait, il est vrai qu'al-Qaïda et Daech ne sont pas tendres avec le mode de gouvernance de ces pétromonarchies d'obédience wahhabo-salafiste mais ces deux mouvements terroristes wahhabites sont loin d'être eux-mêmes exemplaires du point de vue d'une éthique islamique qui, faisant place à l'*ijtihâd*, se veut conforme à une conception dynamique de la charia. La liste de leurs contrevenances est bien trop longue pour être recensée. Il suffit de jeter un coup d'œil à la Déclaration islamique des droits de l'homme pour se rendre compte de l'ampleur des écarts entre des normes islamiques plus ou moins progressistes conformes aux « exigences de l'époque » (*moqtaziyât-e zamân*)[1] et les normes mises en pratique par ces mouvements terroristes. Les médias exposant à longueur de journée les horreurs commises par les différents groupes terroristes se réclamant de l'islam, nous nous contenterons de citer ci-après quelques exemples de défaillances au regard des normes définies dans la Déclaration islamique des droits de l'homme.

D'après l'art. 2 - a) La vie est un don de Dieu ; elle est garantie à chaque être humain. d) L'intégrité physique est garantie ; personne n'a le droit de la violer. Selon l'art. 6 - a) La femme est l'égale de l'homme dans la dignité humaine ; ses droits sont équivalents à ses devoirs. Elle a une personnalité civile, une responsabilité financière indépendante, et le droit de conserver son nom patronyme et ses liens de famille. D'après l'art. 11 - a) L'individu est né libre ; nul n'a le droit de l'humilier, de l'opprimer ou de l'exploiter. Il ne peut y avoir d'autre soumission qu'à Dieu

---

[1] Cf. Azadeh Niknam, « Le statut de la charia en Iran : de l'islamisme au postislamisme », in revue *Esprit*, août-septembre 2001.

le Tout-Puissant. Selon l'art. 20 - Nul ne peut arrêter un individu, restreindre sa liberté, l'exiler ou lui infliger une peine sans raison légale. Nul ne peut l'exposer à la torture physique ou morale ou à tout autre traitement humiliant, brutal ou contraire à la dignité humaine. *Last but not least*, d'après l'art. 22 - d) Est interdit l'appel à la haine nationale ou religieuse et tout ce qui constitue une incitation à toute forme de discrimination raciale.

Il est vrai que les deux groupes terroristes d'obédience wahhabite, al-Qaïda et Daech, n'ont pas encore eu le privilège d'accéder durablement au pouvoir pour faire étalage des incuries dont ils sont capables du point de vue de la gestion islamique de l'économie.

L'on peut néanmoins se faire une idée quant à savoir si l'islam est en mesure d'apporter une réponse idoine à la crise du capitalisme mondial comme Michel Foucault a pu l'espérer. Il n'est évidemment pas question ici d'entrer dans les méandres de la finance islamique et de ses dérives et encore moins de solutionner la question de l'économie islamique[1]. Il n'en demeure pas moins que l'on peut sans grand risque d'erreur faire état de deux réalités en la matière, l'une d'ordre théorique et l'autre pratique. La première est que, comme il a déjà été signalé précédemment, l'islam n'a pas eu à l'origine pour vocation de résoudre les problèmes du capitalisme mondial et depuis, aucun individu n'a à ce jour apporté de réponse islamique idoine apte à remédier à la crise du capitalisme mondial. La seconde, d'ordre pratique, est qu'aucun pays islamique n'a, à ce jour, été en mesure de faire état d'une économie islamique performante et éthiquement plus ou moins irréprochable.

La question se pose alors de savoir ce qu'on entend par une économie performante et éthiquement plus ou moins

---

[1] Cf. supra chapitre 7.

irréprochable. Dans une perspective humaniste, une économie éthiquement plus ou moins irréprochable est une économie qui permet d'assurer la couverture des coûts humains de la vie aux moindres coûts humains du travail. D'un point de vue idéal, une telle économie est exempte de toute corruption, du chômage, de la pauvreté, des injustices socio-économiques qui sont le lot de la majorité des sociétés actuelles. Il est à noter que, pour ce qui est des travaux théoriques ayant pour objet l'humanisation de l'économie, ce sont des économistes de confession chrétienne tels que François Perroux et Henri Bartoli qui, durant la décennie 80, ont été les plus prolifiques. L'économie humaniste a été définie par F. Perroux comme l'économie de la Ressource Humaine qui a pour objet de promouvoir « l'économie de tout l'homme et de tous les hommes ». Etant entendu que le terme homme doit être compris au sens d'être humain.

D'après Olivier Roy[1], « il n'y a pas de modèle politique concret propre à l'islamisme, encore moins d'économie ». Les travaux théoriques concernant l'économie islamique laissent à désirer et, d'un point de vue pratique, l'économie islamique a été un échec. Est-ce à dire que l'islam ne peut, à l'instar du christianisme, servir de base à l'élaboration de projets alternatifs : projet de société viable et projet d'économie viable ? Il est plus facile de mobiliser pour détruire que pour construire. Les mouvements antisystèmes ne peuvent avoir pour seul but la destruction, pour perdurer ils devraient d'abord et surtout être porteurs d'un projet constructif. Etymologiquement, le mot djihad signifie « effort », du point de vue religieux il a été défini comme un « effort dans le chemin vers Dieu ».

---

[1] Olivier Roy, *L'échec de l'islam politique*, Paris, Seuil, 2015 pour la postface inédite.

Pourquoi doit-on, à longueur de journée, entendre parler du « djihad » en termes de violence ? Pourquoi n'entend-on pas parler du grand « djihad », qui est un « effort sur soi » pour le plus grand bien de tout un chacun ? En effet, si l'on considère que le petit « djihad » signifie « guerre sainte » et que, par définition, toute guerre relève du domaine de la violence, peu importe qu'elle soit légitimée par le droit ou la religion, le petit djihad peut alors être assimilé à de la violence. Mais si d'aucuns mettent l'accent sur la légitimité ou la légitimation du petit djihad au nom de l'islam, il n'en demeure pas moins que la même religion prône le grand « djihad » qui est le combat contre soi-même et ses grands démons. Or, celui qui a réalisé le grand djihad et qui a ainsi acquis la maîtrise de ses émotions, de ses passions et de ses pulsions ne peut s'adonner à des actes de violence nihilistes.

D'après Scott Atran[1], « les tenants du Califat pur (Daech) sont violemment opposés à l'idée d'un « grand djihad » qui désigne une bataille spirituelle intérieure. Selon eux, il s'agit d'une conception erronée du djihad, portée par l'hérésie soufie, qui s'est développée à la fin du califat abbasside, et qui a corrompu la pureté du califat arabe pour le précipiter vers sa chute ». Qui veut noyer son chien l'accuse de la rage. Le chien étant considéré comme « impropre » par certains musulmans, ceux-ci n'ont même pas besoin de l'accuser de rage pour le noyer en toute impunité. Daech fait la guerre à toutes les religions et veut par tous les moyens provoquer une guerre civile au sein des sociétés occidentales d'abord et des autres sociétés ensuite. Un mouvement terroriste mû par la haine et l'idée de vengeance n'a évidemment pas pour vocation de promouvoir un concept qui a pour objet la maîtrise des émotions, des pulsions et des passions qui, *in fine*, aurait pour conséquence de pacifier les relations entre êtres

---

[1] Scott Atran *L'Etat islamique est une révolution*, op.cit.

humains. Reste à savoir pourquoi les médias, qui ne cessent de faire étalage des méfaits du petit djihad, ne font pas un tant soit peu la propagande du grand djihad.

A l'instar des chrétiens engagés, pour certains musulmans, l'islam ne peut qu'être politique. Dire que l'islam est politique, cela ne signifie pas que le pouvoir et les privilèges doivent être accaparés par des religieux. Mais que la religion doit être au service de la libération de l'être humain. Dire que l'islam est politique, c'est dire que le social et l'économique ne peuvent être séparés ; que le social ne se limite pas à un système palliatif destiné à corriger a posteriori les dégâts humains causés par une économie prédatrice. Dire que l'islam est politique, c'est invoquer le grand djihad pour réguler et assainir l'économie via un double combat individuel et collectif, c'est s'engager dans un double mouvement de résistance et d'engagement. Résistance individuelle à la violence, à l'aliénation ; résistance collective à la violence sociale, économique et politique ; engagement individuel et collectif dans des projets constructifs initiateurs de paix durable via le grand djihad qui permet à tout un chacun, par un combat intérieur contre ses propres pulsions destructrices, d'acquérir une paix intérieure, condition nécessaire à un engagement collectif harmonieux.

Dire que l'islam est politique, cela ne signifie pas que les acteurs religieux doivent monopoliser les rênes du pouvoir. La prise du pouvoir temporel par les acteurs religieux a pour conséquence un affaiblissement inexorable de leur pouvoir spirituel pouvant aller jusqu'à la perte de toute autorité morale pour la religion concernée, pour la simple raison que, d'une façon générale, le pouvoir corrompt et que le pouvoir absolu corrompt absolument, sans parler du fait que vouloir satisfaire les êtres humains sur Terre n'est pas tâche facile, autant aller sur la Lune ou sur Mars pour y bâtir une nouvelle cité. Pour l'heure, seule la démocratie, avec son système de mise à mort virtuelle du chef qu'est l'alternance, permet d'assurer une gouvernance plus ou moins acceptable à l'ère des NTIC et plus spécifiquement du numérique.

L'être humain étant mortel et aucune génération n'étant éternelle, le message religieux ne peut être compris et vécu de façon parfaitement identique partout et en tout temps. Il est à noter, à cet égard, que, si l'expression italienne *« Traduttore, traditore »* est vraie, ce n'est pas uniquement parce qu'il faut tenir compte du contexte spatio-temporel de l'émission et de la réception du message, mais c'est aussi en raison des caractéristiques et conditionnements propres à l'écrivain et au traducteur. Il y a l'intention de l'émetteur du message, le message effectivement émis et la lecture du message réceptionné. On ne peut en effet ignorer que ce qui est transmis à travers le temps a d'abord été communiqué à un moment donné et dans un lieu et milieu spécifiques. De sorte qu'on peut dire que « lire c'est trahir ». Dans son livre sur *Les limites de l'interprétation*, Umberto Eco écrit : « aucun texte ne peut être interprété selon l'utopie d'un sens autorisé défini, original et final. Le langage dit toujours quelque chose de plus que son inaccessible sens littéral, lequel est déjà perdu dès le début de l'émission textuelle ».

Selon le théologien Soheib Bencheikh[1] : « Ceux qui appellent aujourd'hui à l'application de la Charia dans les pays musulmans, sans la moindre réforme, ni révision, ni mise à jour de ce droit, font que cet appel devient folie, pathologie, voire même crime. Mais on peut appeler à l'application de la Charia sans l'imposer, parce que l'islam initialement est un message qui se propose et non pas un ordre qui s'impose. Rien n'empêche d'enlever, de revisiter, relire avec une intelligence neuve, mais d'abord

---

[1] *Islam : foi et valeurs par Soheib Bencheikh*, op.cit.

désacraliser les œuvres de nos ancêtres, il n'y a que la révélation qui est sacrée, le reste ne l'est pas. Il faut rendre hommage à ce droit, même figé, parce qu'il s'est voulu être justice de son siècle. Mais si la justice d'un siècle stagne, elle devient injustice pour un autre siècle. Un progrès d'un siècle, s'il stagne, devient tout simplement une régression pour un autre siècle.»

Pour perdurer, la religion doit être vivante afin que le message transmis puisse être réapproprié et vécu authentiquement par les individus. L'histoire nous enseigne que, pour pouvoir être suffisamment puissant pour influer sur le cours de l'histoire, le pouvoir religieux doit se tenir à l'abri et au-dessus des contingences temporelles. Aucun pouvoir temporel ne pouvant perdurer, la religion, en devenant instrument du pouvoir temporel, ne peut que se pervertir et se vouer à périr. Pour perdurer, la religion doit opter pour le domaine de l'intemporel et limiter son rôle dans la sphère du temporel à des fonctions de régulation en vue de garantir des lignes d'orientation éthique et morale.

L'on se doit à cet égard de saluer les avancées dans la diplomatie du Saint-Siège qui, nous limitant ici au cas de sa sainteté le pape François, n'a cessé de faire preuve d'intelligence et de sagesse pour intervenir juste à temps et à-propos et à des moments fatidiques pour influencer et guider l'action (questions de la pauvreté, des réfugiés, des guerres d'ingérence, du terrorisme, de l'environnement, etc.). Il est à noter à cet égard le caractère révolutionnaire de l'encyclique sur l'écologie humaine[1]. Il s'agit de réhabiliter l'idée du progrès qui, ainsi qu'il a été signalé plus haut, se doit d'être multidimensionnelle pour être effective. Le progrès doit viser les institutions et les individus, le matériel et le spirituel, l'infrastructure et la

---

[1] Lettre Encyclique Laudato Si' du Saint-Père François sur la sauvegarde de la maison commune, Vatican, le 24 mai 2015.

superstructure. Il faut savoir miser tout autant sur la morale, l'éthique et le spirituel que sur l'économie, la science et la technique.

Moins audibles que l'Eglise catholique, les représentants de l'islam sunnite et chiite se font entendre par intermittence. Il est à noter, à cet égard, le rôle joué par les présidents Mohammed Khatami d'abord et Hassan Rohani ensuite dans la réorientation de la diplomatie de la République islamique d'Iran vers le dialogue et la détente durant la période de leur présidence[1]. Qui plus est, c'est sur proposition de M. Khatami, alors Président de la République islamique d'Iran, que l'année 2001 a été proclamée année des Nations Unies pour le dialogue entre les civilisations. C'est à son instigation qu'en novembre de la même année, l'Assemblée générale des Nations Unies a adopté le Programme mondial pour le dialogue entre les civilisations[2]. Il est vrai que le Concile Vatican II avait déjà ouvert la porte de l'altérité et du dialogue interreligieux[3]. Si la porte de l'*ijtihâd* n'a *de facto* pas été fermée depuis plus de mille ans chez les chiites --ceux qui pratiquent l'*ijtihâd* étant appelés les « *mujahid* » à ne pas confondre avec les « djihadistes »[4]--, il n'en va pas de même chez les sunnites d'obédience wahhabo-salafiste qui, en prétendant qu'il ne faut pas interpréter, ne font en fait pas autre chose que ce qu'ils interdisent. Sauf que, n'en déplaise aux wahhabo-salafistes, la pratique de l'interprétation semble avoir été réhabilitée par certains oulémas de l'islam sunnite et ceci ne

---

[1] Cf. entre autres, le Rapport d'information N° 457 du Sénat français sur sa mission en Iran du 14 au 21 avril 2000.
[2] UNESCO, *Investir dans la diversité culturelle et le dialogue interculturel*, Paris, 2010, p.62.
[3] Concile Vatican II : Déclaration sur les relations de l'Eglise avec les religions non chrétiennes – Nostra aetate, Rome, à Saint-Pierre, le 28 octobre 1965.
[4] Mohamad Ali Amir-Moezzi et Christian Jambet, *Qu'est-ce que le shî'isme ?*, Paris, Cerf, 2014, p.121 et p.188 (notes de bas de page).

se limite pas aux imams opérant en Occident. En effet, dans un entretien accordé au magazine *Le Point*, Cheikh Ahmed al-Tayeb, grand imam d'Al-Azhar, l'institution la plus ancienne et la plus respectée de l'islam sunnite, déclare : « Je suis de très près chaque doctrine qui ne rend pas compte de la vraie religion afin de protéger celle-ci ainsi que la société, de toute déviation. Al-Azhar, sa mosquée et son université, travaille jour et nuit pour promouvoir les vraies valeurs de l'islam. Nos oulémas reflètent l'opinion de la société et traduisent sa réalité. Ils font de leur mieux en matière d'*ijtihâd* (interprétation) pour répondre aux problèmes auxquels sont confrontés les gens dans le monde d'aujourd'hui. Car, à Al-Azhar, nous croyons en la nécessité de renouveler le discours religieux en se fondant sur la méthode du juste milieu (entre tradition et modernité). C'est la meilleure façon de lutter contre le fanatisme salafiste »[1].

Il est vrai que l'université d'Al-Azhar n'a pas toujours été à l'abri de l'idéologie salafiste[2]. Selon Olivier Roy[3] : « A la fin du XIXe siècle, un courant fondé par Jamaluddin Afghani et repris par Mohammed Abdouh lança un mouvement de réforme (salafisme) ». Ainsi qu'il a déjà été signalé, ce courant de pensée promu par Afghani avait pour objet de défendre l'esprit de l'islam des ancêtres et non point de restaurer des pratiques en vogue à l'époque des « salafres » (les pieux ancêtres). C'est ainsi que Mohammed Abdouh, « azharite » (diplômé de l'université d'Al-Azhar), qui a promu le renouveau de la pensée islamique pour lutter contre le colonialisme, la corruption et la polygamie, s'est

---

[1] « Interview avec le grand imam d'Al-Azhar : Ahmed al Tayeb, l'homme qui détient les clés de l'Islam », le magazine *Le Point* du jeudi 16 juin 2016 n° 2284.
[2] Ainsi qu'il a déjà été signalé, le wahhabisme s'est aujourd'hui paré du qualificatif de salafisme pour mieux se faire accepter.
[3] Olivier Roy, *Généalogie de l'islam*, *op.cit.*, p.37.

opposé à ceux qui voulaient mettre un terme à l'*ijtihad* et, contrairement à tout esprit archaïque, a, en tant que mufti, promulgué des *fatwas* permettant de moderniser les pratiques sociales de l'époque. Donc, si, du fait de l'invasion de l'Egypte par les Anglais et de l'affaiblissement de l'Empire ottoman, c'est l'islam salafiste qui avait été promu au Caire au tournant du XXe siècle, pour l'heure, c'est bien une vision éclairée de l'islam qui semble être officiellement soutenue et promue par les oulémas d'Al-Azhar. Dans le même esprit d'ouverture, s'inscrit la rencontre du pape François et des représentants de l'islam chiite et sunnite dont le grand imam d'Al-Azhar, l'institution la plus ancienne et la plus respectée de l'islam sunnite. Autant d'actes destinés à pacifier les esprits par la diplomatie, que l'on peut qualifier, sans risque d'erreur, de grand djihad.

Conclusion

Dévoyé il y a maintenant près de quarante ans, le terme islamisme, qui a depuis lors acquis le statut de concept, continue à faire l'objet de débats quant à sa définition. Interrogé dans le cadre des grands entretiens du quotidien *Le Figaro*[1], l'islamologue Adrien Candiard déclare : « On a forgé ce mot pour désigner des formes jugées extrémistes de l'islam, et la distinction entre islam et islamisme sert couramment — à juste titre — à éviter de faire porter le poids de la violence et du terrorisme à des millions de musulmans qui n'ont rien demandé ». Cette réponse peut en surprendre plus d'un. En fait, la réalité c'est la réalité. Mais si nous voyons tous la même chose, chacun la voit à sa façon et, pour se rassurer, croit que l'autre la voit pareillement. Le problème concernant la distinction entre islam et islamisme, comme nous avons pu le constater, est que, pour une large fraction de la population occidentale aujourd'hui, « l'islamisme, c'est l'islam et l'islam est le problème ». Et le dévoiement du terme islamisme par les islamologues occidentaux y est pour quelque chose. Si, pour comprendre le monde, nous avons besoin de concepts, nous ne devons pas perdre de vue que ces concepts, produits de la société, acquièrent une vie propre et déterminent celle-ci à leur tour. Contingents à l'origine, les concepts s'imposent à nous comme une nécessité, de sorte qu'ils finissent par figer notre perception de la réalité, et plus important encore, ils contribuent à déterminer son évolution. Abstraction contingente pour la génération des

---

[1] Eugénie Bastié (FigaroVox/grand entretien - L'islamologue Adrien Candiard), « Le salafisme fantasme l'islam originel contre la tradition musulmane », *FigaroVox*, le 29/08/2016.

soixante-huitards, le nouveau concept d'« islamisme » est devenu une réalité pour la génération actuelle.

Si bien nommer les choses ne permet pas toujours de résoudre nos problèmes, mal les nommer ajoute sans conteste au malheur des hommes. Pour l'heure, « islamisme » et « djihadisme » font partie des trous noirs de la géopolitique mondiale. Si l'on accepte comme vrai que l'islam est par essence une religion violente et belliciste, on n'a pas d'autre choix que de vouloir extirper le mal que représente cette religion de la terre. Or, l'humanité ne peut se résoudre à la guerre des religions pour la très simple raison que, nous limitant au monde des monothéistes, en admettant qu'on en ait fini avec l'islam, il faudra alors en finir avec le catholicisme, puis viendra le tour du judaïsme et ainsi de suite.

Si l'on rejette la position des essentialistes pour qui il n'existe qu'un seul islam violent par essence et que l'on adhère à l'idée selon laquelle, il existe deux conceptions de l'islam : un bon et un mauvais. Dans ce cas, on a, d'une part, un bon islam : l'islam pacifiste et, d'autre part, un mauvais islam : l'islam violent et belliciste que l'on identifie par les qualificatifs d'islamisme et de djihadisme, le djihadisme n'étant autre que l'islamisme en action, en situation extrême et désespérée. Alors, dans ce cas, nous n'avons d'autre choix que de mener une guerre permanente contre le terrorisme islamiste : al-Qaïda affaibli, on doit combattre le nouveau monstre dénommé Daech. Dissous car devenu infréquentable, al-Nosra renaît de ses cendres sous le patronyme de Fatah al-Cham, etc.

Le spectre de la guerre étant présent dans les deux cas de figure, la posture belliqueuse s'impose. Il ne reste plus qu'à créer des chaires de « djihadologie » et le formatage des esprits sera ainsi assuré.

Comme on a pu le constater tout au long de cet essai, « islamisme » et « djihadisme » sont des trous noirs de la géopolitique mondiale qui parasitent et empêchent d'appréhender la réalité des problèmes et donc la recherche de véritables solutions à nos problèmes. Ainsi qu'il a été illustré, le salafisme djihadiste est un avatar de la politique de wahhabisation de l'islam mondial. Le problème est que la doctrine wahhabo-salafiste s'est propagée, non parce qu'elle représente l'islam vrai, mais parce que ses promoteurs avaient les moyens financiers de mener à bien leur politique d'acculturation religieuse, et aussi et surtout, parce qu'ils ont bénéficié du soutien indéfectible du monde occidental durant la guerre froide. Soutien qui perdure depuis.

Si dans le présent essai nous avons longuement examiné les effets délétères de l'islam politique, nous n'avons pour autant ignoré que, face à la réalité du « wahhabo-néoconservatisme », les critiques attitrés de l'« islamo-gauchisme » font preuve d'une cécité totale si ce n'est d'un mutisme. A croire qu'une réalité qui n'est pas nommé n'existe pas et ce qui est nommé est nécessairement vrai. En effet, à regarder de près on constate que l'élite intellectuelle et politique néo-réactionnaire qui se complait à critiquer la gauche néo-progressiste en l'affublant du qualificatif d'« islamo-gauchiste » n'est pas exempte de toute affiliation. Il est vrai que ce sont les vainqueurs qui ont de tous temps écrit sinon tenu les rênes de l'histoire.

A l'ère planétaire, les problèmes sont multiples, imbriqués, multidimensionnels et complexes à bien des égards : il y a les problèmes économiques (il s'agit d'offrir des emplois décents à la jeunesse, à défaut d'instituer un revenu minimum universel) ; il y a les problèmes d'ordre politique (il faudrait, par exemple, en Irak, cesser de jouer sur le marqueur religieux pour la participation au pouvoir, etc.), il y a les problèmes socio-culturels (ex. la problématique de la reconnaissance, à l'ère de la mondialisation et du cosmopolitisme *de facto*), les problèmes d'ordre écologique (ex. les risque de famine consécutifs aux changements climatiques), les problèmes d'accès aux ressources naturelles (ex. les problèmes d'accès à l'eau ont été considérés comme un des facteurs déclencheurs de la guerre civile en Syrie), les problèmes sont aussi et surtout d'ordre idéologique. Les idéologies sont un rouage important des sociétés car elles font office de « système immunitaire symbolique ». Aujourd'hui, c'est le vide idéologique créé par la mort des religions séculières (l'humanisme des Lumières, le communisme, le maoïsme, le nationalisme, le droits-de-l'hommisme, le tiers-mondisme, etc.) qu'il s'agit de combler. Car nonobstant la disqualification du terme « idéologie » par la pensée marxiste[1], que ceci nous plaise ou pas, les sociétés ont besoin d'idéologie pour permettre aux individus de supporter leur condition, de faire corps et demeurer soudées pour le meilleur et pour le pire des mondes. Les sociétés qui pensent s'être débarrassées des idéologies se trompent.

Comme nous avons pu le constater tout au long de cet essai, ce ne sont pas seulement les musulmans qui ont cru voir dans l'islam une solution à leurs problèmes temporels. Car, si Afghani a été le premier intellectuel oriental à croire au 19e siècle que l'islam en tant qu'idéologie pouvait aider les musulmans à s'arracher de leur sous-développement pour autant qu'il soit respecté dans son esprit et non dans sa forme, Foucault a été le premier intellectuel occidental à avoir cru y trouver la troisième voie : une alternative au capitalisme et au communisme. Aujourd'hui, l'islam est sciemment instrumentalisé à des fins politiques et géopolitiques. L'instrumentalisation de l'islam à des fins

---

[1] Cf. supra, chapitre 5.

idéologiques dans les pays musulmans se fait sur la base du slogan « l'islam est la solution » (exemple, la contestation contre le régime du Chah en Iran s'est cristallisée dans la figure de l'ayatollah Khomeiny promu au rang d'imam[1]) et dans les pays occidentaux sur la base du slogan « l'islam est le problème » (instrumentalisation de l'islam aux fins de la construction de la nouvelle figure de l'ennemi comme facteur de cohésion sociale et/ou d'accession au pouvoir).

Il est certain que, sans la révolution iranienne de 1979 et l'appropriation politique de l'élan mystique et spirituel du peuple iranien par le courant activiste du clergé chiite, l'islam dit politique n'aurait probablement pas connu les beaux jours qui ont été les siens. Mais l'on ne peut pour autant ignorer l'apport des islamologues français en la matière. En effet, ainsi que nous avons pu le relater, porteurs malgré eux de l'esprit de l'idéologie marxiste, les islamologues de la génération soixante-huitarde ont joué le rôle d'intellectuels organiques sans frontières, pour l'« islam politique », triomphe de la révolution iranienne. Le dévoiement du terme islamisme et sa consécration comme concept politique ont non seulement fini au fil du temps par banaliser l'instrumentalisation de l'islam par les acteurs politiques, mais ont conféré aussi à ce phénomène une légitimité épistémologique, ce qui est pire encore. Avatar de la recherche en sciences politique des islamologues, à l'instar de l'« Orient » des orientalistes des premières heures, ce concept, qui n'avait pas lieu d'être, disparaîtra. Mais, entre-temps, combien de malentendus et de dégâts occasionnés !

---

[1] Selon l'eschatologie shiite, le 12e imâm, le Mahdi, est actuellement caché ; il doit se manifester comme Sauveur à la Fin du Temps. D'après la dernière lettre de l'imâm caché, quiconque prétendrait le contraire devrait être considéré comme un imposteur. Cf. Mohamad Ali Amir-Moezzi et Christian Jambet, *Qu'est-ce que le shî'isme ?*, Paris, Cerf, 2014, p.111 et p.116.

Quelques remarques au terme de notre réflexion autour de la problématique de l'instrumentalisation de l'islam à la lumière des bouleversements géopolitiques majeurs qui ont marqué le Moyen-Orient depuis la révolution iranienne de 1979. Certaines de ces remarques relèvent de l'ordre du simple constat et/ou de l'intime conviction, d'autres sont de l'ordre de la proposition, à moins qu'elles ne soient du ressort de l'espérance que d'aucuns qualifieront de vœux pieux, mais, comme qui ne tente rien n'a rien, nous les formulerons quand même envers et contre le chaos apparent de la géopolitique mondiale.

La première remarque concerne le salafisme djihadiste, avatar du processus de wahhabisation de l'islam mondial, elle a été pertinemment formulée par l'ancien directeur de l'Ecole de guerre. Fin 2016, dans une émission portant sur les opérations militaires menées au Proche-Orient, le général Vincent Desportes faisait état d'un simple constat qui a le mérite de la clarté : « La coalition livre une bataille en Irak et Syrie. La victoire sur Daech ne signifie pas que la coalition a gagné. Pour gagner la guerre, il faudrait pouvoir éradiquer le wahhabisme, sinon une entité nouvelle naîtra des cendres de Daech »[1]. Daech n'est pas le véritable danger, car, concernant cette entité terroriste à proprement parler, il est plausible de conclure sur un mode consensuel en paraphrasant l'islamologue Olivier Roy : « le pire ennemi de Daech n'est autre que Daech et la meilleure façon de l'affaiblir est de cesser de lui conférer le prestige d'ennemi mondial numéro un ».

La seconde remarque, présentée en termes d'intime conviction, concerne la religion. Au terme de notre réflexion, nous avons acquis la conviction que les sociétés ne peuvent pas se passer de religion ni l'individu de religiosité. La première raison tient au fait que les êtres

---

[1] Régis Debray, « Guerre et opérations militaires », *Les discussions du soir*, France Culture, le 28 octobre 2016.

humains ont tout autant besoin de croire que de savoir. Et tant que la science n'aura pas pu prouver ce qui advient après la mort, la religion demeure la meilleure réponse que l'homme ait pu donner à son besoin de transcendance et à son problème d'angoisse de la mort, d'abord et surtout, et du sens de la vie ensuite[1]. La religion garde son pouvoir d'attractivité auprès des individus en raison du fait qu'elle constitue un pari : les hommes aiment parier et ils sont d'autant plus enclins à parier qu'ils sont assurés de ne rien perdre, d'où le fameux pari de Pascal.

Les sociétés ne peuvent se passer de religions pour la très bonne et irréfutable raison que celles-ci apportent, chacune à sa façon, des réponses aux énigmes cachées de la condition humaine. Et l'individu, qu'il soit croyant, agnostique ou athée, à un moment ou l'autre de sa vie, est inévitablement envahi par ce "sentiment océanique" qui l'amène, parfois malgré lui, à faire l'expérience de la religiosité comme une vérité singulière. Vérité singulière, car, si l'expérience religieuse est commune à tous les êtres humains, l'interprétation de cette expérience est différente d'un être humain à un autre. C'est pourquoi on peut être enclin à donner raison à Karlfried Graf Dürckheim, psychothérapeute et philosophe allemand initié à l'école du Zen Rinzai, lorsqu'il déclare que « les religions séparent les hommes, mais les expériences religieuses les unissent. Et le travail œcuménique ne donnera rien tant qu'on essaiera de rapprocher les paroles dans lesquelles on a durci les interprétations de l'expérience. C'est seulement lorsqu'on

---

[1] N'en déplaise à Marx, dépouillée de son « voile mystique et nébuleux », la vie ordinaire peut paraître insipide et de peu d'intérêt à pas mal d'individus. C'est pourquoi il serait bon que chaque personne puisse vivre sa vie et pour ce faire dispose des « capabilités » (cf. A. Sen) qui lui permettent de décider librement de ce qu'est la vie bonne pour elle.

se rendra compte du fonds commun des expériences que l'œcuménisme fera un pas. Le reste est effort perdu ».

Interrogé sur la question du « dialogue entre civilisations » vingt ans avant la publication du fameux *The Clash of Civilisations* de Huntington, Henry Corbin[1] rappelle que « le terme « civilisation » est un terme abstrait. Ce ne sont pas les « civilisations » comme telles (les « universaux ») qui peuvent dialoguer. Seuls, leurs porte-parole peuvent être les partenaires réels du dialogue ». Henry Corbin identifie alors deux nihilismes symétriques : « celui d'une théologie affirmative (kataphatique) érigeant d'emblée son dogme en absolu, au-delà duquel rien ne serait à chercher » et « celui d'une théologie négative (apophatique) qui n'aspirerait qu'à une indétermination de l'Absolu, et qui perdrait de vue qu'il est le *nihil a quo omnia prodedunt* (le Trésor caché du *hadîth*) ». Et il précise que seule une théologie capable de déjouer les pièges des deux nihilismes est en mesure de créer les conditions d'un réel dialogue[2].

Nous avons longuement disserté de la question de la pratique de l'interprétation (*ijtihad*) chez les musulmans de différentes obédiences, et de la position des mouvements wahhabo-salafistes en la matière. Ainsi qu'il a été signalé, l'intolérance des wahhabo-salafistes à l'égard de la libre interprétation des textes révélés s'avère d'autant plus illégitime, qu'eux-mêmes, en refusant d'admettre que l'an 2018 n'est pas l'an 622, s'adonnent de ce fait même à une interprétation des textes sacrés et finissent, sous couvert du postulat de « parole incréée », à obliger les autres à se

---

[1] Henry Corbin, *Le paradoxe du monothéisme*, Paris, L'Herne, 1981, cf. « De la théologie apophatique comme antidote du nihilisme », pp. 211-256.
[2] Nous sommes confrontés à l'aporie : « individualisme absolu et universalisme absolu » en relation à Dieu, qui a été mise en lumière par Ernst Troeltsch. Cf. Louis Dumont, *Essai sur l'individualisme*, Paris, Seuil, 1985, p.43.

conformer à leur interprétation de la parole divine, non conscients qu'en la décontextualisant, ils en falsifient l'esprit. Si la familiarisation avec la doctrine wahhabite, à l'instar de la connaissance d'autres doctrines, peut s'avérer utile, la politique de wahhabisation de l'islam mondial des acteurs étatiques et autres est en revanche incompatible avec l'esprit de coexistence pacifique et de tolérance, sous-jacent au principe de non-ingérence inscrit dans la Charte des Nations Unies.

Ceci dit, comment venir maintenant à bout des mouvements terroristes djihadistes ? Comment freiner la montée des mouvements populistes islamophobes ? Comme tout un chacun le sait, on n'a pas affaire à un enjeu mais à des enjeux globaux qui sont tout à la fois, diplomatiques, économiques, sociaux, culturels, écologiques, politiques et religieux. Le monde s'est complexifié pour le meilleur et pour le pire. Le terrorisme nous a fait comprendre qu'il n'y a pas d'un côté les puissants et de l'autre les exclus, les laissés-pour-compte du capitalisme sauvage mondialisé. Chaque individu compte, qu'il soit multimilliardaire ou pauvre. Selon le psychanalyste Fethi Benslama, « Ce mouvement qu'on appelle islamisme s'est présenté comme celui qui pourrait être le porte-parole des masses pauvres »[1]. D'apparence inoffensive, la propagande des wahhabo-salafistes et des laissés-pour-compte de la géopolitique (les ex-partisans de Saddam Hussein) est assassine et lourde de conséquences.

Le concept d'islamisme ayant fait l'objet d'un examen tous azimuts dans les chapitres précédents, il n'est pas question d'entamer ici un nouveau débat d'ordre théorique. Rappelons simplement que le mouvement appelé « islamisme » constitué pour l'heure par al-Qaïda, le

---

[1] Nicolas Dutent, « Ces intellectuels qui tissent un islam progressiste », le quotidien *L'Humanité*, du jeudi 12 février 2015.

Groupe Etat islamique (Daech), al-Qaïda au Maghreb islamique (Aqmi), al-Qaïda dans la péninsule arabique (Aqpa), Boko Haram, Ansar al-Charia, Ansar Dine, pour nous limiter aux plus connus, n'a pas pour objectif de fournir une alternative au capitalisme sauvage, mais d'instaurer un régime fondé sur une interprétation littéraliste de la Charia. Comme nous avons eu l'occasion de l'illustrer tout au long du présent essai, le dévoiement du terme islamisme, promu au rang de concept politique, a pour conséquence de légitimer ces entités terroristes, prétendues porte-parole des masses pauvres, qui, *de facto*, ne font qu'ajouter du malheur au malheur des pauvres d'abord et du reste du monde ensuite.

L'Acte constitutif de l'UNESCO stipule que « les guerres prenant naissance dans l'esprit des hommes, c'est dans l'esprit des hommes que doivent être élevées les défenses de la paix ». Si nous voulons contribuer à l'essor de la paix, nous devrions apprendre à nommer juste pour pouvoir parler vrai et agir bien. Au terme de cette réflexion, l'on est en droit d'espérer que l'abandon du concept d'« islamisme » puisse œuvrer à la pacification des esprits et contribuer peu ou prou à assainir le climat dans la sphère de la géopolitique.

Il est évidemment salutaire de vouloir promouvoir la contribution des valeurs islamiques à l'humanisation de la mondialisation comme des acteurs de confession chrétienne (agents économiques et économistes, etc.) ont œuvré à l'humanisation de l'économie capitaliste en Europe[1], en Amérique latine[2] et en Afrique[3]. L'Occident n'est plus en mesure d'imposer sa vision au reste du monde et n'a pas

---

[1] Partis politiques de tendance démocrate chrétienne.
[2] La théologie de la libération.
[3] Il est à noter que, pour ce qui concerne ces derniers, qu'ils soient partisans ou opposants de la « théologie de libération », personne n'a ressenti le besoin de dévoyer le terme christianisme.

intérêt à le faire pour la très simple raison que l'Orient est en Occident comme l'Occident est en Orient[1]. Nous avons tous le droit et le devoir de participer à la construction du monde. Chaque acteur a son rôle à jouer.

Nous en arrivons à formuler quatre propositions et/ou vœux concernant le système des Nations Unies. Les deux premiers intéressent l'ONU, les deux autres l'UNESCO. La première concerne la quête de solutions aux problèmes du mal-développement mondial, afin que les acteurs cessent d'instrumentaliser la religion à des fins politiques et géopolitiques, la communauté mondiale aura alors franchi un grand pas parmi tant d'autres pour approcher de l'idée de paix mondiale. Tant qu'Israël n'aura pas compris qu'il est de son intérêt de jouer la carte du processus de paix, il n'y a pas beaucoup d'espoir de voir la situation s'améliorer dans cette zone de turbulences du monde qu'est le Moyen-Orient ; l'ONU devrait réfléchir pour voir s'il n'est pas possible de convaincre les responsables politiques d'agir dans l'intérêt de leur peuple et, pour ce faire, de tenter de relancer le processus de paix avant qu'il ne soit trop tard. Les responsables religieux de confessions juive et musulmane ont un rôle majeur à jouer en œuvrant à la communion des esprits. En cherchant à regarder ensemble la lumière divine, de plus en plus de musulmans et de juifs apprendront à « la » déceler dans le regard de l'un et de l'autre et leurs leaders politiques finiront par accepter de mener à terme le processus de paix entamé.

La seconde proposition, intimement liée à la précédente, concerne la fonction proactive de l'ONU ; elle consiste dans la mobilisation de l'ensemble des agences du système des

---

[1] En réalité, l'Orient et l'Occident sont deux pôles de notre conscience. Cf. entre autres, Henry Corbin, *Philosophie iranienne et philosophie comparée*, Téhéran, Académie Impériale Iranienne de Philosophie, 1977, p.24.

Nations Unies en vue de l'exploration d'idées-phares chargées d'espérance et susceptibles de mobiliser la jeunesse mondiale. Promue par des idéaux humanistes, l'action de l'ONU, en faveur des exclus des bienfaits de la mondialisation, ne peut être réduite à l'aide humanitaire, à la lutte contre l'extrême pauvreté et la faim. Si le « revenu universel » et le « développement durable » sont des concepts utiles, voire même indispensables, au regard de l'objectif « accroître la sécurité et le bien-être de tous », ils ne suffisent pas, tels quels, à mobiliser toutes les composantes de la communauté mondiale[1].

Nous avons fait allusion à la récupération par les mouvements wahhabo-salafistes des laissés-pour-compte de la civilisation capitaliste mondialisée. Quoi de plus naturel de la part de mouvements radicaux ? Le problème est que, si des jeunes Occidentaux, qui ne sont pas tous issus de familles de confession musulmane mais pour moitié des convertis de fraîche date, décident de rentrer dans la logique de la violence des mouvements radicaux de l'islam wahhabo-salafiste, ceci tient au fait qu'il n'y a pas d'« idéologie alternative » adaptée à leur cas et à leurs besoins spécifiques qui comporte une dimension utopique de plus grande justice internationale. Pour ce qui est des jeunes des pays pauvres asphyxiés par leur dette et pourvus d'Etats défaillants, l'on ne peut ignorer que c'est dans les

---

[1] Le revenu universel ne fait pas partie des objectifs de l'ONU mais le développement durable, oui. C'est en janvier 2016 que les 17 Objectifs de développement durable (ODD) du Programme de développement durable à l'horizon 2030 de l'ONU sont entrés en vigueur. Si, comme à l'accoutumée, les Objectifs de développement durable (ODD) ne sont pas juridiquement contraignants, il est attendu des gouvernements qu'ils prennent eux-mêmes les choses en main et mettent en place des cadres nationaux pour les atteindre. Comme si la volonté nationale de changer les choses de la part d'un pays pauvre, embourbé parfois dans le chaos de la misère et de la guerre civile, suffisait.

angles morts de la diplomatie internationale, que constituent ces pays défavorisés stratégiquement inintéressants pour les grandes puissances, que les organisations terroristes recrutent leurs combattants.

S'il est communément admis aujourd'hui de considérer l'être humain comme une fin et jamais simplement comme un moyen, l'on ne peut admettre dès lors qu'il soit traité comme moins qu'un moyen. La communauté internationale, qui est à l'affût de la moindre parcelle de terre pour l'accaparer en vue de son exploitation présente ou future, ne peut plus s'offrir le luxe de laisser en friche sa ressource ultime. A l'ère des autoroutes de l'information et de la mondialisation tous azimuts, aucun groupe d'individus ne peut impunément être laissé en marge de l'aventure du village planétaire.

La troisième proposition, qui se présente comme un vœu, est formulée à la lumière de la crise du multilatéralisme en général, et de l'UNESCO en particulier. Nous estimons que l'UNESCO devrait avoir le courage d'une réforme simple mais d'envergure qui consiste à revenir à la modalité de désignation des membres de ses organes directeurs qui fut sienne avant 1993[1]. La composition des organes directeurs de l'UNESCO, avant l'amendement de son Acte constitutif en 1991, lui permettait d'assurer un juste équilibre entre une représentation westphalienne pour la Conférence général (Etats membres) et post-westphalienne pour le Conseil exécutif (personnalité nommément désignées et siégeant à titre personnel). Il importe donc de relancer la pratique selon laquelle les membres du conseil

---

[1] C'est à l'instigation du Japon que l'article V de l'Acte constitutif de l'UNESCO a été amendé, en 1991, pour ce qui est de la qualité des membres du Conseil exécutif de cette Organisation. Depuis 1993, date d'entrée en vigueur de l'amendement, le Conseil est composé d'Etats membres et non plus de personnes (26C/Rés., 19.3).

exécutif de cette organisation de coopération intellectuelle – conscience de l'humanité, selon les termes de Gandhi – soient constitués par des sommités nommées en fonction de leur qualité intellectuelle et morale, et de leur compétence, afin que, l'UNESCO, définie par sa nouvelle directrice générale[1] comme une « intelligence collective en action », puisse être pleinement reconnue comme « conscience de l'humanité ».

Nous en arrivons maintenant à la quatrième et dernière proposition. L'UNESCO a rempli son mandat en matière d'alphabétisation, ses apports dans le domaine de l'éducation sont immenses mais sa tâche demeure inachevée. Car, si les institutions perdurent, avec chaque être humain, c'est l'éternel recommencement. Partout, l'être humain est tout autant mû par la raison que par ses passions et ses émotions. Si les religions partout et de tout temps ont eu pour tâche d'apprendre à dominer les passions et à gérer les émotions, sécularisation oblige, c'est la psychanalyse qui a peu ou prou pris le relais dans certaines contrées. Comme tout le monde n'a pas recours à la psychanalyse, le besoin se fait de plus en plus sentir de parer à cette carence. Les institutions scolaires ne peuvent plus se limiter à la transmission du savoir et au développement de l'esprit critique et de compétition.

L'UNESCO ayant pour mandat de construire la paix dans l'esprit des hommes et des femmes, il y aurait lieu de travailler pour voir s'il n'est pas opportun de promouvoir la thématique de la gestion des émotions au service de la paix au niveau scolaire. Et, pour ce qui concerne l'enseignement supérieur, les experts devraient examiner s'il n'y pas lieu d'envisager la gestion des conflits comme une thématique obligatoire, quel que soit le cursus d'éducation supérieure. Ceci afin que les laissés-pour-compte de la civilisation

---

[1] Audrey Azoulay.

capitaliste mondialisée ne se fassent plus exploser de désespoir et que les détenteurs de pouvoir engagés dans l'« aventure de l'occidentalisation du monde »[1] œuvrent à sa régénérescence à la faveur d'une humanité réconciliée avec son Orient.

---

[1] Si, à la lumière des indicateurs du développement humain, l'on ne peut ignorer, ainsi que Louis Dumont se plaît à le croire, que la plus effective tentative d'« humanisation du monde est sortie à la longue d'une religion qui le subordonnait le plus strictement à une valeur transcendante », l'on doit tout autant admettre que les nations les plus avancés sur les plans scientifico-technique, économique, social, politique et culturel sont capables des pires atrocités. cf. Louis Dumont, op. cit., p.71 et Ninou Garabaghi, « Les Organisations internationales et le progrès du genre humain : Quel avenir pour la culture de la paix et l'éthique de la non-violence ? », revue de l'Académie de géopolitique de Paris : « *Géostratégique* » N° 44, en avril 2015.

Bibliographie[1]

Abderrahim Kader, *Daech, histoire, enjeux et pratiques de l'Etat islamique*, Paris, Eyrolles, 2016.
Amir-Moezzi Mohamad Ali et Jambet Christian, *Qu'est-ce que le shî'isme ?*, Paris, Cerf, 2014.
Appadurai Arjun, *Géographie de la colère. La violence à l'âge de la globalisation*, Payot, Paris, 2007.
Atran Scott, *L'État islamique est une révolution*, Paris, Les liens qui libèrent, 2016.
Barnavi Elie, *Les religions meurtrières*, Paris, Flammarion, 2014.
Bencheikh Ghaleb, *Le Coran : Une synthèse d'introduction et de référence*, Paris, Eyrolles, 2015.
Birnbaum Jean, *Un silence religieux*, Paris, Seuil, 2016.
Boudon Raymond, *Déclin de la morale, Déclin des valeurs*, Paris, PUF, 2002.
Boudon Raymond, *Croire et savoir – Penser le politique, le moral, et le religieux*, Paris, PUF, 2012.
Burgat François, *Comprendre l'Islam politique*, Paris, La Découverte, 2016.
Bzezinski Zbigniew, *Power and principle – Memoirs of the National Security Adviser*, New York, Farrar Straus & Giroux (T), 1983.
Chaliand Gérard, *Pourquoi perd-on la guerre ? Un nouvel art occidental*, Paris, Odile Jacob, 2016.

---

[1] Cette bibliographie sélective se limite aux livres. Le nombre d'articles, d'études, d'enquêtes, etc. qui ont servi de base à la rédaction du présent essai a été estimé beaucoup trop important pour être recensé en fin d'ouvrage.

Châtelet François (dir.), *Histoire des idéologies*, Hachette, Paris, 1978, tome 1 : « Les mondes divins jusqu'au VIIIe siècle de notre ère ».

Chatelus Michel, *Stratégies pour le Moyen-Orient*, Paris, Calmann-Lévy, 1974.

Chomsky Noam, *Un monde complètement surréel*, Paris, Lux éditeur, 2004.

Chomsky Noam, *Les Etats manqués. Abus de puissance et déficit démocratique*, Paris, Fayard, 2007.

Corbin Henry, *En Islam iranien – Aspects spirituels et philosophiques*, Paris, Gallimard, 1971, tome II : « Sohrawardî et les platoniciens de Perse ».

Corbin Henry, *Philosophie iranienne et philosophie comparée*, Téhéran, Académie Impériale Iranienne de Philosophie, 1977.

Corbin Henry, *Le paradoxe du monothéisme*, Paris, L'Herne, 1981.

Cresswell Robert, *Eléments d'ethnologie*, Paris, Armand Colin, 1975.

Debray Régis, *Un mythe contemporain : le Dialogue des civilisations*, Paris, Ed. du CNRS, 2007.

Debray Régis, *Allons aux faits – Croyances religieuses, réalités historiques*, Paris, Coédition Gallimard/France Culture, 2016.

Debray Régis, Girard Renaud, *Que reste-t-il de l'Occident ?*, Paris, Grasset, 2014.

Debray Régis et Geffré Claude, *Avec ou sans Dieu ? : le philosophe et le théologien*, Paris, Fayard, 2006.

Delmas-Marty Mireille, *Les forces imaginantes du droit (I) : Le relatif et l'universel*, Paris, Seuil, 2004.

Delmas-Marty Mireille, *Les forces imaginantes du droit (IV) : La communauté des valeurs*, Seuil, 2011.

Dumont Louis, *Essai sur l'individualisme – une perspective anthropologique sur l'idéologie moderne*, Paris, Seuil, 1985.

Durand Pascal et Sindaco Sarah (dir.), *Le discours "néo-réactionnaire" – Transgressions conservatrices*, Paris, Ed. CNRS, 2015
Eco Umberto, *Construire l'ennemi... et autres écrits occasionnels*, Paris, Grasset, 2014.
Eco Umberto, *Cinq questions de morale*, Paris, Grasset, 2000.
Eco Umberto, *Les limites de l'interprétation*, Paris, Grasset, 1992.
Etienne Bruno, *Islamisme radical*, Paris, Hachette, 1987.
Etienne Bruno, *Islam, les questions qui fâchent*, Paris, Bayard, 2003
Fahmy Mansou, *La condition de la femme dans l'Islam*, Ed. Allia, 2004 (1913, date de la 1ère éd.).
Fillon François, *Vaincre le totalitarisme islamique*, Paris, Albin Michel, 2016.
Fottorino Eric (dir.), *Qui est Daech ? Comprendre le nouveau terrorisme*, Paris, Ed. Le1, 2015.
Foucault Michel, *Dits et Écrits II, 1976-1988*, Paris, Gallimard, 2001.
Garabaghi Ninou, *Les espaces de la diversité culturelle*, Paris, Karthala, 2010.
Geoffroy Eric, *L'islam sera spirituel ou ne sera plus*, Paris, Seuil, 2009.
Gharabaghi Abbass, *Vérité sur la crise iranienne*, Paris, La Pensée Universelle, 1985.
Guidère Mathieu, *Etat du monde arabe*, Louvain-la-Neuve, de Boeck, 2015.
Harari Yuval Noah, *Homo deus : Une brève histoire du futur*, Paris, Albin Michel, 2017.
Houellebecq Michel, *Soumission*, Paris, Flammarion, 2015.
Huntington Samuel P., *Le choc des civilisations*, Paris, Odile Jacob, 1997.
Kepel Gilles, *Jihad – Expansion et déclin de l'Islamisme*, Paris, Gallimard, 2000.

Khankan Sherin, *La femme est l'avenir de l'islam*, Paris, Stock, 2017.
Lamchichi Abderrahim, *Jihâd : un concept polysémique et autres essais*, Paris, L'Harmattan, 2006.
Laurens Henry (en collaboration avec Mireille Delmas-Marty), *Terrorismes : histoire et droit*, Paris, CNRS Éditions, 2010.
Lewis Bernard, *Islam*, Paris, Quarto Gallimard, 2005.
Mervin Sabrina, *Histoire de l'Islam – fondements et doctrines*, Paris, Flammarion - Champs Histoire, 2010.
Meyer David (direction), Simoens Yves, Bencheikh Soheib, *Les versets douloureux. Bible, Évangile et Coran entre conflit et dialogue*, (Préface d'Alexandre Adler), Bruxelles, Ed. Lessius, 2007.
Michel Patrick (dir.), *Religion et démocratie*, Paris, Albin Michel, 1997.
Mohammed Marwan et Hajjat Abdellali, *Islamophobie : comment les élites françaises fabriquent le problème musulman*, Paris, La Découverte, 2013.
Moïsi Dominique (François Boisivon pour la traduction), *La géopolitique de l'émotion : Comment les cultures de peur, d'humiliation et d'espoir façonnent le monde*, Paris, Flammarion, 2011.
Morin Edgar, *La méthode 3. La Connaissance de la connaissance*, Paris, Seuil, 1986.
Morin Edgar, *La méthode 4. Les idées*, Paris, Seuil, 1991.
Morin Edgar, *Vers l'abîme ?*, Paris, Carnets de l'Herne, 2007.
Naraghi Ehsan, *L'Orient et la crise de l'Occident*, Paris, Ed. Entente, 1977.
Nietzsche Friedrich, *La généalogie de la morale*, Paris, Flammarion, 1996.
Oubrou Tareq (entretiens avec Michael Privot et Cédric Baylocq), *Profession Imâm*, Albin Michel, 2015.
Oubrou Tareq (entretien avec Samuel Lieven), *Intégration,*

laïcité, violences - Un Imâm en colère, Paris, Albin Michel, 2015.
Raimbaud Michel, *Tempête sur le Grand Moyen-Orient*, Paris, Ellipses, 2015.
Ricœur Paul, *L'idéologie et l'Utopie*, Paris, Seuil, 1997.
Rodinson Maxime, *Islam et capitalisme*, Paris, Seuil, 1966.
Roy Olivier, *Généalogie de l'islamisme*, Paris, Hachette, 2001.
Roy Olivier, *Islam mondialisé*, Paris, Seuil, 2004.
Roy Olivier, *Echec de l'Islam politique* (1992), Paris, Seuil, 2015 (postface inédite).
Roy Olivier (entretiens avec Jean-Louis Schlegel), *En quête de l'Orient perdu*, Paris, Seuil, 2014.
Saïd Edward W., *L'Orientalisme, L'Orient créé par l'Occident*, Paris, Seuil, 2005.
Sand Shlomo, *La fin de l'intellectuel français ? De Zola à Houellebecq*, La Découverte, 2016.
Sansal Boualem, *2084 : La fin du monde*, Gallimard, 20 août 2015.
Sen Amartya, *Identité et violence – l'illusion du destin*, Odile Jacob, Paris 2007.
Sloterdijk Peter, *Tu dois changer ta vie*, Paris, Libella-Maren Sell Editions, 2011.
Tchouang Tseu, *Œuvre complète*, Paris, Gallimard/UNESCO, 1969.
Vallet Odon, *Petit lexique des idées fausses sur les religions*, Paris, Albin Michel, 2016.
Voltaire, *Essai sur les mœurs et l'esprit des nations (1756)*, Paris, Garnier, éd. Louis Moland, 1878.
Weisskopf Walter A., *Aliénation, idéologie et répression*, Paris, PUF, 1976.

Table des matières

Introduction 9

CHAPITRE 1. Processus de dévoiement du terme « islamisme » : la question de la légitimité du nouveau concept d'« islamisme » 13

CHAPITRE 2. Echec des politiques du développement : de la montée de l'islam spirituel et mystique à la construction de l'« islam politique » 25

CHAPITRE 3. De l'ingérence à l'instrumentalisation des religions à des fins géostratégiques et politiciennes : la construction et l'expansion du djihadisme 37

CHAPITRE 4. Des entités en mal de reconnaissance internationale : "Etat palestinien", "Etat islamique", l'imposture de l'Histoire 63

CHAPITRE 5. Des orientalistes aux islamologues : pour en finir avec le concept d'« islamisme », trou noir de la géopolitique mondiale 85

CHAPITRE 6. De la peur du djihadisme à la haine de l'islam : construction de la figure du nouvel ennemi 135

CHAPITRE 7. Le salafisme djihadiste avatar de la politique de wahhabisation de l'islam mondial 163

CHAPITRE 8. Du mythe de l'islamisation du monde : montée en force et/ou en visibilité de l'islam 193

CHAPITRE 9. De la crise des idéologies politiques à l'échec de l'« islamisme » comme idéologie alternative 227

Conclusion 247

Bibliographie 263

SOCIOLOGIE ET QUESTIONS DE SOCIÉTÉ
AUX ÉDITIONS L'HARMATTAN

Dernières parutions

DES MANIÈRES D'EXISTER ET DE SE DÉPLACER
Les rythmes de vie des citadins – Éclairages anthropologiques
Bonnet Michel
Cet essai développe une série de réflexions sur le rôle et la place du temps existentiel dans la formation des modes de vie et des mobilités des citadins. L'ouvrage entremêle le parcours autobiographique de l'auteur à soixante-dix entretiens pour esquisser un panorama des rythmes de vie des citadins, de catégories sociales et d'âges variés.
(Coll. Logiques sociales, 20.00 euros, 184 p.)
ISBN : 978-2-343-12808-5, ISBN EBOOK : 978-2-14-005299-6

L'ENNEMI AU CŒUR DU POLITIQUE
Beauchard Jacques - Préface de Jean-Pierre Raffarin
Géopolitique, attentats, violences intérieures, guerres : tout concourt à démontrer combien l'identification de l'ennemi pose de plus en plus question ; tout en donnant lieu à des manipulations multiples. Le terrorisme djihadiste et les métamorphoses de la guerre post-westphalienne obligent ainsi à redéfinir les contenus et la construction de cette identification complexe tout en repensant les origines de l'État. L'enjeu de ce livre est donc de prévenir la violence et la guerre par la compréhension de la politique et de l'institution du politique, tout en traitant les caractères les plus cruels de l'actualité.
(17.50 euros, 164 p.)
ISBN : 978-2-343-13425-3, ISBN EBOOK : 978-2-14-005257-6

À LA RECHERCHE DE LA REPRÉSENTATION PERDUE
Nouvelles voies à l'ouest, à l'est et au sud
Petia Gueorguieva et Antony Todorov (dir.)
La crise de la représentation politique qui se manifeste dans de nombreux pays démocratiques à travers l'abstention, le rejet des élites et la crise des partis politiques établis, est un champ d'analyse qui attire l'attention des spécialistes depuis longtemps. Mais doit-on parler d'une crise de la représentation ou d'une crise de la démocratie représentative ? Les « solutions » des alternatives populistes sont-elles démocratiques et viables ? Quelles réponses peut-on apporter à la crise des partis et de nos systèmes démocratiques ?
(Coll. Politique Comparée, 26.50 euros, 254 p.)
ISBN : 978-2-343-13564-9, ISBN EBOOK : 978-2-14-005295-8

DE LA DOMINATION COLONIALE AU REJET DES MIGRANTS
De l'indigène à l'immigré – Essais politiques
Jouard Michel
L'Europe se déchire aujourd'hui face à l'arrivée de migrants venus d'Afrique et du Moyen-Orient. Cet ouvrage dénonce les motivations coloniales, à peine remises en cause, par les puissances mondiales, poussant les ex-colonies à toujours garder un lien de subordination. L'ouvrage met en lumière le refus de ces mêmes grandes nations d'assumer leurs responsabilités historiques et géopolitiques.
(Coll. Questions contemporaines, 22.50 euros, 222 p.)
ISBN : 978-2-343-12146-8, ISBN EBOOK : 978-2-14-005396-2

ÉVALUATION DES POLITIQUES DE SÉCURITÉ ROUTIÈRE
Nouvelles technologies, enjeux économiques et communication
Gilles Blanchard et Laurent Carnis (coordinateurs scientifiques)
En 2016, 3477 personnes ont perdu la vie dans un accident de la route en France métropolitaine marquant une quasi-stabilisation après deux années d'augmentation. Ainsi, la poursuite de la baisse de l'insécurité routière devient de plus en plus difficile. Il ne s'agit plus seulement d'améliorer les infrastructures et les véhicules, mais aussi de changer profondément les comportements tout en accompagnant de nouvelles dynamiques de mobilité. Cet ouvrage rassemble 12 contributions originales, invitant à la fois le lecteur averti et le grand public à porter un regard différent sur la sécurité et ses enjeux.
(20.50 euros, 190 p.)
ISBN : 978-2-343-13181-8, ISBN EBOOK : 978-2-14-005339-9

LA FRANCE ET LES QUESTIONS INTERNATIONALES
Bilan des années 2010 et perspectives
Sous la direction de Daniel Lagot
Quel bilan tirer de l'action de la France dans les années 2010 ? Quand donner priorité à la négociation ou à la guerre ? Quelle politique de défense, en fonction de quels objectifs et de quelles menaces ? L'arme nucléaire, danger pour l'humanité ou instrument de paix entre les mains de certains pays ? Autant de questions d'actualité, parmi d'autres, débattues ici par les porte-parole des principaux candidats à l'élection présidentielle et reprises dans divers articles et analyses complémentaires.
(14.50 euros, 130 p.)
ISBN : 978-2-343-13214-3, ISBN EBOOK : 978-2-14-005063-3

SOCIÉTÉS SÉCURITAIRES OU SOCIÉTÉS DE CONFIANCE ?
Pelletan Jacques
Il s'agit de tenter de lire les sociétés contemporaines à travers le prisme de la sécurité. Le rapport au risque s'est profondément modifié récemment, le monde est devenu plus incertain : violence, terrorisme, mais aussi insécurité économique, l'inquiétude se remet à régner. Est-il possible dans un monde occidental, régi par la crainte et le conservatisme, de restaurer la confiance face aux risques futurs ?
(20.00 euros, 192 p.)
ISBN : 978-2-343-13074-3, ISBN EBOOK : 978-2-14-005028-2

SURVEILLER ET PRÉVENIR – L'ère de la pénalité prédictive
Bourgoin Nicolas
La guerre contre le terrorisme lancée par les États occidentaux après les attentats du 11 septembre a bouleversé le droit pénal et favorisé la montée en force d'un modèle prédictif. Celui-ci conduit à la mise en place de dispositifs destinés à repérer les signes avant-coureurs de passage à l'acte violent. Ainsi, la doctrine du pré-terrorisme instituant une présomption de culpabilité à rebours du droit pénal classique est l'alibi d'un contrôle social toujours plus étroit exercé sur les citoyens. Mais à terme, ne risque-t-elle pas de menacer les fondements mêmes de la démocratie en remettant en cause l'équilibre des pouvoirs ?
(Coll. Logiques sociales, 26.00 euros, 264 p.)
ISBN : 978-2-343-13155-9, ISBN EBOOK : 978-2-14-005138-8

ESSENTIEL DE LA GÉOPOLITIQUE
En collaboration avec Aimé Kayemba Cimanga
et Jean Rostand Kalombo Ntambue – Préface de Félicien Lukiana Mabondo
Ce livre met à la disposition du public certaines notions fondamentales qui peuvent favoriser la compréhension de la géopolitique. Au centre des préoccupations des auteurs se trouvent l'État, les enjeux majeurs, les rivalités, les crises, la paix, les défis à relever, le sport etc. Pour bien comprendre ces dimensions, les auteurs ont effectué une plongée assez détaillée dans les diverses théories géopolitiques, lesquelles marquent la volonté des entités politiques d'agir, de mener la guerre ou d'appliquer leur politique expansionniste, de s'enrichir, de négocier, de faire des alliances, etc.
(Coll. Notes de cours, 21.50 euros, 212 p.)
ISBN : 978-2-343-12615-9, ISBN EBOOK : 978-2-14-005156-2

RÉFORMER LA GOUVERNANCE MONDIALE
Shanda Tonme Jean-Claude
Tous les enjeux évoqués dans ce livre rentrent dans le décor planté par deux déclarations de Kim Dae Jung II, président de la Corée, et Kako Nubukpo, un esprit libre de la société civile. Tous deux expriment une même démarche et mettent en exergue une égale expression, à la fois défi et volonté de correction de la gouvernance mondiale. C'est le socle révélateur d'une extraordinaire remise en cause, qui fait la substance de cet ouvrage.
(27.00 euros, 264 p.)
ISBN : 978-2-343-13374-4, ISBN EBOOK : 978-2-14-005054-1

L'ÉGALITÉ
Réalité, rêve ou utopie ?
Makunga Lendo
Somme-nous égaux ? Non, disent certains, car l'égalité est impossible et n'est pas naturelle, c'est un rêve, une utopie. D'autres pensent que oui, mais tout dépend de quoi on parle. D'autres encore répondent par oui et par non. Aussi, l'égalité est-elle source de polémiques, parce qu'il existe différentes compréhensions de ce concept. Il n'empêche que c'est le deuxième pilier de la devise de la France : Liberté, Égalité, Fraternité. L'égalité est la base sur laquelle s'est construite la

démocratie. Que veut dire ce principe républicain pour la France et pour le monde aujourd'hui ?
(26.00 euros, 260 p.)
ISBN : 978-2-343-13196-2, ISBN EBOOK : 978-2-14-004947-7

HUMANITARIAN ACTION AND TERRORISM
Perceptions from the Muslim World
Macdonald Ingrid A.
The proliferation of designated terrorist groups in majority-Muslim countries has greatly affected civilian populations. Humanitarian workers and assets have been targeted by terrorist groups through the use of violent methods. This study addresses the state of observable perceptions towards humanitarian organisations and workers among local populations. It concludes by proposing the 'humanitarian cube', which is a dynamic new tool for strengthening understanding and actions in response to local perceptions.
(29.00 euros, 288 p.)
ISBN : 978-2-343-13570-0, ISBN EBOOK : 978-2-14-005199-9

L'ORDRE SOCIAL
Giraud Claude
L'ordre social est la part absente de nombreux traités de sociologie, car il cumule les paradoxes. Il est un ordre pensé sans lieu, sans hiérarchie, sans souveraineté, alors même qu'il est un ordre, celui des interdépendances, des fonctionnalités, des solidarités sans conscience. C'est un ordre de normes où les rapports de puissance et de pouvoir sont en concurrence. Cet ouvrage interroge l'ordre social comme catégorie conceptuelle qui rend compte d'une réalité complexe, changeante et persistante.
(Coll. Logiques sociales, 17.50 euros, 158 p.)
ISBN : 978-2-343-13288-4, ISBN EBOOK : 978-2-14-004966-8

CATÉGORISER L'AUTRE
Aires anglophone et lusophone
Sous la direction de Michel Prum
Catégoriser l'Autre, du point de vue de l'ethnicité, c'est se placer soi-même au dessus de toute catégorisation et assigner celui-ci à des particularismes dont on serait exempt. Le «majoritaire» se donne le droit de nommer et de classer les «minoritaires». Neuf contributions constituent cet ouvrage et proposent un tour d'horizon de ce qu'on a appelé la «race» dans l'aire anglophone et lusophone.
(Coll. Racisme et eugénisme, 21.50 euros, 212 p.)
ISBN : 978-2-343-13300-3, ISBN EBOOK : 978-2-14-005029-9

L'HARMATTAN ITALIA
Via Degli Artisti 15; 10124 Torino
harmattan.italia@gmail.com

L'HARMATTAN HONGRIE
Könyvesbolt ; Kossuth L. u. 14-16
1053 Budapest

L'HARMATTAN KINSHASA
185, avenue Nyangwe
Commune de Lingwala
Kinshasa, R.D. Congo
(00243) 998697603 ou (00243) 999229662

L'HARMATTAN GUINÉE
Almamya Rue KA 028, en face
du restaurant Le Cèdre
OKB agency BP 3470 Conakry
(00224) 657 20 85 08 / 664 28 91 96
harmattanguinee@yahoo.fr

L'HARMATTAN CONGO
67, av. E. P. Lumumba
Bât. – Congo Pharmacie (Bib. Nat.)
BP2874 Brazzaville
harmattan.congo@yahoo.fr

L'HARMATTAN MALI
Rue 73, Porte 536, Niamakoro,
Cité Unicef, Bamako
Tél. 00 (223) 20205724 / +(223) 76378082
poudiougopaul@yahoo.fr
pp.harmattan@gmail.com

L'HARMATTAN CAMEROUN
TSINGA/FECAFOOT
BP 11486 Yaoundé
699198028/675441949
harmattancam@yahoo.com

L'HARMATTAN CÔTE D'IVOIRE
Résidence Karl / cité des arts
Abidjan-Cocody 03 BP 1588 Abidjan 03
(00225) 05 77 87 31
etien_nda@yahoo.fr

L'HARMATTAN BURKINA
Penou Achille Some
Ouagadougou
(+226) 70 26 88 27

L'HARMATTAN SÉNÉGAL
10 VDN en face Mermoz, après le pont de Fann
BP 45034 Dakar Fann
33 825 98 58 / 33 860 9858
senharmattan@gmail.com / senlibraire@gmail.com
www.harmattansenegal.com